T0257925

Essential Topics in Ecological Water Quality

Essential Topics in Ecological Water Quality

Edited by **Herbert Lotus**

New York

Published by Callisto Reference,
106 Park Avenue, Suite 200,
New York, NY 10016, USA
www.callistoreference.com

Essential Topics in Ecological Water Quality
Edited by Herbert Lotus

© 2015 Callisto Reference

International Standard Book Number: 978-1-63239-322-7 (Hardback)

Printed in the United States of America.

Contents

Preface   IX

Water Treatment Technologies and Water Reuse   1

Chapter 1   **Water Reuse and Sustainability**   3
Rouzbeh Nazari, Saeid Eslamian and Reza Khanbilvardi

Chapter 2   **Evaluation of the Removal of Chlorine,
THM and Natural Organic Matter from Drinking
Water Using Microfiltration Membranes
and Activated Carbon in a Gravitational System**   17
Flávia Vieira da Silva-Medeiros, Flávia Sayuri Arakawa,
Gilselaine Afonso Lovato, Célia Regina Granhen Tavares,
Maria Teresa Pessoa Sousa de Amorim, Miria Hespanhol
Miranda Reis and Rosângela Bergamasco

Chapter 3   **Application of Hybrid Process of Coagulation/ Flocculation
and Membrane Filtration to Water Treatment**   31
Rosângela Bergamasco, Angélica Marquetotti Salcedo Vieira,
Letícia Nishi, Álvaro Alberto de Araújo
and Gabriel Francisco da Silva

Chapter 4   *In situ* **Remediation Technologies
Associated with Sanitation Improvement:
An Opportunity for Water Quality
Recovering in Developing Countries**   55
Davi Gasparini Fernandes Cunha, Maria do Carmo Calijuri,
Doron Grull, Pedro Caetano Sanches Mancuso
and Daniel R. Thévenot

Chapter 5   **Water Quality Improvement Through an Integrated
Approach to Point and Non-Point Sources Pollution
and Management of River Floodplain Wetlands**   73
Edyta Kiedrzyńska and Maciej Zalewski

Chapter 6   Water Quality in the Agronomic Context:
             Flood Irrigation Impacts on Summer In-Stream Temperature
             Extremes in the Interior Pacific Northwest (USA)   91
             Chad S. Boyd, Tony J. Svejcar and Jose J. Zamora

Chapter 7   The Effect of Wastes Discharge
             on the Quality of Samaru Stream, Zaria, Nigeria   107
             Y.O. Yusuf and M.I. Shuaib

Chapter 8   Elimination of Phenols on a Porous Material   121
             Bachir Meghzili, Medjram Mohamed Salah,
             Boussaa Zehou El-Fala Mohamed
             and Michel Soulard

Chapter 9   Impact of Agricultural Contaminants
             in Surface Water Quality:
             A Case Study from SW China   135
             Binghui He and Tian Guo

Chapter 10   Effects of Discharge Characteristics on Aqueous Pollutant
             Concentration at Jebel Ali Harbor, Dubai-UAE   149
             Munjed A. Maraqa, Ayub Ali, Hassan D. Imran,
             Waleed Hamza and Saed Al Awadi

Chapter 11   Water Quality in Hydroelectric Sites   167
             Florentina Bunea, Diana Maria Bucur,
             Gabriela Elena Dumitran and Gabriel Dan Ciocan

Chapter 12   Removal Capability of Carbon-Soil-Aquifer
             Filtering System in Water Microbiological Pollutants   185
             W.B. Wan Nik, M.M. Rahman, M.F. Ahmad,
             J. Ahmad and A. M Yusof

Chapter 13   An Overview of the Persistent
             Organic Pollutants in the Freshwater System   201
             M. Mosharraf Hossain, K. M. Nazmul Islam
             and Ismail M. M. Rahman

Chapter 14   Fluxes in Suspended Sediment
             Concentration and Total Dissolved Solids
             Upstream of the Galma Dam, Zaria, Nigeria   217
             Y.O. Yusuf, E.O. Iguisi and A.M. Falade

Chapter 15   **Rainwater Harvesting Systems in Australia**   **233**
            M. van der Sterren, A. Rahman and G.R. Dennis

            Permissions

            List of Contributors

Preface

The pollution of aquatic environment reduces possible uses of water, especially those that require high quality standards i.e. for drinking purposes. This book discusses different problems regarding the quality of water and also focuses on the current techniques of water treatment. The primary emphasis is on sustainable choices of water usage that prevent water quality difficulties and aim at the protection of available water resources, and the development of the aquatic ecosystems. The book contains different procedures employed for water treatment technologies.

Significant researches are present in this book. Intensive efforts have been employed by authors to make this book an outstanding discourse. This book contains the enlightening chapters which have been written on the basis of significant researches done by the experts.

Finally, I would also like to thank all the members involved in this book for being a team and meeting all the deadlines for the submission of their respective works. I would also like to thank my friends and family for being supportive in my efforts.

Editor

Water Treatment Technologies and Water Reuse

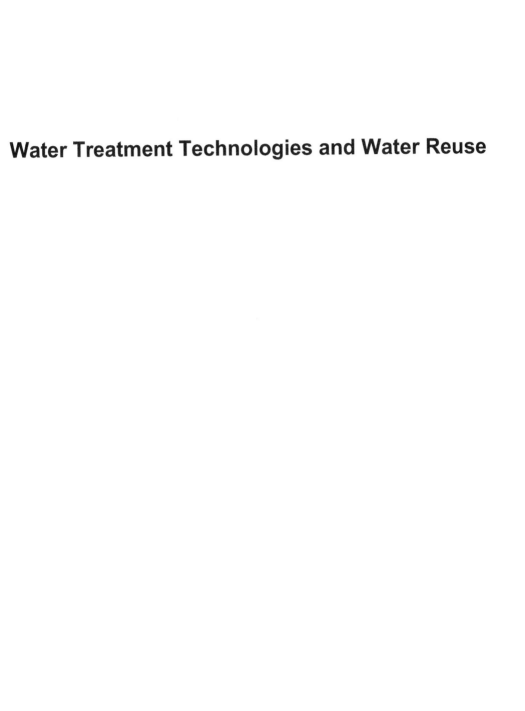

Water Reuse and Sustainability

Rouzbeh Nazari[1], Saeid Eslamian[2] and Reza Khanbilvardi[1]
[1]City University of New York
[2]Isfahan University of Technology,
[1]USA
[2]Iran

1. Introduction

Water reuse simply is the use of reclaimed water for a direct beneficial purpose in various sectors from home to industry and agriculture. For a number of semi-arid regions and islands, water reuse provides a major portion of the irrigation water. In addition, the reuse of treated wastewater for irrigation and industrial purposes can be used as strategy to release freshwater for domestic use, and to improve the quality of river waters used for abstraction of drinking water. Specific water reuse applications meet the water quality objectives. Water quality standards and guidelines which are related to irrigation and industrial water reuse are described in this chapter. Other reuse consumptions such as urban, recreational and environmental are also discussed.

Water quality is the most important issue in water reuse systems in ensuring sustainable and successful wastewater reuse applications. The main water quality factors that determine the suitability of recycled water for irrigation are pathogen content, salinity, specific ion toxicity, trace elements, and nutrients. It will be introduced the important criteria for evaluation water quality and World Health Organization guidelines (WHO, 1989) and the United States Environmental Protection Agency guidelines (USEPA, 1992, 2004) which are the two main guidelines that frequently used in many countries around the world. Finally, it will be discussed briefly about different treatment method selections; the degree of treatment required and the extent of monitoring necessary which depend on the specific application. Wastewater reuse can be applied for various beneficial purposes such as agricultural irrigation, industrial processes, groundwater recharge, and even for potable water supply after extended treatment. Water reuse allows the communities to become less dependent on groundwater and surface water resources and can decrease the diversion of water from sensitive ecosystems. Additionally, water reuse may reduce the nutrient loads from wastewater discharges into waterways, thereby reducing and preventing pollution. This "new" water resource may also be used to replenish overdrawn water resources and rejuvenate or reestablish those previously destroyed. Most common types of wastewater reuses are summarized in Table 1.

2. Agriculture reuse

The reuse of wastewater has been successful for irrigation of a wide array of crops, and increases in crop yields from 10-30% have been reported (Asano, 1998, 2004). For a number of semi-arid regions and islands, water recycling provides a major portion of the irrigation water. In addition, the reuse of treated wastewater for irrigation and industrial purposes can be used as strategy to release freshwater for domestic use, and to improve the quality of river waters used for abstraction of drinking water by reducing disposal of effluent into rivers (USEPA, 2003). By knowing that water for agriculture is critical for food security and also by understanding that agriculture remains the largest water user, with about 70% of the world's freshwater consumption, it can be understood that how important it is to have new source of water available for this section. According to recent Food and Agriculture Organization data (FAO Website), only 30 to 40% of the world's food comes from irrigated land comprising 17% of the total cultivated land. One of the broad strategies to address this challenge for satisfying irrigation demand under conditions of increasing water scarcity in both developed and emerging countries is to conserve water and improve the efficiency of water use through better water management and policy reforms. In this context, water reuse becomes a vital alternative resource and key element of the integrated water resource management at the catchment scale (Asano and Levine, 1996; Lazarova, 2000, 2001).

However, despite widespread irrigation with reclaimed wastewater, water-reuse programs are still faced with a number of technical, economic, social, regulatory, and institutional challenges. Some of the water-quality concerns and evaluation of long-term environmental, agronomic, and health impacts remain unanswered. But water quality is the most important issue in water reuse systems so to ensure sustainable and successful wastewater reuse applications, the potential public health risk associated with wastewater reuse should be evaluated and also the specific water reuse applications should meet water quality objectives. Water quality of the effluent which is going to be used as reuse water, is the most important issue related to water reuse systems that determines the acceptability and safety of the use of recycled water for a given reuse application. The options for sustainable reuse projects are related to the quality of the effluent, and the environmental risk associated with land application for a variety of crops and activities and irrigation type and even the quality standard can vary during irrigation and non-irrigation period (Eslamian et al., 2010). It might be higher during interim periods when irrigation is not practiced to ensure a relatively safe discharge to receiving water bodies. The main water quality factors that determine the suitability of recycled water for irrigation are pathogen content, salinity, sodicity (levels of sodium that affect soil stability), specific ion toxicity, trace elements, and nutrients. All modes of irrigation may be applied depending on the specific situation. If applicable, drip irrigation provides the highest level of health protection, as well as water conservation potential (Valentina and Akica, 2005). The most important criteria for evaluation of the treated wastewater are as follows (Kretzschmar et al., 2002):

-   Salinity (especially important in arid zones)
-   Heavy metals and harmful organic substances
-   Pathogenic germs

Table 1 presents the most important water quality parameters and their significance in the case of municipal wastewater reuse.

The goal of each water reuse project is to protect public health without necessarily discouraging wastewater reclamation and reuse. The guidelines or standards required removing health risks from the use of wastewater and the amount and type of wastewater treatment needed to meet the guidelines are both contentious issues. The cost of treating wastewater to high microbiological standards can be so prohibitive that the use of untreated wastewater is allowed to occur unregulated.

Types of Reuse	Treatment	Reclaimed Water Quality	Reclaimed Water Monitoring	Setback Distances
Urban Reuse Landscape irrigation, vehicle washing, toilet flushing, fire protection, commercial air conditioners, and other uses with similar access or exposure to the water	Secondary[1] Filtration[2] Disinfection[3]	pH = 6–9 <10 mg/L biochemical oxygen demand (BOD) < 2 turbidity units (NTU)[5] No detectable fecal coliform/100 mL[4] 1 mg/L chlorine (Cl₂) residual (min.)	pH – weekly BOD – weekly Turbidity – continuous Coliform – daily Cl₂ residual – continuous	50 feet (15 m) to potable water supply wells
Agricultural Reuse For Non-Food Crops Pasture for milking animals; fodder, fiber and seed crops	Secondary Disinfection	pH = 6–9 < 30 mg/L BOD < 30 mg/L total suspended solids (TSS) < 200 fecal coliform/100 mL[5] 1 mg/L Cl₂ residual (min.)	pH – weekly BOD – weekly TSS – daily Coliform – daily Cl₂ residual – continuous	300 feet (90 m) to potable water supply wells
Indirect Potable Reuse Groundwater recharge by spreading into potable aquifers	Site specific Secondary and disinfection (min.) May also need filtration and/or advanced wastewater treatment	Site specific meet drinking water standards after percolation through vadose zone.	pH – daily Turbidity – continuous Coliform – daily Cl₂ residual – continuous Drinking water standards– quarterly Other – depends on Constituent	100 feet (30 m) to areas accessible to the public (if spray irrigation) site specific

[1] Secondary treatment processes include activated sludge processes, trickling filters, rotating biological contactors, and many stabilization pond systems. Secondary treatment should produce effluent in which both the BOD and TSS do no exceed 30 mg/L.
[2] Filtration means passing the effluent through natural undisturbed soil or filter media such as sand and anthracite.
[3] Disinfection means the destruction, inactivation or removal of pathogenic microorganisms. It may be accomplished by chlorination, or other chemical disinfectants, UV radiation or other processes.
[4] The number of fecal coliform organisms should not exceed 14/100 mL in any sample.
[5] The number of fecal coliform organisms should not exceed 800/100 mL in any sample.

Table 1. Reuse Chart (USEPA, 2004)

Parameter	Significance	Approximate Range in Treated Wastewater
Total Suspended solids (TSS)	TSS can lead to sludge deposits and anaerobic conditions. Excessive amounts caused clogging of irrigation systems. Measures of particles in wastewater can be related to microbial contamination, turbidity. Can interfere with disinfection effectiveness	< 1 to 30 mg/l
Organic indicators TOC Degradable Organics (COD, BOD)	Measure of organic carbon. Their biological decomposition can lead to depletion of oxygen. For irrigation only excessive amounts cause problems. Low to moderate concentrations are beneficial.	1 – 20 mg/l 10 – 30 mg/l
Nutrients N,P,K	When discharged into the aquatic environment they lead to eutrophication. In irrigation, they are beneficial, nutrient source. Nitrate in excessive amounts, however, may lead to groundwater contamination.	N: 10 to 30 mg/l P: 0.1 to 30 mg/l
Stable organics (e.g. phenols, pesticides, chlorinated hydrocarbons)	Some are toxic in the environment, accumulation processes in the soil.	
pH	Affects metal solubility and alkalinity and structure of soil, and plant growth.	
Heavy metals (Cd, Zn, Ni, etc.)	Accumulation processes in the soil, toxicity for plants	
Pathogenic organisms	Measure of microbial health risks due to enteric viruses, pathogenic bacteria and protozoa	Coliform organisms: < 1 to 104 /100 ml other pathogens: Controlled by treatment technology
Dissolved Inorganics (TDS, EC, SAR)	Excessive salinity may damage crops. Chloride, Sodium and Boron are toxic to some crops, extensive sodium may cause permeability problems	

Table 2. Water quality parameters for wastewater reuse and their significance (Asano, 1998)

Regulatory approaches stipulate water quality standards in conjunction with requirements for treatment, sampling and monitoring. These standards or guidelines are highly dependent on the kind of water use. Obviously, the landscape and forest irrigation has the lowest requirements concerning the treatment of effluent, compared to the potable reuse. But, the requirements of irrigation of limited crops (crops that need further processing) are not high and therefore it is applicable in economic terms.

The greatest health concern when using recycled water for irrigation is related to pathogens that could be present (Kretschmer et al., 2000). It is widely known that it is not practical to establish the presence or absence of all pathogenic organisms in wastewater or recycled water in a timely fashion. For this reason, the indicator organism, E-coli, was established many years ago to allow monitoring of a limited number of microbiological constituents.

Standards for wastewater reuse in many countries have been influenced by the WHO Health Guidelines (1989) (Table 3) and the USEPA Guidelines (2004) (Table 4). The Guidelines are set to minimize exposure to workers, crop handlers, field workers and consumers, and recommend treatment options to meet the guideline values. WHO's 1989 Guidelines which seems somehow old and there are no any newer WHO guidelines; for the safe use of wastewater in agriculture take into account all available epidemiological and microbiological data. The fecal coliform guideline (e.g. =1000 FC/100 ml for food crops eaten raw) was intended to protect against risks from bacterial infections, and the newly introduced intestinal nematode egg guideline was intended to protect against helminthes infections (and also serve as indicator organisms for all of the large settable pathogens, including amoebic cysts). The exposed group that each guideline was intended to protect and the wastewater treatment expected to achieve the required microbiological guideline was clearly stated. Waste stabilization ponds were advocated as being both effective at the removal of pathogens and the most cost-effective treatment technology in many circumstances.

Category	Reuse Conditions	Exposed Group	Intestinal Nematode (arithmetic mean no. eggs per liter)	Fecal Coliforms (geometric mean no. per 100 ml)	Wastewater treatment expected to achieve the required microbiological guideline
A	Irrigation of crops likely to be eaten uncooked, sports fields, public parks	Workers, consumers, publics	≤ 1	≤ 1000	A series of stabilization ponds designed to achieve the microbiological quality indicated, or equivalent treatment
B	Irrigation of cereal crops, industrial crops, fodder crops, pasture and trees	Workers	≤ 1	No standard recommendation	Retention in stabilization pond for 8-10 days or equivalent helminth and fecal coliform removal
C	Localized irrigation of crops in category B if exposure to workers and the public does not occur	None	Not applicable	Not applicable	Pretreatment as required by irrigation technology, but not less than primary sedimentation

Table 3. WHO (2001) guideline for use of treated wastewater in agriculture

Reuse Type	Treatment	Water Quality	Setbacks	Monitoring
.... Public Contact				
Irrigation for public areas: * Parks * Cemetery * Golf Courses * Other landscapes **Agricultural irrigation for:** * Food crops that will not be commercially processed * Any crops eaten raw	Secondary Filtration and Disinfection	* pH 6-9 * ≤ 10 mg/L BOD * ≤ 2 NTU * No detectable fecal coliforms/100 mL * at least 1 mg/L residual chlorine	* 50 feet to potable water	* Weekly: pH, BOD * Monthly: Coliforms * Continuously: Turbidity, Chlorine Residue
.... Limited or No Public Contact...				
Irrigation of restricted access areas: * Sod farms * Silvicultures * Other areas with limited or no public access **Agricultural irrigation for:** * Food crops that will be commercially processed * Non-food crops and pastures	Secondary Disinfection	* pH 6-9 * ≤ 30 mg/L BOD * ≤ 30 mg/L TSS * ≤ 200 fecal coliforms/100 mL * at least 1 mg/L residual chlorine	* 300 feet to potable water * 100 feet to areas accessible to public (if spray irrigation is used)	* Weekly: pH, BOD * Monthly: Coliforms and TSS * Continuously: Chlorine Residue

Table 4. USEPA (2004) guideline for agricultural reuse of wastewater

In contrast, USEPA (2004) has recommended the use of much stricter guidelines for wastewater use in the USA. The USEPA (2004) has established guidelines to encourage states to develop their own regulations. The primary purpose of federal guidelines and state regulations is to protect human health and water quality. To reduce disease risks to acceptable levels, reclaimed water must meet certain disinfection standards by either reducing the concentrations of constituents that may affect public health and/or limiting human contact with reclaimed water. The elements of the guidelines applicable to reuse in agriculture are summarized in Table 4. For irrigation of crops likely to be eaten uncooked, no detectable fecal coliform/100 ml are allowed (compared to 1000 FC/100ml for WHO), and for irrigation of commercially processed crops, fodder crops, etc, the guideline is 200 FC/100 ml.

Much wastewater reuse in agriculture is indirect and that is, the wastewater is predisposed into rivers and the contaminated river water is used later on for irrigation. However, international guidelines for the microbiological quality of irrigation water used on a particular crop do not exist (Ayers and Westcott, 1985). The United States Environmental Protection Agency (USEPA) recommended that the acceptable guideline for irrigation with natural surface water, including river water containing wastewater discharges, be set at 1000 FC/10 ml (USEPA, 1981). This standard has been adopted in some other countries as an irrigation water quality standard, for example, Chile, in 1978 (Ayers and Westcott, 1985). This standard is also consistent with guidelines for unrestricted irrigation. FAO has now recommended that the WHO (1989) Guidelines be used interim irrigation water standards, until more epidemiological information is available. Eslamian and Tarkesh-Isfahani (2010b) evaluate the most efficient irrigation systems in wastewater reuse.

3. Industrial reuse

Reuse of reclaimed water for industrial proposes is developed in many industries of United State of America, Europeans and other developed countries. Reclaimed water reuse is one of the strategies for sustainable management. Industrial reuse has increased substantially since the early 1990s for many of the same reasons urban reuse has gained popularity, including water shortages and increased populations, particularly in drought areas, and legislation regarding water conservation and environmental compliance. Utility power plants are ideal facilities for reuse due to their large water requirements for cooling, ash sluicing, rad-waste dilution, and flue gas scrubber requirements (Metcalf and Eddy, 2003, 2007). Petroleum refineries, chemical plants, and metal working facilities are among other industrial facilities benefiting from reclaimed water not only for cooling, but for processing needs as well. For the majority of industries, cooling water is the largest use of reclaimed water because advancements in water treatment technologies have allowed industries to successfully use lesser quality waters. These advancements have enabled better control of deposits, corrosion, and biological problems often associated with the use of reclaimed water in a concentrated cooling water system. The most frequent water quality problems in cooling water systems are corrosion, biological growth, and scaling. These problems arise from contaminants in potable water as well as in reclaimed water, but the concentrations of some contaminants in reclaimed water may be higher than in potable water (EPA, 1981).

Industrial reuse can be explained and defined for a number of industries in the world, but if the most industrial water consumption, cooling towers, is considered to this subject, the industrial reuse is defined for each industry and it can be defined as a quality standard for reclaimed water reuse. Eslamian and Tarkesh-Isfahani (2010a) evaluate the urban reclaimed water for industrial reuses in North Isfahan, Iran. Based on this and other research projects results on eight various industries, and case studies, articles and books, reclaimed water quality parameter limitation for use in cooling towers are defined and shown in Table 5.

Parameter	Measured Standard Method	Unit	Selected Range of Concentration for IOR Consumed
Electrical conductivity (EC)	Platinum Electrode, number 2510 B of Standard methods	μmhos/cm	500-600
Hardness (as $CaCo_3$)	EDTA Titrimetric, number 2340 C of standard methods	mg\L	150-250
Alkalinity	Titrimetric, number 2320 B of standard methods	mg\L	100-150
Chloride	Argentometric, number 4500-Cl- B of standard methods	mg\L	175-250
Orthophosphate (PO4)	Vanadomolybdophosphoric Acid Colorimetric, number 4500-P C of	mg\L	0-1
polyphosphate	Vanadomolybdophosphoric Acid Colorimetric, number 4500-P C of standard methods	mg\L	Good
NO_2^-	Colorimetric, number 4500-NO_2^- B of standard methods	mg\L	<1
NO_3^-	Ultraviolet Spectrophotometric, number 4500-NO_3^- B of standard methods	mg\L	<5
NH_3	Nesslerization, number D1426 of ASTM	mg\L	<1
TSS	Gravimetric, number 2540 D of standard methods	mg\L	5-10

Turbidity	Nephelometric, number 2130 B of standard methods	NTU	<2
TDS	Platinum Electrode, number 2510 B of Standard methods	mg\L	250-500
Ca	EDTA Titrimetric, number 3500-Ca B of standard methods	mg\L	50-75
BOD$_5$	Respirometric, number 5210 D of standard methods	mg\L	0-5
COD	Closed Reflux-Titrimetric, number 5220 C of standard methods	mg\L	20-40
pH	Electrometric, number 4500-H$^+$ B of standard methods	-	6-8
SO$_4^{2-}$	Gravimetric, number 4500-SO$_4^{2-}$ C of standard methods	mg\L	0-250
Na$^+$ (as NaCl)	Direct Air-Acetylene Flame Atomic Absorption Spectrometric, number 3111 B of standard methods	mg\L	150.200
Cr	number 3111 B of standard methods	mg\L	<0.5
Cu	number 3111 B of standard methods	mg\L	<0.05-1
Se	number 3111 B of standard methods	mg\L	<1
Mn	number 3111 B of standard methods	mg\L	<0.3-1
Pb	number 3111 B of standard methods	mg\L	<1
Zn	number 3111 B of standard methods	mg\L	<1
Mg	number 3111 B of standard methods	mg\L	<20-30
Co	number 3111 B of standard methods	mg\L	<1
Cd	number 3111 B of standard methods	mg\L	<0.1-1
Fe	number 3111 B of standard methods	mg\L	0.1-0.3
Sr	number 3111 B of standard methods	mg\L	<1
As	number 3111 B of standard methods	mg\L	<1
Hg	number 3111 B of standard methods	mg\L	<1
SiO$_2$	Molybdosilicate Colorimetric, number 4500-SiO$_2$ C of standard methods	mg\L	<10-20
Oil and Greece	Partition- Gravimetric, number 5520 B of standard methods	mg\L	<1
Total Chlorine Residual	DPD Colorimetric, number 4500-Cl G of standard methods	mg\L	<4
Total coliform (as log)	-	Log MPN/100ml	<2.2 MPN/100ml
Fecal coliform (as log)	-	Log MPN/100ml	<2.2 MPN/100ml
SRB (sulfate reducing bacteria)	-	MPN/100ml	nil
PAHs	Gas Chromatographic-Flame Ionization Detector	µg/L	nil
THMs	Gas Chromatographic-Mass Spectrometry	µg/L	nil
MTBE	Gas Chromatographic-Flame Ionization Detector	µg/L	nil
OCP-Pesticide	Gas Chromatographic-Electron Capture Detector	µg/L	nil
OPP-Pesticide	Gas Chromatographic-Nitrogen Phosphorous Detector	µg/L	nil
2,4-D	High Performance Liquid Chromatographic	µg/L	nil

Table 5. Range of water quality parameters for reuse of reclaimed water in cooling towers

4. Urban reuse

Urban reuse systems are a crucial part of water recycling since it can provide the reclaimed water for various non-drinking purposes such as Irrigation of public parks and recreation centers, athletic fields, school yards and playing fields, highway medians and shoulders, and landscaped areas surrounding public buildings and facilities, Irrigation of landscaped areas surrounding single-family and multi-family residences, general wash down, and other maintenance activities. Urban reuse can be expanded to cover commercial uses such as vehicle washing facilities, laundry facilities, window washing and mixing water for pesticides, herbicides, liquid fertilizers, toilet and urinal flushing in commercial and industrial buildings. Reclaimed water can also help with human health and safety in dust control and concrete production for construction projects and control the expansion of suspended particles in the air and provide water for fire hydrants. A 2-year field demonstration/research garden compared the impacts of irrigation with reclaimed versus potable water for landscape plants, soils, and irrigation components. The comparison showed few significant differences; however, landscape plants grew faster with reclaimed water (Lindsey et al., 1996). But such results are not a given. Elevated chlorides in the reclaimed water provided by the City of St. Petersburg have limited the foliage that can be irrigated (Johnson, 1998). Dual distribution systems could be used to deliver the reclaimed water to customers through a parallel network of distribution completely separated and marked to distinguish from the community's drinking water line. Design considerations for urban water reuse systems should include two major components: water reclamation facilities and reclaimed water distribution system, including storage and pumping facilities. The reclaimed water distribution system has the potential to become a third water utility, along with drinking water and wastewater. Reclaimed water systems are operated, maintained, and managed in a manner similar to the drinking water system. One of the oldest municipal dual distribution systems in the U.S., in St. Petersburg, Florida, has been in operation since 1977. The system provides reclaimed water for a mix of residential properties, commercial developments, industrial parks, a resource recovery power plant, a baseball stadium, and schools. The City of Pomona, California, first began distributing reclaimed water in 1973 to California Polytechnic University and has since added two paper mills, roadway landscaping, a regional park and a landfill with an energy recovery facility. As part of planning of an urban reuse system, communities have the option of choosing continuous or interruptible reclaimed water system. In general, an interruptible source of reclaimed water can be used as long as reclaimed water will not be used as the only source of fire protection. For example, the City of St. Petersburg, Florida, decided that an interruptible source of reclaimed water would be acceptable, and that reclaimed water would provide only backup for fire protection. If a community determines that a non-interruptible source of reclaimed water is needed, then reliability, equal to that of a potable water system, must be provided to ensure a continuous flow of reclaimed water. This reliability could be ensured through a municipality having more than one water reclamation plant to supply the reclaimed water system, as well as additional storage to provide reclaimed water in the case of a plant upset. However, providing the reliability to produce a non-interruptible supply of reclaimed water will have an associated cost increase. In some cases, such as the City of Burbank, California, reclaimed water storage tanks are the only source of water serving an isolated fire system that is kept separate from the potable fire service. Retrofitting a developed urban area with a reclaimed water distribution system can be expensive. In some cases, however, the benefits of conserving potable water may justify the cost.

5. Environmental and recreational reuses

Water reuse provides a dependable, locally-controlled water supply and tremendous environmental benefits. Environmental reuse includes creating artificial wetlands, enhancing natural wetlands and sustaining stream flows. Uses of reclaimed water for recreational purposes range from landscape impoundments, water hazards on golf courses, to full-scale development of water-based recreational impoundments, incidental contact (fishing and boating) and full body contact (swimming and wading). As with any form of reuse, the development of recreational and environmental water reuse projects will be a function of a water demand coupled with a cost-effective source of suitable quality reclaimed water. In California, approximately 10 percent (47.6 mgd) (2080 l/s) of the total reclaimed water use within the state was associated with recreational and environmental reuse in 2000 (Leverenz et al., 2002). In Florida, approximately 6 percent (35 mgd or 1530 l/s) of the reclaimed water currently produced is being used for environmental enhancements, all for wetland enhancement and restoration (Florida Department of Environmental Protection, 2002). In Florida, from 1986 to 2001, there was a 53 percent increase (18.5 mgd to 35 mgd or 810 l/s to 1530 l/s) in the reuse flow used for environmental enhancements (wetland enhancement and restoration). Two examples of large-scale environmental and recreational reuse projects are the City of West Palm Beach, Florida, wetlands-based water reclamation project and the Eastern Municipal Water District multipurpose constructed wetlands in Riverside County, California. Other applications of environmental and recreational water reuse include creation of natural and man-made wetlands, recreational and aesthetic impoundments and stream augmentation. The objectives of these reuse projects are typically to create an environment in which wildlife can thrive and develop an area of enhanced recreational or aesthetic value to the community through the use of reclaimed water. Other benefits of environmental reuse include decreasing wastewater discharges and reducing and preventing pollution. Recycled water can also be used to create or enhance wetlands and riparian habitats.

6. Economic considerations

One the major aspects of water reuse is the socio economic impacts assessment of implementation of such resources. Wastewater can decrease impacts of water shortage in arid and semi-arid regions of the world and promote means of sustainable development in the world. However, this will be highly dependent of environmentally sound implementation and management for reuse systems. Poor planning and management could leave significant damages on health and environment by contaminating valuable drinking water supplies and bring unwanted socio economic losses. Economic sustainability and public reception depend on the usage of reclaimed water. Most researches and surveys (Angelakis et al., 2001; Mantovani et al., 2001), have concluded that the best practices are those that substitute reclaimed water in lieu of potable water for use in irrigation, environmental restoration, cleaning, toilet flushing, and industrial uses. The main benefits of using reclaimed water in these situations are conservation of water resources and pollution reduction. Treating and reusing wastewater is economically reasonable in terms of increasing the water availability and the benefits of saving the environment from discharge of wastewater into other systems and controlling the spread of contamination into water and soil. Demand for municipal, industrial and agriculture is on the rise and are expected to reach 37, 23, and 340 bcm; respectively. Provided the low consumptive use of the municipal and industrial sectors, most of the appropriated water can be recovered (Kretschmer et al., 2003). In the agricultural sector, the large size of withdrawals encourages the collection and reuse of irrigation water. Wastewater is already in use around the world. In China, Chile and Mexico, extensive agriculture lands around are irrigated by wastewater (Sadik et al., 1994; Xie et al., 1993). Arab regions have also practicing wastewater reuse. About 7 bcm of wastewater was reused in 1996 out of 191 bcm the total withdrawal that year; this implies less than 4% recovery. Reused agriculture drainage was about 5 bcm out of 168 bcm withdrawn for that sector, less than 3% recovery and 2 bcm of municipal and industrial wastewater out of 23 bcm withdrawn, about 9% recovery (El-Ghamam, 1997). Wastewater is a source real economic activity involving local and federal government along with private industries. Various entities invest in getting rid of it or suffer the environmental damage. Either practice has a pervasive impact on public health and the sustainability of development. If wastewater is properly treated and reused, solves two major of saving local and regional environment and resolving water shortage. Over all the economic viability of water reuse has to be studied individually and the required treatment and cost efficiency, depend on type of pollutants, concentration and type of reuse.

7. Public health concerns and acceptance

The major emphasis of wastewater reclamation and reuse has been on non-drinking applications so far, such as agricultural and landscape irrigation, industrial cooling and in-building applications, such as toilet flushing, in large commercial buildings. Indirect or direct potable reuse raises more public concern because of real or perceived perception of aesthetics and long-term health concerns. Regardless, the value of water reuse is weighed within a context of larger public issues of necessity and opportunity and will not be implemented until two major problems of public health concerns and public acceptance is resolved. Each of these problems involves various issues from scientific concerns to human psychology. In the case of public health concerns, which are extremely viable concerns, presence of pathogenic organism and inorganic micro pollutants should be carefully examined for their short and long term impacts. Pathogens could impose serious threat to human health. They are found in water as bacteria, protozoa, helminthes and ruses which some of them can be easily detected and removed (Dishman et al., 1989). However, others are more difficult to detect and removed and there are not enough studies to assign a safe concentration limits to them. Furthermore, the risk of viral infections and waterborne diseases in general is still an unresolved issue. The inorganic pollutants of concerns in water reuse are nitrates, other nitrogen compounds and heavy metals which are easy to detect and remove. Organic micro pollutants also represent a large problem in direct potable reuse mainly because of lack of sufficient information on the health significance of many the known or suspected carcinogenic, mutagenic, allergenic, teratogenic organic compounds found in water (Crook, 1985). It is also necessary to mention that there are thousands of organic compounds in water that are awaiting discovery (Golden, 1984; Dishman et. al., 1989). The second problem that the potable use of reclaimed water has to facing is public acceptance. This a major obstacle to reuse and it roots in educational and psychological barriers which have to overcome in order to obtain public support. Numerous researches have highlighted the fact that public is not welcoming in this regard and most of the polls revealed major opposition to direct potable use (Gallup, 1973; Kasperson et al., 1974; Carley, 1985). The general feeling about use of wastewater for drinking purposes is negative, regardless of the degree of treatment and these feelings embody the psychological factors in the public's rejection of direct potable use of reclaimed water.

8. Conclusions

The world's population is on the rise and is expected to increase dramatically between now and the year 2020 (United Nations, 2006). This growth will put more pressure on our already scarce and damaged water resources. Communities around the world will be faced with an increased level of wastewater production with no use. Water reclamation and reuse can offer significant help for conserving and extending available water supplies. Water reuse may also present communities with an alternate wastewater disposal method as well as providing pollution abatement by diverting effluent discharge away from sensitive surface waters. However, water reuse has its own advantages and disadvantages which have been summarized in Table 6.

Advantages	Disadvantages
This technology reduces the demands on drinkable sources of freshwater.	If implemented on a large scale, revenues to water supply and wastewater utilities may fall as discharge of wastewaters is reduced.
It may reduce the need for large wastewater treatment systems, if significant portions of the waste stream are reused or recycled.	Reuse of wastewater may be seasonal in nature, resulting in the overloading of treatment and disposal facilities during the rainy season.
The technology may diminish the volume of wastewater discharged, resulting in a beneficial impact on the aquatic environment.	Application of untreated wastewater as irrigation water or as injected recharge water may result in groundwater contamination.
Capital costs are low to medium for most systems and are recoverable in a very short time; this excludes systems designed for direct reuse of sewage water.	Health problems, such as water-borne diseases and skin irritations, may occur in people coming into direct contact with reused wastewater
Operation and maintenance are relatively simple except in direct reuse systems where more extensive technology and quality control are required	In some cases, reuse of wastewater is not economically feasible because of the requirement for an additional distribution system.
Provision of nutrient-rich wastewaters can increase agricultural production in water-poor areas.	Gases, such as sulfuric acid, produced during the treatment process can result in chronic health problems.

Table 6. Advantages and disadvantages of water reuse

9. References

Angelakis, A., Thairs, T. and Lazarova, V., 2001, Water Reuse in EU Countries: Necessity of Establishing EU-Guidelines, State of the Art Review, Report of the EUREAU Water Reuse Group EU2-01-26, 52p.

Asano, T., and Levine, A.D., 1996, Wastewater reclamation, recycling and reuse: past, present, and future, Water Science and Technology, 33(10-11), 1-14.

Asano, T., 1998, Wastewater Reclamation and Reuse, Water Quality Management Library, Vol. 10, Technomic Publishing Company, Lancaster, Pennsylvania.

Asano, T., 2004, Water and Wastewater Reuse, An Environmentally Sound Approach for Sustainable Urban Water Management (http://www.unep.or.jp/)

Ayers, R.S., and Westcott, D.W., 1985, Water Quality for Agriculture, Food and Agricultural Organization of the United Nations, FAO Irrigation and Drainage, Paper 29, Rome, Italy.

Carley, R. L., 1985, Wastwater reuse and Public Opinion, Journal of the American Water Works Association, 77(7), 72.

Crook, J., 1985, Water Reuse in California, Journal of the American Water Works Association, 77(7), 61.

Dishman, M., Sherrard, J., and Rebhun, M., 1989, Gaining Support for Direct Potable Water Reuse, Journal of Professional Issues in Engineering, 115 (2), 154-161.

El-Ghamam, A. R. I. H., 1997, The Future of Agriculture and Food Production in Saudi Arabia: A Briefing, Country Report.

Eslamian, S., Tarkesh-Isfahani, S. and Malekpour, I., 2010, Investigating heavy metals concentration of a wastewater treatment plant for agricultural and landscape reuses, Dryland Hydrology: Global Challenges Local Solutions, September 1-4, Westin La Paloma,Tucson, AZ, USA.

Eslamian, S., and Tarkesh-Isfahani, S., 2010a, Evaluating the urban reclaimed water for industrial reuses in North Isfahan, Iran, The 4th International Symposium on Water Resources and Sustainable Development, Algiers, Algeria.

Eslamian, S. and Tarkesh-Isfahani, S., 2010b, Evaluating the most efficient irrigation systems in wastewater reuse, Pakistan Agriculture: Challenges and Opportunities, Kashmir, Pakistan.

Florida Department of Environmental Protection (2002), (dep.state.fl.us/water/ wetlands/).

Food and Agriculture Organization (http://www.fao.org/corp/statistics/en/).

Gallup, G.J. ,1973, Water Quality and Public Opinion, Journal of the American Water Works Association, 65(8), 513.

Johnson, W.D., 1998, Innovative Augmentation of a Community's Water Supply – The St. Petersburg, Florida Experience, Proceedings of the Water Environment Federation, 71st Annual Conference and Exposition, October 3-7, Orlando, Florida.

Kasperson, R. E. et al., 1974, Community adoption water reuse system in the United States, U. S. Office of Water Resources Research.

Kretzschmar, R., 2002, Best Management Practices for Florida Marinas, Florida Department of Environmental Quality, Florida.

Kretschmer, N., Ribbe, L., and Gaese, H., 2000, Wastewater Reuse for Agriculture, Technology Resource Management and Development, Scientific Contributions for Sustainable Development, Special Issue, Water Management, Vol. 2, Cologne.

Lazarova, V., 2000, Wastewater Disinfection: Assessment of the Available Technologies for Water Reclamation, in Water Conservation, Vol. 3: Water Management, Purification and Conservation in Arid Climates, Goosen, M.F.A. and Shayya, W.H., eds., Technomic Publishing Co. Inc., 171.

Lazarova, V., 2001, Role of Water Reuse in Enhancing Integrated Water Resource Management, Final Report of the EU Project CatchWater, EU Commission.

Leverenz, H., Tchobanoglous, G. and Darby, J. L., 2002, Review of Technologies for the Onsite Treatment of Wastewater in California, Report No. 02-2, Prepared for California State Water Resources Control Board, Center for Environmental and Water Resources Engineering, UC Davis, Davis, CA.

Lindsey, P.R., Waters, K., Fell, G. and Setka Harivandi, A., 1996, The Design and Construction of a Demonstration/Research Garden Comparing the Impact of Recycled vs. Potable Irrigation Water on Landscape Plants, Soils and Irrigation Components, Water Reuse Conference Proceeding, American Water Works Association, Denver, Colorado.

Mantovani, P., Asano, T., Chang, A. and Okun, D.A., 2001, Management Practices for Nonpotable Water Reuse, WERF, Project Report 97-IRM-6.

Metcalf and Eddy, 2003, Wastewater Engineering: Treatment and Reuse, 4rd ed., Mc-Graw Hill Inc., New York.

Metcalf and Eddy, 2007, Water Reuse: Issues, Technologies, and Applications, McGraw-Hill, New York.

Sadik, A., and Shawki B., 1994, The Water Problems of the Arab World: Management of Scarce Resources." In Rogers, Peter, and Peter Lydon (eds.), Water in the Arab World: Perspectives and Prognoses, Ch 1, 1-37, the American University in Cairo Press, Cairo, Egypt.

United Nations, 2006, World Population, the World at Six Billion and World Prospects: The 2006 Revision, Department of Social and Economic Affairs, Population Division, New York.

USEPA, 1981, Process Design Manual: Land Treatment of Municipal Wastewater, EPA Center for Environmental Research Information, EPA 625/1-81-013, Cincinnati, Ohio.

USEPA, 1992, Guidelines for Water Reuse, United States Environmental Protection Agency and & USAID (United States Agency for International Development), Cincinnati, OH.

USEPA, 2003, National Primary Drinking Water Standards, United States Environmental Protection Agency, EPA 816-F-03-016, Washington, D.C.

USEPA, 2004, Guidelines for Water Reuse, United States Environmental Protection Agency, EPA 645-R-04-108, Washington, D.C.

Valentina L. and Akica B., 2005, Water Reuse for Irrigation: Agriculture, Landscapes, and Turfgrass, CRC Press.

Virginia Cooperative Extension Materials (http://pubs.ext.vt.edu/452/452-014/452-014.html).

WHO Website (http://www.who.int/bulletin/archives/78%289%291104.pdf)

World Bank 1995, From Scarcity to Security: Averting a Water Crisis in the Middle East and North Africa, Washington D.C.

World Health Organization (WHO), 1989, Health Guideline for the Use of Wastewater in Agriculture and Aquaculture, WHO Technical Report Series, No. 778, Geneva, Switzerland.

World Health Organization (WHO), 2001, Water Quality: Guidelines, Standards and Health Assessment of Risk and Risk Management for Water-related Infectious Diseases, Available from http://www.who.int/water_sanitation_health/dwq/whoiwa/en/

Xie, M., Kuffner, U. and Le Moignee, G., 1993, Using Water Efficiently: Technological Options, World Bank Technical Paper No. 205., Washington D.C.

Evaluation of the Removal of Chlorine, THM and Natural Organic Matter from Drinking Water Using Microfiltration Membranes and Activated Carbon in a Gravitational System

Flávia Vieira da Silva-Medeiros[1], Flávia Sayuri Arakawa[1],
Gilselaine Afonso Lovato[1], Célia Regina Granhen Tavares[1],
Maria Teresa Pessoa Sousa de Amorim[2], Miria Hespanhol Miranda Reis[3]
and Rosângela Bergamasco[1]

[1]Universidade Estadual de Maringá, Maringá, PR, Brazil
[2]Universidade do Minho, Campus Azurém, Guimarães,
[3]Universidade Federal de Uberlândia, Uberlândia, MG,
[1,3]Brazil
[2]Portugal

1. Introduction

Due to its low cost, stability, and effectiveness, adding chlorine to drinking water is one of the most common treatments to ensure its bacteriological quality (Al-Jasser, 2007). Chlorine inactivates various types of micro-organisms and its residual properties help to prevent micro-organism regrowth during water flow in pipes (Connell, 1997).

Beyond off-flavors development due to chlorination by-products, chlorine flavor by itself constitutes one of the major complaints against tap water. In 1996, chlorine taste was the third most reported taste default of tap water in the US (Suffet et al., 1996). Due to the unpleasant taste of tap water, consumers may prefer bottled water as drinking water, even if bottled drinking water consumption would be associated with a higher economic and ecological cost. (Rodriguez et al., 2004) showed that the perception of tap water quality is closely related to the residual chlorine level: people living near a treatment plant who may receive a higher chlorine level in their tap water were generally less satisfied by tap water quality and perceived more risks associated with it than people living far from the plant. It was reported that, in the US, bottled water drinkers have three main categories for decisions: safety of water; healthfulness of the water; and taste of the water (Mackey et al., 2004). Consumers supplied with tap water containing a residual chlorine level greater than 0.24 mg/L Cl_2 were less satisfied with tap water when compared to consumers receiving lower concentrations (Rodriguez et al., 2004). This value almost coincides with the free chlorine residual (0.2 mg/L) that must be maintained in the distribution system, reducing the likelihood of further contamination (Clark & Coyle, 1990). When taken together, this studies underline that the consumers would reject tap water in safe conditions due the chlorine flavor.

The residual chlorine could be removed using an activated carbon filter at the time of consumption. Activated carbon has already been used to remove chlorine excess of tap water used in food industries (Jaguaribe et al., 2005). Due to its well-developed pore structure, activated carbon in either powder or granules has an excellent adsorbent capacity.

Beyond this concern with chlorine flavor, several studies reported that chlorination of organic matter in fresh water resulted in the formation of disinfection by-products (DBPs) (Richardson, 2003; Rook, 1976), especially trihalomethanes (THMs), which remain a human health concern. Trihalomethanes (THMs) are a group of volatile organic compounds (VOCs) classified as disinfection by-products (DBPs). They were first identified by (Rook, 1976) and are formed during the chlorination of water, when chlorine reacts with naturally occurring organic matter: mainly humic and fulvic acids. Their general formula is CHX_3, where X may be any halogen or a combination of halogens. However, generally speaking this term is used to refer only to those compounds containing either chlorine or bromide, because these are the ones most commonly detected in chlorinated water (chloroform, bromodichloromethane, dibromochloromethane and bromoform). Brominated trihalomethanes are formed when hypochlorous acid oxidizes bromide ion present in water to form hypobromous acid, which subsequently reacts with organic materials to form these compounds (Pavon et al., 2008; Richardson, 2003). Iodinated THMs have been identified in chlorinated drinking water; however, they are not widely measured and are not regulated, even though iodinated compounds may be more toxic than brominated and chlorinated compounds (Richardson, 2003).

The International Agency for Research on Cancer (IARC) has classified chloroform and bromodichloromethane as possible carcinogens for humans (Group 2B) based on limited evidence of carcinogenicity in humans but sufficient evidence of carcinogenicity in experimental animals. Dibromochloromethane and bromoform belong to Group 3 (not classifiable as regards their carcinogenicity to humans), based on inadequate carcinogenicity in humans and inadequate or limited carcinogenicity in experimental animals (Pavon et al., 2008; WHO, 2006). In the case of THMs, approximately equal contributions to total exposure come from four sources: the ingestion of drinking water, inhalation of indoor air, inhalation and dermal exposure during showering or bathing, and the ingestion of foods (WHO, 2006). Trihalomethanes have been detected in different aqueous matrixes: tap water, swimming pool water, distilled water, ultrapure water and even in water that has not been subjected to chlorination processes, such as ground water, mineral water, snow, rain water, sea, and river water. However, the concentrations of these compounds in unchlorinated water tend to be much lower than those usually found in tap water. The presence of these levels of THMs may be due to several causes. In cases in which the chloroform > bromodichloromethane > dibromochloromethane > bromoform pattern is conserved, the THMs are likely to have originated from the infiltration of chlorinated water. The sources of chlorinated water to ground water may include the irrigation of lawns, gardens and parks; leaking drinking water distribution and sewer pipes, and industrial spills, among others.

Adsorption in carbonaceous materials, as carbon nanotubes and carbon spheres was reported as an effective technique in removing THM from water (Lu et al., 2005; Morawski et al., 2000). However, the authors did not consider the application of simpler and cheaper technology, as granular activated carbon.

Evaluation of the Removal of Chlorine, THM and Natural Organic Matter from Drinking Water Using Microfiltration
Membranes and Activated Carbon in a Gravitational System

19

(Amy et al., 1990) found that the majority of THM formation potential is presented by small to medium organic compounds with a specific ultraviolet absorbance (SUVA) values less than 3.0. The SUVA parameter represents the ratio UV_{254}/DOC and constitutes an indicator of carbon aromacity in water (Uyak et al., 2008).

Numerous research studies involving microfiltration (MF) and ultrafiltration (UF) of surface waters in rivers and lakes have proved that NOM is the main source of fouling during membrane processes (Uyak et al., 2008). The application of activated carbon (AC) in conjunction with MF/UF membranes is a promising technology for the removal of organic compounds in drinking water treatment, which incorporates the adsorption capabilities of activated carbon and the microorganism and particle removal ability of the MF/UF membranes (Tsujimoto et al., 1998; Yuasa, 1998). (Ravanchi & Kargari, 2009) highlighted the importance to propose innovative integrated membrane processes in order to became this process more commercial.

Moreover, nanofiltration membranes also showed potential in removing THMs from drinking water (Uyak et al., 2008). The application of nanofiltration and reverse osmosis in drinking water treatment is increasing in developed countries (Clever et al., 2000). However, their application in developing countries is still limited due the high costs of the membranes and pumping (Glucina et al., 2000).

Alternatively, microfiltration membrane processes can be designed to operate with gravity as the driving force, in simple systems that could be operated by non-trained people. Besides this, gravitational systems present the following advantages: energy saving, once pumps are not necessary; simpler tubing is required because it operates at low pressures. However, this kind of researched is scarce in the scientific literature (Peter-Varbanets et al., 2009).

In this way, the objective of this publication is to show the efficiency of a microfiltration membrane working alone and associated with activated carbon for removing THM and their precursors from tap water in a gravitational module.

2. Process configuration

2.1 Materials

Acetate cellulose microfiltration membrane (pore diameter=3.0 μm) was purchased from ADVANTEC/MFS (Japan). Commercial Granular Activated Carbon (GAC) 20x40 mesh made from coconut was supplied by BAHIACARBON (Brazil). Table 1 presents the textural characteristics of this activated carbon.

Surface area (m² g⁻¹)	Micropore area (m² g⁻¹)	Total pore volume (cm³ g⁻¹)	Average pore diameter (Å)
715.46	677.77	0.3856	20.04

Table 1. Textural characteristics of the activated carbon used in this work.

Filtration tests were carried out with the membrane working alone and with GAC as a pretreatment. The gravitational module used in this study (Fig. 1) operated exclusively with gravity as driven force, which produced a pressure of approximately 0.36 bar, due the position of the lung tank in relation to the filtration module.

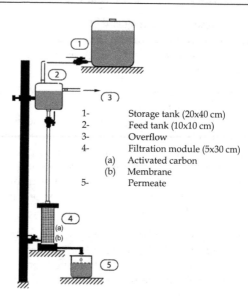

1-	Storage tank (20x40 cm)
2-	Feed tank (10x10 cm)
3-	Overflow
4-	Filtration module (5x30 cm)
(a)	Activated carbon
(b)	Membrane
5-	Permeate

Fig. 1. Scheme of the gravitational filtration system.

The filtration module was composed of two parts: a flat-sheet membrane cell made of stainless steel and an acrylic cartridge to support the activated carbon. The effective membrane was 103 cm² and the activated carbon bed was 20 cm high. In each experiment, a new membrane was used, rinsed with ultrapure water and compacted by filtering ultrapure water during 2 h before starting a filtration test.

2.2 Raw water

The raw water used in this study was prepared using finished water of Pirapó River water (PRW), located in Maringá-Brazil. Quality parameters of this water are summarized in Table 2.

Parameters	Value
DOC (mg/L)	1.89
UV$_{254}$ (L/cm)	0.032
SUVA (L/mg m)	1.69
Color (UC)	1.0
Turbidity (NTU)	0.67
pH	7.26

Table 2. Finished Pirapó River water (PRW) quality parameters

Two different types of water were prepared, and the details of each preparation are given below:

a.   Chlorinated water

Chlorinated water samples were obtained by adding sodium hypoclorite (NaOCl ~2.5%, commercial grade) to tap water (PRW) in order to adjust the chlorine concentration to 2.0

mg/L, the limit regulated by Brazilian Government. The removal of chlorine, DOC, UV_{254} and SUVA were evaluated using the chlorinated water.

b.   Water containing THMs

Trihalometanes mix at 2000 µg/L, produced by Supelco was diluted to 10 mg/L, and this final solution was used to adjust the THMs concentration in tap water (PRW), in order to maintain the total THMs concentration around 120 µg/L, slightly higher than the limit regulated by the Brazilian Government of 100 µg/L.

2.3 Analytical methods

Dissolved organic carbon (DOC) measurements were performed with a HACH DR2010 spectrophotometer, using the low range direct method. Besides, UV_{254} absorbance measurements were conducted in accordance with Standard Methods by a HACH DR 2010 UV spectrophotometer at a wavelength of 254 nm. Water samples for DOC and UV_{254} measurements were first filtered through a pre-washed 0.45 µm membrane filter to remove turbidity, which can interfere in these measurements, and distilled ultra filtered water was used as the background correction on the spectrophotometer.

Chlorine measurements were performed with a HACH DR/2010 spectrophotometer, using the DPD method for free chlorine determination, according to the Standard Methods for the Examination of Water and Wastewater (APHA, 1998). The samples were collected and immediately analyzed, using the HACH DPD free chlorine reagent.

THM concentrations were determined by Gas chromatography–mass spectrometry (GC-MS). For determination of THM was used in a chromatograph GC-MS with mass detector DSQ II with autosampler from Thermo triplus Red Space. The recovery of THMs was optimized using excess of KCl (4g per 12 mL of sample), according to Caro et al. (2007).

2.4 Flux and resistance measurements

Firstly, the permeated flux of deionized water was measured during 120 min in a clean membrane using new and clean membrane.

After the tests with raw water, final flux with deionized water was determined, also during 120 min. At the end of this measurement, the membrane was removed from the experimental module (Fig. 1) and mechanically cleaned in order to remove the cake formed on the membrane surface. After that, a new flux measurement with deionized water was done.

Resistances due to different fouling mechanisms were determined in order to investigate the fouling behavior. Resistances were calculated following the resistance-in-series model adapted from (A.I. Schafer, 2005) as presented in equation (1):

$$J = \frac{\Delta P}{\eta(R_m + R_p + R_c)} \tag{1}$$

where J is permeate flux [kg m^{-2} s^{-1}], ΔP is trans-membrane pressure [kg m^{-1} s^{-2}], η is dynamic viscosity [kg m^{-1} s^{-1}], and R denotes a resistance [m^2 kg^{-1}]: R_m is membrane hydraulic resistance, R_p is resistance due to pore blocking, and R_c is resistance due to cake formation.

Each resistance was experimentally measured in the gravitational system with and without GAC pretreatment. Membrane hydraulic resistance, R_m, was determined measuring the flux of deionized water through a clean membrane sheet. In this case, the others resistances are equals to zero.

The sum of resistances due to pore blocking and cake formation ($R_p + R_c$) was determined measuring the flux of deionized water with the fouled membrane, i.e. with the membrane that was used for raw water filtrations without clean procedures.

After mechanical cleaning, the flux measurements was carried out in order to obtain the resistance due to pore blocking, R_p.

Membrane fouling percentage (%F) was calculated according to equation (2), as proposed by (Balakrishnan et al., 2001). This percentage represents the drop in deionized water flux after the tests of filtration with raw water.

$$\%F = \frac{(J_i - J_f)}{J_i} \times 100 \tag{2}$$

where %F is membrane fouling percentage, and J_i and J_f are deionized water fluxes in clean and fouled membranes, respectively.

3. Process performance

3.1 Flux measurements with raw water

In general, the application of activated carbon as pretreatment to micro/ultrafiltration membranes affects the permeate flux of membranes and improves the removal of several parameters that may cause fouling (Fabris et al., 2007; Kim & Gai, 2008). However, there is a lack of information in the literature about this application in gravitational systems.

The permeate fluxes of the membrane operating alone and with activated carbon as pretreatment were determined during the evaluation of chlorine and THM removals from tap water in the gravitational module. Considering that the observed values of permeate fluxes were almost equals in both assays, Fig. 2 illustrates the average values of both assays during the 480 min of operation. The membrane working alone is referred as M3 and the membrane working with activated carbon as pretreatment is referred as M3C.

The permeate flux increased when using activated carbon as pretreatment. In the first 50 minutes of operation, it is observed a permeate flux between 2000 and 400 kg h^{-1} m^{-2} to the membrane M3 (without GAC pretreatment) and between 3000 and 700 kg h^{-1} m^{-2} to the membrane M3C (with GAC pretreatment). After 50 min of operation, the permeate flux was around 200 and 300 kg h^{-1} m^{-2} to the membranes M3 and M3C, respectively. However, in the last 100 minutes of operation, it is observed that the permeate flux was around 100 kg h^{-1} m^{-2} to the membrane M3, and around 200 kg h^{-1} m^{-2} to the membrane M3C. The better performance of the membrane M3C in comparison to the membrane M3 is probably due to the adsorption of organic matter in the activated carbon surface, which mitigates fouling effects (Kim & Gai, 2008).

Evaluation of the Removal of Chlorine, THM and Natural Organic Matter from Drinking Water Using Microfiltration
Membranes and Activated Carbon in a Gravitational System

23

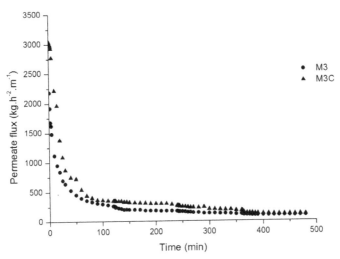

Fig. 2. Permeate fluxes of tap water to the membranes with and without GAC as pretreatment (M3C and M3, respectively).

Fig. 3 presents the results of total resistance against time to the 3.0 μm membrane working alone and associated with activated carbon, calculated using Equation 1. Total resistance is referred as the sum of R_m, R_p, and R_c.

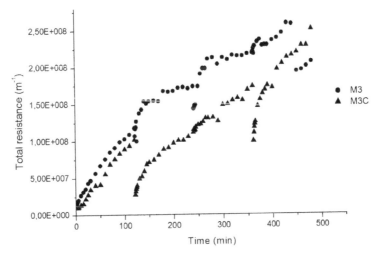

Fig. 3. Total resistance in tap water assays.

It is possible to notice from Fig. 3 that the application of GAC as pretreatment reduced the total resistance of the membrane during the operation with tap water. This decrease in the total resistance is probable the main cause of the flux increase observed during the filtration with GAC as pretreatment.

3.2 DOC and UV$_{254}$ rejection capacity

Figs. 4 and 5 present the reduction in DOC and UV$_{254}$ compounds, respectively, after microfiltrations of tap water with and without GAC pretreatment.

DOC rejection by the process with the membrane working alone (M3) was between 50 and 70% (Fig. 4) during the complete experiment. The range presented by UV$_{254}$ rejection was between 35 and 55% (Fig. 5). These results are comparable to previous results found to nanofiltration membranes (Alborzfar et al., 1998), with the advantage of using microfiltration and a gravitational system, much more cheaper than a nanofiltration system using pumps.

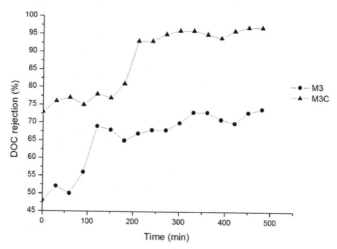

Fig. 4. DOC rejection in tap water after microfiltrations with the membrane working alone (M3) and with GAC as pretreatment (M3C).

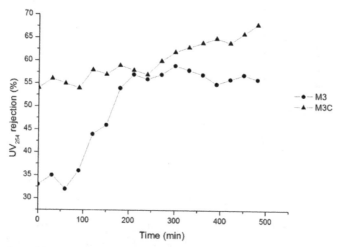

Fig. 5. UV$_{254}$ rejection of in tap water after microfiltrations with the membrane working alone (M3) and with GAC as pretreatment (M3C).

Evaluation of the Removal of Chlorine, THM and Natural Organic Matter from Drinking Water Using Microfiltration
Membranes and Activated Carbon in a Gravitational System

25

Besides this, M3C presented a removal of 70 to 95% to DOC (Fig. 4) and a removal of 55 to 70% to UV254 (Fig. 5). The application of activated carbon as pretreatment significantly increased the removal of dissolved organic matter, as reported before (Bao et al., 1999).

3.3 Chlorine rejection capacity of the evaluated systems

Fig. 6 shows the rejection performances of free chlorine, in the initial concentration of 2 mg/L, by the two studied systems, M3 and M3C, in the gravitational module.

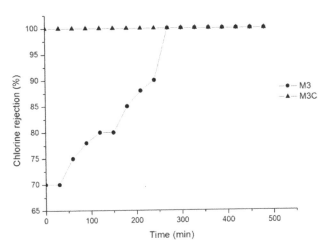

Fig. 6. Chlorine rejection in tap water after microfiltrations with the membrane working alone (M3) and with GAC as pretreatment (M3C).

Chlorine rejection increased gradually to the membrane working alone (M3), from 70% in the first minutes of operation to 100% after 270 min. It probably happened due to a formation of a cake fouling layer on the surface of the membrane, which improved the removal rate of contaminants (Kim & Gai, 2008).

In contrast to M3, the chlorine rejection of the system using activated carbon as pretreatment (M3C) was of 100% during the complete experiment. (Jaguaribe et al., 2005) reported that, due to its well-developed pore structure, coconut shells activated carbon (working alone) reduces around 40% of free chlorine presented in water.

3.4 THM rejection capacity of the evaluated systems

Total THM and its four compounds were chosen to evaluate the change in permeate concentration over the course of filtration tests using the membrane working alone (M3) and using activated carbon as pretreatment to the membrane (M3C). Fig. 7 illustrates the total THM rejection capacity of M3 and M3C processes.

Total THM rejection of M3 process was approximately equals to 74% during the complete experiment, which could be considered a suitable result, since similar results were obtained when using nanofiltration membranes at higher pressures (Uyak et al., 2008), and considering the fact that M3 is a microfiltration membrane operating only by gravity.

Moreover, M3C process presented a removal of 93 to 99% of THM compounds due to the considerable adsorption capacity of activated carbon toward various pollutants, especially THM, as reported before (Razvigorova et al., 1998).

Fig. 8 shows the rejection performances of three species of THM by the two studied systems, M3 and M3C during the 480 min of filtration. The THM species are chloroform (CFM), bromodichloromethane (BDCM) and dibromochloromethane (DBCM). Since bromoform (BFM) was not detected in the tested water, hence, three species were taken in account.

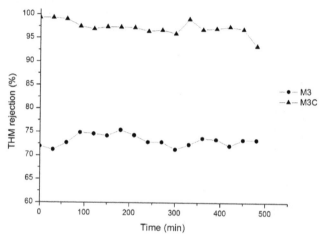

Fig. 7. Total THM rejection in tap water after microfiltrations with the membrane working alone (M3) and with GAC as pretreatment (M3C).

The selected microfiltration membrane working alone (M3) was effective in removing the THM compounds during the 480 min of filtration. It was depicted in Fig. 8 that the rejection efficiencies of CFM were found to be around 65 and 70% during the 480 min of operation. Moreover, the rejection efficiencies of BDCM and DBCM were practically 100%. As observed in previous similar studies (Uyak et al., 2008), the removal efficiency of M3 process was increased with increasing the molecular weight of THM species. As bromine atom replace the chlorine atoms, greatly increasing the molecular weight, resulted in higher removal efficiency. The higher removal efficiency of BDCM and DBCM was attributed to higher molecular weight and brominating characteristics (Uyak et al., 2008).

For the system using activated carbon as pretreatment to the microfiltration membrane (M3C), Fig. 8 illustrates the removal of CFM to be around 99% in the first minutes of operation and around 94% after 100 minutes. The rejection rates of BDCM and DBCM were also around 100%. The application of activated carbon as pretreatment enhanced the THM compounds removal, and this fact is especially notable to CFM, once the rejection rates of BDCM and DBCM were considerably high also when the gravitational module operated only with the microfiltration membrane (M3). It was found in the literature that magnitude of adsorption of these chlorinated compounds was in the following order: BDCM>DBCM>CFM. The molecules containing bromine were adsorbed with highest efficiency compared to the remainder of lower radius (chlorine). It means that not only the size of pores determines the adsorption but also surface chemical character may influence it (Razvigorova et al., 1998).

Evaluation of the Removal of Chlorine, THM and Natural Organic Matter from Drinking Water Using Microfiltration
Membranes and Activated Carbon in a Gravitational System

27

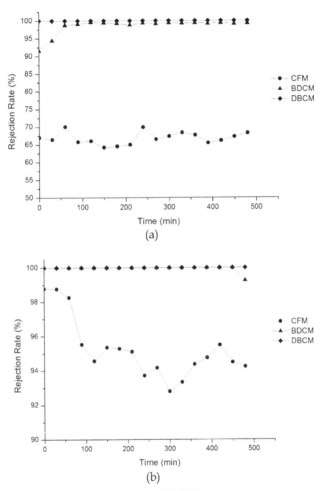

Fig. 8. THM compounds rejection of (a) M3 and (b) M3C processes.

3.5 Flux and resistance measurements

Fig. 9 illustrates the permeate flux values measured with deionized water to the membrane with and without the GAC pretreatment before and after the filtrations with tap water.

The association with activated carbon significantly improved the performance of the 3.0 μm membrane in relation to the initial and final flux values. The initial flux to the carbon+membrane system was approximately 2700 kg m^{-2} h^{-1}, while the membrane working alone presented initial flux around 2000 kg m^{-2} h^{-1}.

Considering final flux values, the membrane associated with GAC achieved values around 222 kg m^{-2} h^{-1} and the membrane working alone presented values not higher than 60 kg m^{-2} h^{-1}. These results could be related to the adsorption of organic matter by the activated carbon (Choo & Kang, 2010).

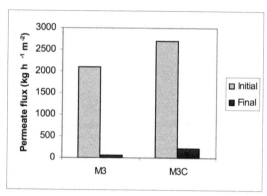

Fig. 9. Initial and final flux values of deionized water to the membrane with and without GAC as pretreatment (M3C and M3, respectively).

Table 3 shows the results of resistances and fouling percentage to the 3.0 μm membrane working alone and associated with activated carbon.

Process	R_m x10^8 (m^{-1})	R_c x10^8 (m^{-1})	R_p x10^8 (m^{-1})	R_t x10^8 (m^{-1})	%F
M3	0.141	1.31	4.17	5.621	97.3
M3C	-	1.13	0.156	1.395	91.9

Table 3. Results of resistance and fouling percentage.

It was observed a considerable decrease in the R_c, R_p, and R_t values when GAC was applied as pretreatment, indicating that the activated carbon reduced the fouling due the cake formation (R_c) and pore blocking (R_p). These results are in accordance with the previous reports (Kim & Gai, 2008). The hydraulic membrane resistance (R_m) was not determined to the membrane working with GAC as pretreatment because it is considered an inherent property of the membrane.

It was also observed a decrease in the fouling percentage with the application of activated carbon as pretreatment to the 3.0 μm membrane, which could also be expected, according to previous reports in the literature (Choo & Kang, 2010).

4. Conclusions

Microfiltration and its association with activated carbon are technologies that have potential for use in drinking water treatment. The conclusions that can be drawn from the results of this experimental investigation are as follows:

- Experimental results show that the microfiltration membrane evaluated in this study was effective in removing DOC, UV$_{254}$, chlorine, CFM, BDCM, and DBCM compounds. Further, brominated THM compounds were removed more significantly than chlorinated THM ones. The higher removal efficiency of DBCM was attributed to higher molecular weight and brominating characteristics.
- The application of activated carbon as pretreatment increased the permeate flux and also increased the rejection efficiency of DOC, UV$_{254}$, chlorine, CFM, BDCM, and DBCM compounds. It was attributed to the chemical interaction that probably happens between activated carbon surface and the studied compounds.

Evaluation of the Removal of Chlorine, THM and Natural Organic Matter from Drinking Water Using Microfiltration
Membranes and Activated Carbon in a Gravitational System

29

- The system using activated carbon and microfiltration membrane (M3C) applied in a gravitational module is a promising alternative to improve drinking water quality, due to its efficiency, simplicity and low cost.

5. Acknowledgments

The authors acknowledge the financial support of the CNPq and the FAPEMIG.

6. References

Al-Jasser, A. O. (2007). Chlorine decay in drinking-water transmission and distribution systems: Pipe service age effect. *Water Research*, Vol. 41, No. 2, (Jan), pp. 387-396, ISSN 0043-1354.

Alborzfar, M.;Jonsson, G. & Gron, C. (1998). Removal of natural organic matter from two types of humic ground waters by nanofiltration. *Water Research*, Vol. 32, No. 10, (Oct), pp. 2983-2994, ISSN 0043-1354.

Amy, G. L.;Alleman, B. C. & Cluff, C. B. (1990). Removal of Dissolved Organic-Matter by Nanofiltration. *Journal of Environmental Engineering-Asce*, Vol. 116, No. 1, (Jan-Feb), pp. 200-205, ISSN 0733-9372.

APHA. American Public Health Association (1998). Standard Methods for the Examination of Water and Wastewater, 20th ed. Washington.

Balakrishnan, M.;Dua, M. & Khairnar, P. N. (2001). Significance of membrane type and feed stream in the ultrafiltration of sugarcane juice. *Separation Science and Technology*, Vol. 36, No. 4, pp. 619-637, ISSN 0149-6395.

Bao, M. L.;Griffini, O.;Santianni, D.;Barbieri, K.;Burrini, D. & Pantani, F. (1999). Removal of bromate ion from water using granular activated carbon. *Water Research*, Vol. 33, No. 13, (Sep), pp. 2959-2970, ISSN 0043-1354.

Caro J., Serrano A., Gallego M. (2007), Sensitive headspace gas chromatography-mass spectrometry determination of trihalomethanes in urine. *Journal of Chromatography B, Analytical Technologies in the Biomedical and Life Sciences*, Vol. 848, No. 2, (Nov 7), pp. 277-282, ISSN: 1570-0232.

Choo, K. H. & Kang, S. K. (2010). Why does a mineral oxide adsorbent control fouling better than powdered activated carbon in hybrid ultrafiltration water treatment? *Journal of Membrane Science*, Vol. 355, No. 1-2, (Jun 15), pp. 69-77, ISSN 0376-7388.

Clark, R. M. & Coyle, J. A. (1990). Measuring and Modeling Variations in Distribution-System Water-Quality. *Journal American Water Works Association*, Vol. 82, No. 8, (Aug), pp. 46-53, ISSN 0003-150X.

Clever, M.;Jordt, F.;Knauf, R.;Rabiger, N.;Rudebusch, M. & Hilker-Scheibel, R. (2000). Process water production from river water by ultrafiltration and reverse osmosis. *Desalination*, Vol. 131, No. 1-3, (Dec 20), pp. 325-336, ISSN 0011-9164.

Connell, G. F. (Ed.). 1997. *The Chlorination/Chloramination Handbook*. Denver.

Fabris, R.;Lee, E. K.;Chow, C. W. K.;Chen, V. & Drikas, M. (2007). Pre-treatments to reduce fouling of low pressure micro-filtration (MF) membranes. *Journal of Membrane Science*, Vol. 289, No. 1-2, (Feb 15), pp. 231-240, ISSN 0376-7388.

Glucina, K.;Alvarez, A. & Laine, J. M. (2000). Assessment of an integrated membrane system for surface water treatment. *Desalination*, Vol. 132, No. 1-3, (Dec 20), pp. 73-82, ISSN 0011-9164.

Jaguaribe, E. F.;Medeiros, L. L.;Barreto, M. C. S. & Araujo, L. P. (2005). The performance of activated carbons from sugarcane bagasse, babassu, and coconut shells in removing residual chlorine. *Brazilian Journal of Chemical Engineering*, Vol. 22, No. 1, (Jan-Mar), pp. 41-47, ISSN 0104-6632.

Kim, H. S. & Gai, X. J. (2008). The role of powdered activated carbon in enhancing the performance of membrane systems for water treatment. *Desalination*, Vol. 225, No. 1-3, (May 1), pp. 288-300, ISSN 0011-9164.

Lu, C. S.;Chung, Y. L. & Chang, K. F. (2005). Adsorption of trihalomethanes from water with carbon nanotubes. *Water Research*, Vol. 39, No. 6, (Mar), pp. 1183-1189, ISSN 0043-1354.

Mackey, E. D.;Baribeau, H.;Crozes, G. F.;Suffet, I. H. & Piriou, P. (2004). Public thresholds for chlorinous flavors in US tap water. *Water Science and Technology*, Vol. 49, No. 9, pp. 335-340, ISSN 0273-1223.

Morawski, A. W.;Kalenczuk, R. & Inagaki, M. (2000). Adsorption of trihalomethanes (THMs) onto carbon spheres. *Desalination*, Vol. 130, No. 2, (Nov 1), pp. 107-112, ISSN 0011-9164.

Pavon, J. L. P.;Martin, S. H.;Pinto, C. G. & Cordero, B. M. (2008). Determination of trihalomethanes in water samples: A review. *Analytica Chimica Acta*, Vol. 629, No. 1-2, (Nov 23), pp. 6-23, ISSN 0003-2670.

Peter-Varbanets, M.;Zurbrugg, C.;Swartz, C. & Pronk, W. (2009). Decentralized systems for potable water and the potential of membrane technology. *Water Research*, Vol. 43, No. 2, (Feb), pp. 245-265, ISSN 0043-1354.

Ravanchi, M. T. & Kargari, A. (2009). *New Advances in Membrane Technology*. InTech, ISBN 978-953-307-009-4, Vienna, Austria.

Razvigorova, M.;Budinova, T.;Petrov, N. & Minkova, V. (1998). Purification of water by activated carbons from apricot stones, lignites and anthracite. *Water Research*, Vol. 32, No. 7, pp. 2135-2139, ISSN 0043-1354.

Richardson, S. D. (2003). Disinfection by-products and other emerging contaminants in drinking water. *Trac-Trends in Analytical Chemistry*, Vol. 22, No. 10, (Nov), pp. 666-684, ISSN 0165-9936.

Rodriguez, M. J.;Turgeon, S.;Theriault, M. & Levallois, P. (2004). Perception of drinking water in the Quebec City region (Canada): the influence of water quality and consumer location in the distribution system. *Journal of Environmental Management*, Vol. 70, No. 4, (Apr), pp. 363-373, ISSN 0301-4797.

Rook, J. J. (1976). Haloforms in Drinking-Water. *Journal American Water Works Association*, Vol. 68, No. 3, pp. 168-172, ISSN 0003-150X.

Schafer, A.; Fane, A. & Waite, T. (2004). *Nanofiltration – Principles and Applications*, Elsevier, ISBN 978-1-85617-405-3, Oxford.

Suffet, I. H. M.;Corado, A.;Chou, D.;McGuire, M. J. & Butterworth, S. (1996). AWWA taste and odor survey. *Journal American Water Works Association*, Vol. 88, No. 4, (Apr), pp. 168-180, ISSN 0003-150X.

Tsujimoto, W.;Kimura, H.;Izu, T. & Irie, T. (1998). Membrane filtration and pre-treatment by GAC. *Desalination*, Vol. 119, No. 1-3, (Sep 20), pp. 323-326, ISSN 0011-9164.

Uyak, V.;Koyuncu, I.;Oktem, I.;Cakmakci, M. & Toroz, I. (2008). Removal of trihalomethanes from drinking water by nanofiltration membranes. *Journal of Hazardous Materials*, Vol. 152, No. 2, (Apr 1), pp. 789-794, ISSN 0304-3894.

WHO. 2006. Guidelines for Drinking-Water Quality. Geneva.

Yuasa, A. (1998). Drinking water production by coagulation-microfiltration and adsorption-ultrafiltration. *Water Science and Technology*, Vol. 37, No. 10, pp. 135-146, ISSN 0273-1223.

Application of Hybrid Process of Coagulation/ Flocculation and Membrane Filtration to Water Treatment

Rosângela Bergamasco[1], Angélica Marquetotti Salcedo Vieira[1],
Letícia Nishi[1], Álvaro Alberto de Araújo[2] and Gabriel Francisco da Silva[2]
[1]*Universidade Estadual de Maringá,*
[2]*Universidade Federal de Sergipe,*
Brazil

1. Introduction

Nowadays, the concern about contamination of aquatic environments has increased, especially when water is used for human consumption (Madrona et al., 2010). Contamination of water resources, especially in areas with inadequate sanitation and water supply, has become a risk factor for public health, with water playing a role as a vehicle for transmission of biological agents (viruses, bacteria, and parasites) as well as a source of contamination by chemicals (industrial effluents). Among the waterborne diseases, enteric diseases are most frequent. Approximately 19% of waterborne gastroenteritis outbreaks in the United States are attributed to protozoan parasites (Lindquist, 1999), particularly *Giardia* and *Cryptosporidium* species, due to their wide distribution in the environment, high incidence in the population, and resistance to conventional water treatment (Iacovski et al., 2004).

Giardia duodenalis is a flagellated protozoan parasite that causes giardiasis. The cysts are transmitted by fecal-oral route in a direct (person to person) or indirect way (contaminated water and food). In humans it can cause from self-limited to chronic enteritis, debilitating diarrhea with steatorrhea and weight loss (Rey, 2001). *Cryptosporidium* is a genus of protozoan parasites with several species that have been treated as a public health problem, especially after the advent of HIV/AIDS (Fayer, 2004). In immune depressed individuals it causes more serious infection; in some cases, the patient can die if untreated. It is considered an opportunistic parasite (Rey, 2001).

Despite regulations and control measures turning to be more and more stringent, outbreaks of waterborne *Cryptosporidium* spp. and *Giardia* spp. have been reported worldwide (United States Environmental Protection Agency [USEPA], 1996; Centers for Disease Control and Prevention [CDC], 2006). Therefore, the treatment applied to the collected water must ensure that it is free of pathogens and chemicals that pose health risks, when distributed by the water supply system. Furthermore, physicochemical parameters must meet the drinking water standards required by the laws of each country (Bergamasco et al., 2011). Thus, there is great importance in either the development of more sophisticated treatments or the improvement of the current ones.

In addition to these microorganisms, many other impurities can harm human health if not reduced or eliminated. They do not approach each other naturally during the coagulation/ flocculation process, being necessary the presence of a coagulant agent. Coagulation and flocculation processes are essential parts of water treatment, and the clarification of water using coagulants is practiced since ancient times (International Water Association [IWA], 2010). Alum has been the most widely used coagulant because of its proven performance, cost effectiveness, relative easy handling, and availability. Recently, much attention has been drawn on the extensive use of alum. According to Driscoll & Letterman (1995), the utilization of alum has raised a public health concern because of the large amount of sludge produced during the treatment and the high level of aluminum that remains in the treated water. McLachlan (1995) discovered that the intake of a large quantity of alum salt may cause Alzheimer disease.

Among the new techniques for drinking water treatment is the use of natural coagulants, aiming at a better quality of treated water by reducing the use of chemicals and also due to others advantages of natural coagulants. The biopolymers may be of great interest since they are natural low-cost products, characterized by their environmentally friendly behavior. Advantages of natural coagulant/flocculants have led some countries, like Japan, China, India, and United states, to adopt the use of natural polymers in surface water treatment do produce drinking water (Kawamura, 1991). The application of these coagulants in the coagulation/flocculation process has been successfully performed to remove turbidity, color, and natural organic matter (NOM) from natural water in order to produce clean potable water.

Thus, considering the coagulation step, the use of natural polyelectrolytes such as chitosan and *Moringa oleifera* (moringa) could be an option with many advantages over chemical agents, particularly the biodegradability, low toxicity, low residual sludge production, and less risks to health. Polyelectrolytes such as chitosan have a large number of surface charges that increase the efficiency of the coagulation process. Regarding the moringa, many researchers are currently seeking to identify the compound responsible for the coagulating characteristic of the seed of this plant, although there are still no definitive conclusions. It is known that it is a protein or polypeptide with coagulant properties (Ndabigengesere et al., 1995; Okuda et al., 2001; Ghebremichael et al., 2005). According to Davino (1976) the mechanism of coagulation/flocculation caused by the protein found in the seed of *Moringa oleifera* Lam resembles the mechanism caused by polyelectrolytes.

The use of coagulants for drinking water treatment, in spite of being efficient in the removal of most contaminants, is not able to generate water of high potability standards, which leads to the necessity of the simultaneous use of other techniques. Membrane filtration technique is already widely recognized and can be implemented in combination with coagulation processes.

This way, this chapter will look at the use of alternative techniques for water treatment based on the use of natural coagulants (chitosan and moringa seeds) associated with the membrane filtration process (micro and ultrafiltration) to obtain drinking water for human consumption.

1.1 Chitosan

Chitosan is a linear copolymer of d-glucosamine and N-acetyl-dglucosamine produced by the deacetylation of chitin, a natural polymer of major importance and the second most abundant natural polymer in the world, after cellulose (Rinaudo, 2006). It is described as a cationic polyelectrolyte and is expected to coagulate negatively charged suspended particles found in natural waters with increased turbidity (Divakaran & Pillai, 2002). Chitosan has been investigated as coagulant/flocculant for the removal of impurities from natural water and wastewater (Kawamura, 1991; Klopotek et al., 1994) because it can be conditioned and used for pollutant complexation in different forms, from water-soluble forms to solid forms (Renault et al., 2009). Some of the applications proposed for chitosan are: (i) removal of turbidity (Divakaran & Pillai, 2002; Bergamasco et al., 2011), natural organic matter (NOM) and color (Eikebrokk, 1999); (ii) inactivation of bacteria (Chung et al., 2003); and (iii) metal removal with higher efficiency when associated with ultrafiltration (Verbych et al., 2005).

The use of chitosan as coagulant in the coagulation/flocculation process (CFQ) for surface water treatment was studied by Bergamasco et al. (2011), achieving satisfactory results regarding color and turbidity removal, with values above 87%. The surface water sample in this case was from the Pirapó River at Maringá – PR, Brazil, with initial turbidity of 240 NTU and initial color of 1045 Hu. The chemical oxygen demand (COD) was initially of 19,3 mgO_2/L and it was reduced in 59.9% after treatment with chitosan, showing that the higher coagulation/flocculation of chitosan is related with the compounds that give color and turbidity to surface water.

The efficiency of the use of chitosan is more evident when compared with the process of coagulation/flocculation using aluminum sulfate as coagulant (CFS). The main difference in removal efficiency occurs when evaluating the removal of COD and TDS (Total dissolved solids). Comparing CFQ and CFS processes, COD removal efficiency was 59.9% using CFQ and 38.1% using CFS, and TDS removal was 41.9% for CFQ and 7.5% for CFS. In the same study, Bergamasco et al. (2011) observed that coagulation with aluminum sulfate generated sludge with SVI (sludge volume index) of 38.7 mg/mL, and the sludge formed by coagulation using chitosan showed SVI of 56.8 mg/mL. According to McLachlan (1995), the biggest advantage of chitosan over aluminum sulfate as coagulant is the fact that it is biodegradable, generating an easy to handle organic sludge that can be taken to a common landfill. Furthermore, chitosan improves the sedimentation step, as the flocs are more compact.

In a study by Eikebrokk & Saltnes (2001), it was verified that the fractions of color and organic carbon removal, with chitosan concentration of 4.0 mg/L, were 70 and 30%, respectively. In this work the authors also compared chitosan with metallic coagulants, otaining a reduction of 50% in the generated sludge when using chitosan, solving this way the problem of the concentration of trace metals in treated water. Still, the sludge disposal was simplified due to its biodegradability characteristics.

One can verify that chitosan presents good removal of turbidity and color, and is advantageous in terms of generated sludge, which can be disposed of in ordinary landfills, since it is biodegradable and has no trace metals.

1.2 *Moringa oleifera*

Moringa oleifera (moringa) is a tropical plant belonging to the family Moringaceae (Katayon et al., 2006), a single family of shrubs with 14 known species. Moringa is native of India but is now found throughout the tropics (Bhatia et al., 2007). Moringa seeds contain a non-toxic natural organic polymer which is an active agent with excellent activity and coagulating properties. The tree is generally known in the developing world as a vegetable, a medicinal plant, and a source of vegetable oil (Katayon et al., 2006). Its leaves, flowers, fruits, and roots are used locally as food ingredients. The medicinal and therapeutic properties of moringa have led to its utilization as a cure for different ailments and diseases, physiological disorders, and in Eastern allopathic medicine (Akhtar et al., 2007). Additionally, the coagulant is obtained at extremely low or zero net cost (Ghebremichael et al., 2005).

If moringa is proven to be active, safe, and inexpensive, it is possible to use it widely for drinking water and wastewater treatment. Besides, moringa may yet have financial advantages bringing more economic benefits for the developing countries (Okuda et al., 1999).

The moringa seed has a protein that when solubilized in water is able to promote coagulation and flocculation of compounds that cause color and turbidity in highly turbid water. Several studies have also shown their effective antimicrobial and antifungal capacity, thereby contributing to good water quality at low cost (Chuang et al., 2007; Coelho et al., 2009).

The process used to obtain the coagulant (which would be mainly constituted by the protein in the seed) is usually performed with the solubilization of the protein in water under stirring and filtration, but the use of salts is able to promote a greater solubilization of this protein in the medium, which would favor the coagulation/flocculation process.

Madrona et al. (2010) evaluated the extraction of the coagulant protein in the presence of potassium chloride (KCl) at different concentrations. The authors have shown that greatest coagulation efficiencies are achieved with KCl 1 mol/L, reaching nearly 100% removal of color and turbidity from water with initial turbidity of 850 NTU. In the same study, conducted by Madrona (2010), the author evaluated the effectiveness of other salts in the coagulant extraction process, such as magnesium chloride ($MgCl_2$) and sodium chloride (NaCl), compared with KCl, all at a concentration of 1 mol/L. A protein content of 4499 mg/L was achieved for the extraction using NaCl, and 4818 mg/L using KCl. The process of protein extraction with $MgCl_2$ was able to release only 950 mg/L of protein in solution, which is very close to the concentration released by the extraction with water (873 mg/L).

Moringa has been found to be ineffective as a natural coagulant for low turbidity drinking water but effective for high turbidity water in previous studies (Okuda et al., 2001). This was verified in a study by Nishi (2011). The authors obtained values of color and turbidity removal over 90% when the water to be treated showed high values of initial turbidity, between 350 and 450 NTU. A moringa concentration of 150 mg/L would have been sufficient to achieve this level of removal. The coagulant derived from moringa seeds, as it contains a certain amount of organic matter, can give color and turbidity to the treated water. Thus, when water with low initial turbidity undergoes coagulation/flocculation, depending on the concentration of moringa coagulant used, the effect of turbidity and color removal is not as satisfactory as the level of removal obtained for these parameters using this coagulant in the treatment of high turbidity water. In this case the effect of the coagulant protein overlaps the additional organic load and the removal of the evaluated parameters increases.

An important point to be considered when using moringa as a coagulant is related to the pH of the water to be treated. For chemical coagulants, water pH adjustment is necessary for the flakes to be properly formed. In the case of the moringa, there is no need for this adjustment, and this parameter is not changed after treatment, as evidenced by Vieira et al. (2010). Moringa is an efficient coagulant in a wide pH range (6-8), which is an advantage compared with other coagulants, as the pH adjustment step can be eliminated in the coagulation/flocculation processes.

1.3 Coagulation/flocculation and membrane filtration

In conventional water treatment plants, the coagulation/flocculation process is followed by filtration. However, nowadays, membrane separation has been widely studied for potable water production, since the MF/UF membranes are physical barriers that are able to efficiently remove suspended particles and colloids (Xia et al., 2007; Guo et al., 2009), turbidity, bacteria, algae, parasites, and viruses for clarification and disinfection purposes (Guo et al., 2009), as well as to control trihalomethane precursors (Bottino et al., 2001).

To overcome the problems caused by natural organic matter (NOM) in MF and UF applications, conjunctive use of coagulation and membranes is becoming more attractive for water treatment because the coagulation is an opportunity to join NOM with other particles present in water before NOM reaches the membrane surface. Application of the coagulation and ultrafiltration unit operations contributes to the improvement of treated water quality and the enhancement of the membrane performance. In comparison with conventional processes such as coagulation, flocculation, sedimentation and/or flotation, and rapid or slow sand filtration, MF/UF technology has many advantages such as superior quality of treated water, much greater compactness, easier control of operation and maintenance, use of fewer chemicals, and lower production of sludge (Bergamasco et al., 2009). The combined processes of coagulation/flocculation followed by micro or ultrafiltration usually result in better water quality, since the membrane processes often function as a polishing step in water treatment, being able to remove the impurities which are not removed by coagulation/flocculation processes. In addition, unlike the conventional filtration process, micro/ultrafiltration can retain bacteria and other microorganisms.

Bouchard et al. (2003) studied the processes of microfiltration and coagulation/microfiltration, using ferric chloride and aluminum sulfate as coagulants in the combined process. Microfiltration tests were performed in a mini-module of submerged cross-fiber Zenon membranes, with porosity of 0.1 μm. For the combined process, the results obtained were 60% removal of TOC and a reduction of more than 80% of the compounds that absorb UV at 254nm. The coagulation/microfiltration process was shown to be beneficial, allowing a significant reduction in membrane fouling. These results were already expected, because when coagulation occurs, colloids are destabilized and cluster forming larger flocs, thus contributing to reduce membrane fouling. Comparing the processes of ultrafiltration and coagulation/ultrafiltration, using ferric chloride as a coagulant agent, removal of TOC and compounds that absorb UV at 254nm of 30% and 60%, respectively, was observed for the ultrafiltration process. For the combined coagulation/ultrafiltration, the removal of TOC and compounds that absorb UV at 254nm increased to 60 and 80%, respectively (Bouchard et al., 2003).

Konradt-Moraes (2004) studied the combined process of coagulation/flocculation/ ultrafiltration using ceramic membrane with pore size of 0.05μm and transmembrane pressure of 2 bar. Under optimum conditions of coagulation and flocculation for the biopolymer chitosan, removal of color, turbidity, compounds that absorb UV at 254nm, nitrite, phosphate, total coliforms, and *Escherichia coli* close to 100% and TOC removal of 75% were achieved. Thus, according to Konradt-Moraes (2004), there is great potential in the use of the combined coagulation/flocculation/membrane separation process. However, few research papers have been published to date, and therefore a wide range of study possibilities is open, that can lead to a deeper knowledge of the variables involved in the process as a whole, which in turn allows not only process scale-up, but also its transfer to the companies responsible for public drinking water supply.

2. Experimental results – Case studies

The case studies presented below used the processes of coagulation/flocculation with natural coagulants and membrane filtration for removing color, turbidity, *Giardia* and *Cryptosporidium*, to obtain drinking water for human consumption.

The first case describes the utilization of chitosan as natural coagulant and ceramic ultrafiltration membranes in pilot scale. The second case deals with moringa and polymeric microfiltration membranes in a bench-scale filtration module. Both cases used surface water from the Pirapó River, which serves a population of over 300,000 inhabitants in the city of Maringá, Brazil.

2.1 Process of coagulation/flocculation with chitosan followed by ultrafiltration for surface water treatment

A pilot plant, shown in Figure 1, has been used by the research group headed by the researcher Professor Rosângela Bergamasco, in the Environmental Preservation and Control Laboratory at the State University of Maringá, Brazil. This unit has been the basis for studies of water purification processes that are subsequent to coagulation/flocculation using natural coagulants such as chitosan and *Moringa oleifera*. These studies, based on the treatment of water from the Pirapó River, which is responsible for supplying the city of Maringá - PR, Brazil, have demonstrated the effectiveness of the evaluated coagulants, as well as the applicability of the combined processes, yielding high-quality water within the specifications required by Brazilian law. The surface water characterization is presented in Table 1.

In the study conducted by Bergamasco et al. (2011), the removal of UV-254nm absorbing compounds showed a significant increase when the hybrid process was used, changing from 85.8% with CFQ to 99.4% with CFQ-UF at 2 bar. The results achieved in the filtration with ceramic membranes of stainless steel with Al_2O_3/ZrO_2 (0.1μm) (TAMI, France), at transmembrane pressures of 1 and 2 bar are presented in Table 2. The UV absorbance of organic matter, in the range of 254–280 nm, reflects the presence of unsaturated double bonds and π-π electron interactions such as in aromatic compounds. However, it is known that natural organic matter (NOM) is a mixture of organic compounds called humic materials, but proteins, polysaccharides and other classes of biopolymers also contribute to NOM. This indicates that besides the compounds detected by UV-254nm absorption, other organic compounds may also be present in surface water, and therefore this parameter is not a suitable indicator of NOM removal. Other parameters should be considered for a better understanding of the process.

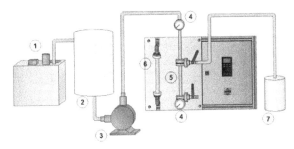

Fig. 1. Schematic diagram of the micro/ultrafiltration experimental unit. 1- thermostatic bath; 2- feed tank; 3- pump; 4- manometers; 5- membrane filtration module; 6- flowmeter (rotameter); 7- permeate.

Water quality parameter	Values
Apparent color (Hu)[1]	1695
True color (Hu)[1]	1045
Turbidity (NTU)[2]	240
Chemical oxigen demand (COD) (mgO$_2$/L)	19.3
Total organic carbon (TOC) (mgC/L)	6.4
UV-254nm absorbing compounds (UV-254nm) (cm^{-1})	0.923
pH	8.17
Total suspended solids (TSS) (mg/L)	1332
Total dissolved solids (TDS) (mg/L)	228
Total coliforms (CFU/100 mL)[3]	3955
Escherichia coli (CFU/100 mL)[3]	800

1. Hu = mg$_{Pt-Co}$/L
2. NTU = Nefelometric turbidity unit
3. CFU/100 mL = Colony forming units per 100 mL of water sample

Table 1. Characterization of the surface water from Pirapó River

Filtration tests were performed to characterize the flow of pure water through the membranes, using deionized water. Flux was calculated using Equation 1, where f$_{permeate}$ is the permeate flux, m is the mass of collected water, ρ25°C is water density at 25°C, Δt is the time interval during which water was collected, and Am is the filtering area of the membrane.

$$f_{permeate} = \frac{m}{\rho 25°C \times \Delta t \times Am} \tag{1}$$

The removal efficiency for each parameter analyzed in the different treatment processes was calculated from Equation 2, where C_i and C_f are the initial and final concentrations, respectively, for each parameter.

$$\% \; Removal \; efficiency = \left(\frac{C_i - C_f}{C_i}\right) \times 100 \tag{2}$$

Deionized water (DW) fluxes were determined before each experiment (J_i) and after the filtration of solutions SW, CFQ and CFS (J_f) to determine the fouling of the membrane. The percentage of fouling (%F) was calculated according to Equation 3, proposed by Balakrishnan et al. (2001), using the steady-state flux values, which assume that the flux tends to constant values. The percentage of fouling represents a decrease in the deionized water flux after tests with contaminated water.

$$\%F = \frac{(J_i - J_f)}{J_i} \times 100 \tag{3}$$

In Equation 3, %F is the percentage of membrane fouling, J_i is the initial water flux obtained in the first filtration with deionized water and J_f is the final water flux obtained by filtration of deionized water after the filtration of surface water.

The parameters apparent color, turbidity, and pH were evaluated according to the Standard Methods (American Public Health Association [APHA], 1995). Turbidity measurements were conducted using a turbidimeter (HACH, 2100P). A digital pH meter (Digimed DM-2) was used for pH measurements. Color measurements were conducted using HACH DR/2010 spectrophotometer – Method 8025. COD values were determined using HACH DR/2010 – Method 10129. TOC was determined using an Aurora 1030C TOC Analyzer with 1088 Rotary TOC Autosampler. Absorbance measurements at 254nm were performed using a Logen Scientific UV-Vis spectrophotometer. UV absorbance at 254nm was also used in this study as an indication of the removal of organics from water. UV absorbance is commonly used as an index of the aromatic level (Kim & Yu, 2005).

Comparing CFQ and CFS processes, the main differences in removal efficiency are observed with respect to COD removal (59.9% using CFQ and 38.1% using CFS) and TDS removal (41.9% for CFQ and 7.5% for CFS). It is also observed in Table 2 that the process of coagulation/flocculation with the natural coagulant chitosan was very effective in removing compounds responsible for color and turbidity, as well as UV-254nm absorbing compounds. Similar results were observed by other authors such as Roussy et al. (2005) and Rizzo et al. (2008).

The working conditions for the experiments presented in Table 2 were as follows: The pH was adjusted to 5.0 and concentration of natural chitosan coagulant was 1.0 mg/L. When using aluminum sulfate as coagulant (15 mg/L), pH was maintained at 7.0. In the rapid mixing step the speed was kept at 120rpm for 2.5 min, whereas the speed used in the slow mixing step was 20rpm for 20 min (Konradt et al., 2008). Temperature was maintained at 25.0±2°C during coagulation/flocculation. All experiments were performed at least in triplicate.

Parameter	CFQ (%)	CFQ – UF (%)		CFS (%)	CFS – UF (%)	
		P=1 bar	P=2 bar		P=1 bar	P=2 bar
Apparent color	98.1	99.4	99.1	99.8	99.2	100
True color	97.1	99.2	99.1	99.3	99.1	99.8
Turbidity	99.3	99.9	99.9	99.2	99.8	99.8
UV-254nm	85.8	91.8	99.4	83.4	96.3	88.5
COD	59.9	90.9	97.4	38.1	89.3	85.1
Total coliforms	62.5*	99.0	99.0	99.0	99.0	99.0
Escherichia coli	99.0	99.0	99.0	99.0	99.0	99.0
TSS	60.3	97.8	88.8	94.8	93.5	97.9
TDS	41.9	40.2	86.1	7.52	49.3	42.9

*reduction of 62.5% in the number of CFU/100mL initially present in the surface water. Source: Bergamasco et al., 2011.

Table 2. Removal efficiency (%) of coagulation/flocculation and coagulation/flocculation-ultrafiltration processes at 1 bar and 2 bar, using chitosan (CFQ) or aluminum sulfate (CFS) for the coagulation of surface water

Coagulation pretreatment allows a higher rejection of organics by microfiltration (MF) and UF and the cut-off criterion due to initial membrane pore size is no longer valid (Schafer et al., 2001). The most consistent system was CFQ-UF at the pressure of 1 bar, achieving removal efficiencies above 90% for all parameters assessed, except for TDS, which had 40.2% removal. But if the use of CFS-UF is considered, TDS removal reached a maximum of 49.3%. Another relevant parameter to be analyzed is COD, whose removal was higher when using CFQ-UF at 2 bar (97.4%) than when using CFS-UF at the same pressure (85.1%). An important point to be evaluated is the type of membrane used for filtration, because depending on the material the membrane is made of, fouling effects may be more or less pronounced, which will result in differences in permeate flux and percent removal of the assessed parameters.

The hybrid process of coagulation with chitosan followed by filtration in 0.1 μm pore size polysulfone membrane, resulted more effective for UV-254nm, TOC, and iron removal. Adding chitosan as coagulant, 70% of UV-254nm absorbing compounds and 47% of TOC (averaged values) were removed (Bergamasco et al., 2009). In this study the authors evaluated water from the Saint-Charles River in Quebec, Canada, with an initial UV-254nm absorbance value of 0.083-0.091.

One can see that the ceramic membranes can be more effective for water treatment than polysulfone membranes. Comparing the results with two membranes of different materials and same porosity, ceramic membranes led to a reduction of 91.8% in UV-254nm absorbance (Bergamasco et al., 2011), while the reduction obtained with polysulfone membranes was 70% (Bergamasco et al., 2009). Chitosan improves the sedimentation step, as the flocs are more compact, which is favorable when hybrid coagulation/flocculation/filtration systems are operated. Another fact that should be taken into consideration when applying UF for water treatment is the occurrence of membrane clogging, which causes a drop in permeate flux and is a result of a set of phenomena related to the solution nature and to the characteristics of the membrane (Bergamasco et al., 2011). This fact can be explained by two mechanisms, commonly attributed to the removal of organic matter by UF: sieve retention and adsorption sequestration. In sieve retention the UF membrane acts as a barrier for particle penetration. The particles are retained on the membrane surface and form a cake that grows in thickness as the filtration progresses. The second mechanism involves the entry and capture of the particles into the membrane matrix (Guo et al., 2009).

Bergamasco et al. (2011), using ceramic membrane in the combined process of coagulation/flocculation with chitosan and ultrafiltration, obtained higher permeate fluxes than with UF of surface water (SW) and CFS-UF under the same pressures. For the pressure of 2 bar the permeate flux of the CFQ-UF process was approximately twice that of the CFS-UF process, thus justifying the use of chitosan as a coagulant prior to the ultrafiltration step for surface water treatment. The authors presented the results obtained by the resistance-in-series model for the different types of resistance observed in the UF step using SW, CFQ, and CFS at pressures of 1 bar and 2 bar, as shown in Table 3. The coagulation/flocculation conditions to obtain water for use in the ultrafiltration experiments were the same as for the experiments presented in Table 2.

	Resistance x10^{11} (m^{-1})*		
	R_f	R_{cp}	R_t
SW			
ΔP=1 bar	1.22	6.79	9.62
ΔP=2 bar	8.11	9.58	20.76
CFQ			
ΔP=1 bar	4.63	0.82	7,06
ΔP=2 bar	9.92	0.12	13.11
CFS			
ΔP=1 bar	0.42	5.78	8.11
ΔP=2 bar	5.89	12.89	21.86

*Rf: fouling resistance; Rcp: concentration polarization resistance; Rt: total resistance.
Source: Bergamasco et al. (2011).

Table 3. Resistances on the membrane during the ultrafiltration process with SW, CFQ, and CFS at 1 bar and 2 bar, using Al2O3/ZrO2 (0.1 µm) ceramic membranes and filtration time of 200 min.

It was observed by means of Table 3, that for the same type of water (without treatment, coagulated with chitosan, or coagulated with aluminum sulfate) the fouling resistance (R_f) due to solute adsorption into membrane pores and walls (Chang et al., 2001) increased with increasing transmembrane pressure, and this can be explained by the higher compression.

This type of resistance can be eliminated only by washing the membrane. It was also observed that R_f was greater for CFQ than for SW and CFS, but concentration polarization resistance (R_{cp}) and R_t were lower for CFQ than for SW and CFS at 1 bar and 2 bar. The floc cake resistance is lower than the resistance due to the unsettled floc and the uncoagulated organics, as reported by Guigui et al. (2002). The use of chitosan as a coagulant can lead to the formation of denser flocs. Thus, its negative impact on the filtration can be explained. The cake is formed by large aggregates, decreasing the average flux through and among these aggregates. However, the performance evaluation of the hybrid systems (CFS-UF and CFQ-UF) showed that the permeate quality was increased when compared with individually operated systems (CFS and CFQ). This is justified by the excellent ability of the UF process to remove particles and colloids.

2.2 Process of coagulation/flocculation with moringa followed by microfiltration for surface water treatment

The other study on water purification processes performed in the Environmental Preservation and Control Laboratory at the State University of Maringá, Brazil, evaluated the coagulation/flocculation process using the natural coagulant *Moringa oleifera*, followed by microfiltration (Nishi, 2011). As mentioned previously, surface water from the Pirapó River was used for this study. Samples of high and low turbidity were mixed to obtain water with different initial turbidity values. The samples used in this study had initial turbidity of 50, 150, 250, 350, and 450 NTU. The prepared samples were artificially contaminated with 10^6 cysts/L of *Giardia* spp. and 10^6 oocysts/L of *Cryptosporidium* spp. obtained from the positive control (suspension of cysts and oocysts) present in the commercial kit Merifluor (Meridian Bioscience, Cincinnati, OH, USA). After being prepared, the samples were subjected to the processes of coagulation/ flocculation with moringa seeds (CFM), microfiltration (MF), and the combined coagulation/flocculation with moringa seeds followed by microfiltration (CFM-MF).

Moringa coagulant solution was prepared and used the same day. Mature moringa seeds from the Federal University of Sergipe (UFS) were used as raw material. The seeds were manually removed from the dry pods and peeled. To prepare the 1% stock solution of moringa (concentration of 10,000 mg/L), 1 g of peeled seeds was crushed and added to 100 mL of distilled water. Subsequently, the solution was stirred for 30 min and vacuum filtered (Cardoso et al., 2008; Madrona et al., 2010). From the 1% stock solution, moringa solutions were prepared with different concentrations: 25, 50, 75, 100, 125, 150, 175, 200, 225, 250, 275, and 300 mg/L. CFM tests were performed on a simple jartest, under the following conditions: rapid mixing speed (RMS) of 100 rpm, coagulation time (CT) of 3 min, slow mixing speed (SMS) of 15 rpm, flocculation time (FT) of 15 min, and settling time (ST) of 60 min (Madrona et al. 2010).

To determine the coagulant concentrations which resulted in the highest removal of the evaluated parameters, 5x12 factorial experiment was applied. For the factor "A" five different levels of initial water turbidity were tested: 50, 150, 250, 350, and 450 NTU. Factor "B" consisted of twelve levels of concentration of moringa coagulant. The parameters analyzed in the experiments – color, turbidity, pH, and removal of *Giardia* and *Cryptosporidium* – were evaluated in triplicate for each combination of the factors "A" and "B". The results were analyzed by ANOVA using the F test and phase contrast (Nkurunziza et al., 2009) to obtain the optimum concentration of each coagulant for each water sample with initial turbidity of 50 to 450 NTU, to be later used in the combined process of coagulation/flocculation/membrane filtration. The software Statistica, version 8.0/2010, was used for the statistical analysis, and p values of less than 0.05 were considered significant.

In this study, the same methodologies described in section 2.1 were used to evaluate the removal efficiency of turbidity and color and the pH of treated water. The concentration of (oo)cysts of *Giardia* and *Cryptosporidium* was assessed by the membrane filtration technique with mechanical extraction and elution (Aldom & Chagla, 1995; Dawson et al., 1993; Franco et al., 2001).

The initial characteristics of water samples used in the study are presented in Table 4.

Turbidity (NTU)	Color (uH)	pH
50	350	7.80
150	902	7.81
250	1000	7.50
350	1849	7.64
450	1885	7.70

Source: Nishi, 2011

Table 4. Water sample parameters before treatment processes.

The results for the removal efficiency of turbidity, color, *Giardia* and *Cryptosporidium*, and pH values for water after treatment with moringa, under the aforementioned conditions, are presented in Table 5. Using moringa as coagulant agent, turbidity removal ranged from 3 to 97.4%. The lowest removal efficiencies, between 3 and 45.6%, were observed for water with low initial turbidity (50 NTU). Removals above 70% were observed for the samples with

Initial turbidity (NTU)	Removal efficiency (%)	Moringa concentration in the water samples (mg/L)											
		25	50	75	100	125	150	175	200	225	250	275	300
50	Turbidity	27.7	23.0	20.7	33.4	33.2	35.6	44.4	45.6	41.6	27.6	10.0	3.0
	Color	0.35	0.11	1.65	3.07	4.00	6.60	22.8	27.6	30.0	26.0	15.4	16.7
	Giardia	6.00	42.3	38.4	69.2	84.6	76.9	82.0	80.0	80.0	76.9	85.0	69.0
	Cryptosporidium	76.0	91.0	90.0	98.0	93.0	91.0	98.0	86.0	91.0	88.0	89.0	83.0
	pH	7.90	8.20	8.10	8.20	8.20	8.10	8.20	7.90	8.00	8.10	8.00	7.90
150	Turbidity	42.0	52.4	69.8	71.0	74.0	75.8	67.7	65.3	69.0	73.6	76.0	72.0
	Color	10.0	47.5	67.0	68.8	70.8	73.5	73.5	71.4	65.0	63.4	66.4	61.6
	Giardia	74.0	97.0	85.0	98.0	94.0	98.0	98.0	98.0	97.0	97.0	91.0	82.0
	Cryptosporidium	42.0	50.0	77.0	81.0	81.0	92.0	92.0	92.0	85.0	92.0	81.0	58.0
	pH	7.93	7.96	7.85	7.74	7.8	7.76	7.70	7.74	7.81	7.74	7.72	7.68
250	Turbidity	68.9	74.8	80.6	93.4	90.1	94.4	93.9	94.2	90.9	94.6	91.8	92.7
	Color	21.8	46.5	46.2	68.4	64.4	80.7	79.0	81.3	78.0	77.8	88.4	81.6
	Giardia	80.0	65.0	65.0	80.0	80.0	95.0	95.0	95.0	95.0	90.0	92.5	90.0
	Cryptosporidium	67.0	61.0	75.0	74,0	86.0	96.0	95.0	92.0	94.0	90.0	87.0	78.0
	pH	7.60	7.70	7.80	7.80	7.60	7.60	7.70	7.80	7.80	7.80	7.70	7.80
350	Turbidity	49.4	62.8	70.8	75.0	82.0	90.0	93.7	95.0	96.0	95.8	96.4	92.5
	Color	47.0	82.3	76.4	94.0	94.0	88.2	97.0	97.0	94.0	88.2	94.0	94.0
	Giardia	47.0	82.3	76.4	94.0	94.0	88.2	97.0	97.0	94.0	88.2	94.0	94.0
	Cryptosporidium	22.0	81.0	68.0	95.0	86.0	81.0	97.0	96.0	92.0	86.0	96.0	92.0
	pH	7.78	7.87	7.75	7.77	7.81	7.74	7.82	7.75	7.73	7.76	7.73	7.78
450	Turbidity	63.0	61.0	75.0	79.9	92.0	94.0	97.2	97.4	97.2	97.0	96.7	94.7
	Color	39.0	47.8	61.6	68.5	88.3	91.5	96.1	96.1	96.4	96.0	95.6	92.8
	Giardia	63.0	92.0	96.0	96.0	92.0	96.0	90.0	97.0	94.0	97.0	93.0	89.2
	Cryptosporidium	45.0	51.0	94.0	85.0	80.0	94.0	86.0	93.0	95.0	98.0	90.0	76.0
	pH	7.80	7.70	7.60	7.60	7.60	7.60	7.00	7.60	7.60	7.60	7.50	7.60

Source: Nishi, 2011.

Table 5. Percentage of removal efficiency of turbidity, color, *Giardia*, and *Cryptosporidium*, and pH values after the process of coagulation/flocculation with moringa.

turbidity of 250, 350, and 450 NTU (Nishi, 2011). The decrease in efficiency of turbidity removal from water with 50 NTU of initial turbidity, after the addition of moringa, can be explained by the increased organic load. This is justifiable as long as moringa is an oilseed which is rich in organic substances such as oil, protein, fat, and vitamins. This increase in turbidity and color in water treated with moringa was also observed in other studies, especially when the water had relatively low initial turbidity and color (Ramos, 2005).

Nkurunziza et al. (2009), using a 3% solution of moringa seeds, prepared with saline water, to treat water from the rivers in the province of Rwanda, obtained removal efficiency of 83.2% in samples with turbidity of 50 NTU and higher values (99.8%) in water with turbidity of 450 NTU. The optimum concentrations found in this study were 150 mg/L for 50 NTU and 125 mg/L for other turbidity levels tested by the researchers. The results of turbidity removal from water with low initial turbidity (50 NTU) were higher than those obtained in the present case study (45.6%) and for water with high initial turbidity (450 NTU) the results were similar (97.4%). The differences may be due to the different preparation procedures of the moringa solution, by aqueous or saline extraction, as well as the different concentrations of the stock solution of moringa. In both studies, the coagulant properties of the moringa appear to be more efficient in water of high initial turbidity, in agreement with other literature reports (Ndabigengesere et al., 1995; Madrona et al., 2010).

Ndabigengesere et al. (1995), applying an aqueous solution of 5% moringa seeds to synthetic turbid water (kaolin added to tap water) with initial turbidity of 426 NTU, obtained removals from 80 to 90% and reached the optimum concentration of 500 mg/L of coagulant solution. This concentration is higher than the optimum concentration for water of 450 NTU obtained in this case study, which was 275 mg/L. This difference between the optimum concentrations of the moringa solution may be due to the different water source: Ndabigengesere et al. (1995) used synthetic turbid water prepared with kaolin, and the present study used surface water. The different efficiencies of turbidity removal and optimum concentrations can be explained by the different compositions of water samples used in the studies (raw water, synthetic turbid water), that is, the substances present in water can influence the action of the coagulant agent and the formation of flocs, as well as by the preparation procedure of the moringa solution (aqueous or saline extraction), evaluated concentrations, and seed quality, among other factors.

Regarding color, the removal ranged from 0.11 to 30% for water with initial turbidity of 50 NTU. The highest removals for this sample were obtained with moringa concentration ranging from 175 to 250 mg/L. For water with higher initial turbidity (150 to 450 NTU), the removal efficiency ranged from 10 to 97%, the highest values being obtained for concentrations of 150 mg/L or higher (Nishi, 2011). It is observed that color removal by moringa is similar to its behavior with respect to turbidity: the lowest values of this parameter are obtained for water with high initial turbidity, which agrees with literature data (Cardoso et al., 2008; Nkurunziza et al., 2009; Madrona et al., 2010).

Concerning the pH of water samples after the coagulation process with different concentrations of moringa, it was observed that the average pH was 7.6, with variation of approximately 10% (Nishi, 2011). There was little variation among the samples regardless of the amount of moringa solution added, which consists of one of the benefits of moringa as a coagulant agent, that is, its addition does not significantly alter the pH of the water (Ndabigengesere et al., 1995; Nkurunziza et al., 2009), unlike the treatment with aluminum sulfate, in which it is necessary to adjust the pH of the water to improve its coagulant action, increasing the amount and cost of chemicals for water treatment.

Considering the removal of *Giardia* cysts and *Cryptosporidium* oocysts, similar behaviors were observed among samples. The best removal of both *Giardia* and *Cryptosporidium* occurred at moringa solution concentrations of 150 mg/L or higher, for all treated water samples (50 to 450 NTU), with average removal efficiency of 93% (1.2 log removal) and 90% (1 log removal), respectively (Nishi, 2011). No studies were found in the literature regarding the removal of these protozoan parasites using moringa as coagulant agent. The high removal obtained can be explained by the coagulant action of moringa, which is based on the presence of cationic proteins in the seeds. These proteins are densely charged cationic dimers with a molecular weight of about 13 kDa, and adsorption and charge neutralization are the main mechanisms of coagulation (Ndabigengesere et al., 1995). Since the zeta potential calculated for (oo)cysts of *Giardia* and *Cryptosporidium* in water at neutral pH are, on average, -17 and -38 mV, respectively (Hsu & Huang, 2002), the mechanism of charge neutralization of the proteins of the natural coagulant could act in the removal of these protozoan parasites.

The removal of protozoan parasites obtained in this study is close to the results of other reports in the literature, using chemical coagulants such as aluminum sulfate and ferric chloride for the removal of these microorganisms (Bustamante et al., 2001; Xagoraraki & Harrington, 2004), and neutralization of charges is also the primary mechanism of coagulation with aluminum sulfate. Brown & Emelko (2009) applied another natural coagulant, chitosan, for the removal of *Cryptosporidium parvum* in pilot-scale treatment of synthetic raw water (dechlorinated tap water with kaolinite-induced turbidity), using concentrations of 0.1, 0.5, and 1.0 mg/L chitosan solution. The authors achieved great reductions in turbidity, but did not obtain good results in *C. parvum* removal, with average values below 10%. A possible explanation for this difference, since chitosan is also a cationic polymer, is the possibility that during the coagulation/flocculation process, the oocysts are also removed by physical entrapment in the flocs, which is another mechanism participating in protozoan removal (Bustamante et al., 2001). Considering that the flocs formed depend on the characteristics of the particles in the water, it can be said that the removal of microorganisms will also depend on these characteristics, as Brown and Emelko (2009) used artificial raw water and in this study natural surface water was used.

Moringa presented good results of color, turbidity, *Giardia* and *Cryptosporidium* removal from all water samples for the coagulation/flocculation process, most notably in samples of high initial turbidity (150, 250, 350, and 450 NTU) and with coagulant concentration of 100 mg/L or higher. The process of coagulation/flocculation with moringa yielded 1.2 log removal for *Giardia* and 1 log removal for *Cryptosporidium*. These removals are in line with the recommendations of the World Health Organization (Lechevallier & Au, 2004). According to Lechevallier & Au (2004), in the conventional water treatment processes, coagulation is a critical step for the removal of pathogenic microorganisms. Coagulation, flocculation, and sedimentation can result in 1-2 log removal of bacteria, viruses and protozoa when properly handled. Also according to the authors, in the case of *Giardia* and *Cryptosporidium*, there is great difficulty in interpreting results in relation to studies on bench scale, as well as on pilot scale, due to the low concentrations at which these protozoa are found and the detection methods, which are still limited.

Statistical analysis showed that there is a relationship of the turbidity, *Giardia* and *Cryptosporidium* removal with the moringa solution concentration and the initial water turbidity. Statistical analysis was applied to obtain the concentration of moringa which showed the best removals of turbidity, *Giardia* and *Cryptosporidium* for each initial turbidity of water samples. It was observed that for the sample with turbidity of 50 NTU, the concentration of the moringa solution showed no statistically significant interaction with the values of turbidity, *Giardia* and *Cryptosporidium* removal. Therefore, it was not possible to obtain the optimum concentration for the water sample with initial turbidity of 50 NTU. For the remaining samples, the moringa solution concentrations which showed the best removal of the evaluated parameters were obtained and are presented in Table 6. It is observed that coagulation/flocculation provided good removal efficiencies of turbidity and color, depending on water characteristics, initial turbidity, and coagulant concentration.

Initial turbidity (NTU)	Optimum concentration (mg/L)
150	250
250	150
350	275
450	275

Source: Nishi, 2011.

Table 6. Moringa concentration which resulted in the best removal of turbidity, color, and (oo)cysts of *Giardia* and *Cryptosporidium*, according to the initial turbidity of the water sample.

After obtaining the optimal concentrations of moringa coagulant for each sample of surface water (Table 6), the samples were subjected to the MF process and to the combined process of coagulation/flocculation with moringa followed by MF (CFM-MF).

The membrane filtration tests were carried out in a bench-scale microfiltration membrane module, using the tangential filtration principle. This module is shown in Figure 2.

Fig. 2. Frontal view of the MF/UF module (Operation manual – MF/UF module): (1) polymeric membranes; (2) pressure gauges; (3) speed controller; (4) feed tank; (5) valve used to collect the permeate; (6) tubing through which the concentrate returned to the feed tank.

The MF membrane was composed of hollow fibers made of polyimide, with porosity of 0.40 μm. The operating pressure was 1.0 bar. To maintain uniformity in the experiments, the initial volume was fixed in 5 L, and the test time was 60 min.

In this study, the same methodologies described in section 2.1 were used to evaluate membrane flux and fouling, as well as to analyze the removal efficiency of turbidity, color, *Giardia* and *Cryptosporidium*, and the pH of treated water.

The results obtained in the processes of microfiltration (MF) and coagulation/flocculation with moringa followed by microfiltration (CFM-MF) are presented below. These results are presented together to show if the pretreatment (coagulation/flocculation with moringa) had differences in relation to the MF process without pretreatment. The removal efficiencies and the pH of the water treated by the MF and CFM-MF processes are presented in Table 7.

Treatment process	Removal efficiency (%)	Initial turbidity (NTU)			
		150	250	350	450
MF	Turbidity	81.09	84.16	76.82	76.33
	Color	78.28	83.45	74.27	72.56
	Giardia	ND	ND	ND	ND
	Cryptosporidium	ND	ND	ND	ND
	pH	7.38	7.85	7.36	7.81
CFM-MF	Turbidity	93.54	92.28	84.78	99.39
	Color	96.15	92.19	88.96	100.0
	Giardia	ND	ND	ND	ND
	Cryptosporidium	ND	ND	ND	ND
	pH	7.33	7.72	7.34	7.51

ND – not detected. Source: Nishi, 2011

Table 7. Removal efficiencies of turbidity, color, *Giardia* and *Cryptosporidium*, and pH values of the water treated by the MF and CFM-MF processes.

It can be observed that the largest color and turbidity removals occurred with the combined CFM-MF process, compared with the MF process without pretreatment. There were no changes in the pH of the treated water. It is clear that the use of coagulation/flocculation with moringa prior to microfiltration improves the quality of treated water (Nishi, 2011).

A few studies were found in the literature regarding the CF/MF process using moringa as a coagulant for surface water treatment. Madrona (2010) evaluated the combined process of coagulation/flocculation with moringa and MF with ceramic membranes, and obtained 97 to 100% removal of turbidity and color in the treatment of surface water from the Pirapó River, in Maringá, Paraná. These results were similar to those obtained in the present study, which used a polymer membrane for the MF process. Parker et al. (1999), using hollow fiber MF membranes with 0.2 μm pores for the treatment of water that had been previously treated in settling tanks, obtained water with turbidity below 0.1 NTU, with average removal of 99.46%, similar to those obtained in this study.

Neither in the microfiltration (MF) process alone, nor in the combined (CFM-MF) processes, (oo)cysts of *Giardia* and *Cryptosporidium* were detected in the filtered water, being below the detection limit (<1 cyst or oocyst/L) (approximately 6 log removal), in agreement with literature data. Jacangelo et al. (1995), studying the application of three MF membranes with pore sizes between 0.08 and 0.22 μm for the treatment of water contaminated with *Giardia* and *Cryptosporidium*, found that the protozoa concentration was below detectable levels in the filtered water (<1 cyst or oocyst/L) from two of the membranes (corresponding to log removal> 4.7 to> 7.0 for *Giardia* and > 4.4 to> 6.9 for *Cryptosporidium*). They also concluded that the level of removal depends on the concentration of protozoa in the water to be treated and on membrane integrity. In another study, MF membranes with average pore size of 0.2 μm resulted in significant removal of particles that were the same size as *Giardia* cysts (5-15 μm). Log removal was, on average, 3.3 to 4.4. The removal of particles that were the same size as *Cryptosporidium* oocysts (2-5 μm) was lower, 2.3 to 3.5 log removal. These removals were obtained according to the concentration of (oo)cysts used for artificial contamination of water and proved to be independent of the membrane flux (114-170 L/hm²) (Karimi et al., 1999).

Thus, one can say that MF may act as a barrier against protozoan (oo)cysts. The coagulation/flocculation with moringa associated with microfiltration resulted in high levels of removal of the evaluated parameters.

Figure 3 shows the permeate flux versus time for the microfiltration of deionized water (DW), raw water without coagulant (SW), and pretreated water (CFM).

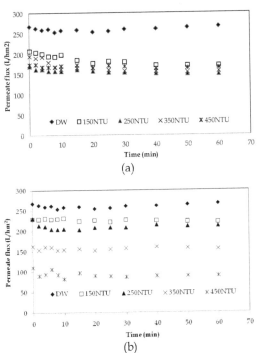

Source: Nishi, 2011.

Fig. 3. Permeate flux with deionized water (DW) and raw water with initial turbidity from 150 to 450 NTU in the MF (a) and CFM-MF (b) processes.

For the MF process with raw water, that is, without previous treatment (coagulation/ flocculation), permeate flux ranged from 157 to 187 L/hm^2 for water samples of turbidity from 150 to 450 NTU. In the combined process (CFM-MF), permeate flux ranged from 157 to 226 L/hm^2 for water samples with initial turbidity of 150 to 350 NTU. Samples of 450 NTU presented the lowest permeate flux, 91 L/hm^2, on average (Nishi, 2011). This may be due to the presence of a greater number of particles that can cause the process of concentration polarization and due to superposition of various fouling mechanisms in the membrane, which may cause the decrease of the permeate flux (Stopka et al., 2001).

The combined processes of coagulation/flocculation/microfiltration showed slightly higher fluxes when compared with the microfiltration process alone. The improvement in permeate flux using coagulation/flocculation prior to microfiltration was also observed in other studies (Katayon et al., 2007; Horčičková et al., 2009).

The percentage of fouling (%F) for the MF process with raw water (SW) and water coagulated/flocculated with moringa (CFM) with initial turbidity from 150 to 450 NTU is shown in Figure 4.

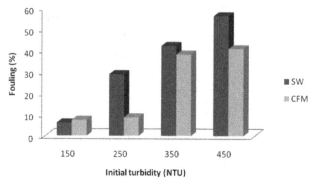

Fig. 4. Percentage of fouling for the MF process with raw water (SW) and water coagulated/flocculated with moringa (CFM) with initial turbidity from 150 to 450 NTU.

It is observed that the MF process with raw water showed higher percentages of fouling, ranging from 6.13 to 56.32% when compared with the combined process of coagulation/ flocculation with moringa followed by MF, which presented percentages of fouling from 7.48 to 40.9% (Nishi , 2011). This reduction in membrane fouling when using the process of coagulation/flocculation as pretreatment was also observed in other studies. Madrona (2010) used coagulation/ flocculation with moringa, followed by MF with ceramic membranes with porosity of 0.1 and 0.2 μm, for the treatment of surface water and observed fouling percentages of around 94% during the filtration of raw water and slightly lower values, around 88%, when water previously coagulated/flocculated with moringa was filtered. Carroll et al. (2000) used polypropylene hollow fiber MF membrane to filter surface water from the Moorabool River, Australia, and observed fouling percentages of 80% for water without pretreatment and 50% for water pretreated by coagulation with alum.

According to Cheryan (1998), the type and extent of fouling depend on the chemical nature of the membrane, the solute, and the solute-membrane interactions, as well as on the porosity of the membrane and the working pressure used in the process.

3. Conclusions

Performance evaluation of the hybrid systems (CFS-UF, CFQ-UF, and CFM-MF) showed that the permeate quality was increased when compared with individually operated systems. This is justified by the excellent ability of the MF/UF process to remove particles and colloids. The results also indicate that when applying CF-MF/UF at optimum conditions, a hygienic barrier effect was achieved for the treatment scheme, in which nearly 100% removal of total coliforms, *E. coli, Giardia* and *Cryptosporidium* was obtained at the end of the process. In addition, the combined processes CFQ-UF, CFM-MF, and CFS-UF produced drinking water in accordance with the legislation.

Given the above considerations, one can say that chitosan and *Moringa oleifera* have a potential application as natural coagulants in CF-MF/UF hybrid processes for treating drinking water with relatively high turbidity. This process can be used reliably to produce drinking water of excellent quality.

4. List of abbreviations

Coagulation/flocculation = CF
Coagulation/flocculation using chitosan as coagulant = CFQ
Coagulation/flocculation using aluminum sulfate as coagulant = CFS
Coagulation/flocculation using chitosan as coagulant followed by ultrafiltration = CFQ-UF
Coagulation/flocculation using aluminum sulfate as coagulant followed by ultrafiltration = CFS-UF
Coagulation/flocculation using moringa as coagulant = CFM
Coagulation/flocculation using moringa as coagulant followed by microfiltration = CFM-MF
Microfiltration = MF
Ultrafiltration = UF
Moringa oleifera = moringa
Natural organic matter = NOM

5. References

Akhtar, M., Moosa Hasany, S., Bhanger, M. I. & Iqbal, S. (2007). Sorption potential of *Moringa oleifera* pods for the removal of organic pollutants from aqueous solutions. *Journal of Hazardous Materials*, Vol. 141, No. 3, pp. 546–556.

Aldom, J.E. & Chagla, A.H. (1995). Recovery of *Cryptosporidium* oocysts from water by a membrane filter dissolution method. *Letters in Applied Microbiology*, Vol. 20, pp. 186-187.

American Public Health Association [APHA]. [1995]. *Standard Methods for the Examination for Water and Wastewater* (19th edition). Byrd Prepess Springfield, ISBN 0875532233, Washington.

Balakrishnan, M., Dua, M. & Khairnar, P.N. (2001). Significance of membrane type and feed stream in the ultrafiltration of sugarcane juice. *Journal of Separation Science and Technology*, Vol. 36, No. 4, pp. 619-637.

Bergamasco, R., Bouchard, C., da Silva, F.V., Reis, M.H.M., Fagundes-Klen, M.R. (2009). An application of chitosan as a coagulant/flocculant in a microfiltration process of natural water. *Desalination*, Vol. 245, pp. 205–213.

Bergamasco, R., Konradt-Moraes, L.C., Vieira, M.F., Fagundes-Klen, M.R. & Vieira, A.M.S. (2011). Performance of a coagulation–ultrafiltration hybrid process for water supply treatment. *Chemical Engineering Journal*, Vol. 166, pp. 483–489.

Bhatia, S., Othman, Z. & Ahmad, A.L. (2007). Pretreatment of palm oil mill effluent (POME) using *Moringa oleifera* seeds as natural coagulant. *Journal of Hazardous Materials*, Vol. 145, No. 1-2, pp. 120-126.

Bottino, A., Capannelli, C., Del Borghi, A., Colombino, M. & Conio, O. (2001). Water treatment for drinking purpose: ceramic microfiltration application. *Desalination*, Vol. 141, pp. 75–79.

Bouchard, C., Laflamme, E., Serodes, J., Ellis, D., Rahni, M. & Rodrigues, M. (2003). Étude en la laboratorie de l'ultrafiltration et de la coagulation-ultrafiltration d'une eau colorée. *Proceedings of 17éme Symposium de l'Est du Canada sur la Recherche Portant sur la Pollution de l'eau.*

Brown, T.J. & Emelko, M.B. (2009). Chitosan and metal salt coagulant impacts on *Cryptosporidium* and microsphere removal by filtration. *Water Research*, Vol. 43, No. 331–338.

Bustamante, H.A., Shanker, S.R., Pashley, R.M. & Karaman, M.E. (2001). Interaction between *Cryptosporidium* oocysts and water treatment coagulants. Water *Research*, Vol. 35, pp. 3179-3189.

Cardoso, K.C., Bergamasco, R., Cossich, E.S. & Konradt-Moraes, L.C. (2008). Otimização dos tempos de mistura e decantação no processo de coagulação/floculação da água bruta por meio da *Moringa oleifera* Lam. *Acta Scientiarum – Technology*, Vol. 30, pp. 193-198.

Carroll, T., King, S., Gray, S. R., Bolto, B. A. & Booker, N. A. (2000). The fouling of microfiltration membranes by nom after coagulation treatment. *Water Research*, Vol. 34, No. 11, pp. 2861 – 2868.

Centers for Disease Control and Prevention [CDC]. (2006). *Surveillance Summaries*, December 22. MMWR, 55 (No. SS-12).

Chang, I.S., Bag, S.O. & Lee, C.H. (2001). Effects of membrane fouling on solute rejection during membrane filtration of activated sludge. *Process Biochemistry*, Vol. 36, pp. 855–860.

Cheryan, M. (1998). *Ultrafiltration and microfiltration handbook.* Technomic Publishing CO, Illinois, Lancaster, USA.

Chuang, P.H., Lee, C.W., Chou, J.Y., Murugan, M., Shieh, B.J. & Chen, H.M. (2007). Anti-fungal activity of crude extracts and essential oil of *Moringa oleifera* Lam. *Bioresource Technology*, Vol. 98, pp. 232–236.

Chung, Y.C., Wang, H.L., Chen, Y.M. & Li, S.L. (2003). Effect of abiotic factors on the antibacterial activity of chitosan against waterborne pathogens. *Bioresource Tech.*, Vol. 88, pp. 179-184.

Coelho, J.S., Santos, N.D.L., Napoleão, T.H., Gomes, F.S., Ferreira, R.S., Zingali, R.B., Coelho, L.C.B.B., Leite, S.P., Navarro, D.M.A.F. & Paiva P.M.G. (2009). Effect of **Moringa** *oleifera* lectin on development and mortality of *Aedes aegypti* larvae. *Chemosphere*, Vol. 77, No. 7, pp. 934-938.

Davino, F. (1976). *Tecnologia de tratamento de água: água na indústria.* Almeida Neves, Rio de Janeiro, Brazil.

Dawson, D.J., Maddocks, M., Roberts, J. & Vidler, J.S. (1993). Evaluation of recovery of *Cryptosporidium parvum* oocysts using membrane filtration. *Letters in Applied Microbiology*, Vol. 17, pp. 276-279.

Divakaran, R. & Pillai, V.N.S. (2002). Flocculation of river silt using chitosan. *Water Research*, Vol. 36, No. 9, pp. 2414-2418.

Driscoll, C.T. & Letterman, R.D. (1995). Factors regulating residual aluminium concentrations in treated waters. *Environmetrics*, Vol. 6, pp. 287-309.

Eikebrokk, B. (1999). Coagulation-direct filtration of soft, low alkalinity humic waters. *Water Science and Technology*, Vol. 40, No. 9, pp. 55–62.

Eikebrokk, B. & Saltnes, T. (2001). Removal of natural organic matter (NOM) using different coagulants and lightweight expanded clay aggregate filters. *Water Science & Technology: Water Supply*, Vol. 1, No. 2, pp. 131–140.

Fayer, R. (2004). *Cryptosporidium*: a water-borne zoonotic parasite. *Veterinary Parasitology*, Vol. 126, pp. 37–56.

Franco, R.M.B., Rocha-Eberhardt, R. & Cantusio Neto, R. (2001). Occurrence of *Cryptosporidium* oocysts and *Giardia* cysts in raw water from the Atibaia river, Campinas, Brazil. *Rev. Inst. Med. Trop. S. Paulo*, Vol. 43, No. 2, pp. 109-111.

Ghebremichael, K.A., Gunaratna, K.R., Henriksson, H., Brumer, H. & Dalhammar, G. (2005). A simple purification and activity assay of the coagulant protein from Moringa *oleifera* seed. *Water Research*, Vol. 39, No. 11, pp. 2338-2344.

Guigui, C., Rouch, J.C., Durand-Bourlier, L., Bonnelye, V. & Aptel, P. (2002). Impact of coagulation conditions on the in-line coagulation/uf process for drinking water production. *Desalination*, Vol. 147, pp. 95-100.

Guo, X., Zhang, Z., Fang, L. & Su, L. (2009). Study on ultrafiltration for surface water by a polyvinylchloride hollow fiber membrane. *Desalination*, Vol. 238, pp. 183–191.

Horčičková, J., Mikulášek, P. & Dvořáková, J. (2009). The effect of pre-treatment on crossflow microfiltration of titanium dioxide dispersions. *Desalination*, Vol. 240, pp. 257-261.

Hsu, B.M. & Huang, C. (2002). Influence of ionic strength and pH on hydrophobicity and zeta potential of *Giardia* and *Cryptosporidium*. *Colloids and Surfaces A: Physicochemical and Engineering Aspects*, Vol. 201, pp. 201-206.

Iacovski, R.B., Barardi, C.R.M. & Simões, C.M.O. (2004). Detection and enumeration of *Cryptosporidium* sp. oocysts in sewage sludge samples from the city of Florianópolis (Brazil) by using immunomagnetic separation combined with indirect immunofluorescence assay. *Waste Manage Res.*, Vol. 22, pp. 171–176.

International Water Association [IWA]. (2010). Coagulation and Flocculation in Water and Wastewater Treatment. In: *Water Wiki*, april 28Th 2011, Available from: http://iwawaterwiki.org/xwiki/bin/view/Articles/CoagulationandFlocculationi nWaterandWastewaterTreatment.

Jacangelo, J. G., Adham, S. S. & Laîné, J-M. (1995). Mechanism of *Cryptosporidium, Giardia* and MS2 virus removal by MF and UF. *Journal of the American Water Works Association*, Vol. 87, No. 9, pp. 107–121.

Karimi, A.A., Vickers, J.C. & Harasick, R.F. (1999). Microfiltration goes Hollywood: the Los Angeles experience. *Journal of the American Water Works Association*, Vol. 91, No. 6, pp. 90–103.

Katayon, S., Noor, M.J.M.M., Asma, M., Ghani, L.A.A., Thamer, A.M., Azni, I., Ahmad, J., Khor, B.C. & Suleyman, A.M. (2006). Effects of storage conditions of *Moringa oleifera* seeds on its performance in coagulation. *Bioresource Technology*, Vol. 97, No. 13, pp. 1455–1460.

Katayon, S., Noor, M.J.M.M., Tat, W.K, Halim, G.A., Thamer, A.M. & Badronisa, Y. (2007). Effect of natural coagulant application on microfiltration performance in treatment of secondary oxidation pond effluent. *Desalination*, Vol. 204, pp. 204-212.

Kawamura, S. (1991). Effectiveness of natural polyelectrolytes in water treatment. *Journal Awa*, Japan, Vol. 79, No. 6, pp. 88-91.

Kim, M.H. & Yu, M.J. (2005). Characterization of NOM in the Han River and evaluation of treatability using UF - NF membrane, Environmental Research, No. 97, pp. 116 - 123.

Klopotek, A.D., Wlaasiuky, D. & Klopotek, B.B. Compounds based on chitosan as coagulants and flocculants, *Proceedings of International Conference on Chitin and Chitosan*, Gydnia, Poland, August, 1994.

Konradt-Moraes, L.C. (2004). Estudo da coagulação-ultrafiltração para produção de água potável. Master of Science Thesis, Universidade Estadual de Maringá – Maringá, PR, Brasil, 135 pp. (in Portuguese).

Konradt, L.C.; Bergamasco, R.; Tavares, C.R.G.; Bongiovani, M.C. & Hennig, D. (2008). Utilization of the coagulation diagram in the evaluated of the natural organic matter (NOM) removal for obtaining potable water, International Journal of Chemical Reactor Engineering, No. 6, pp. 1 - 6.

Lechevallier, M.W. & Au, K. (2004). *Water treatment and pathogen control: process efficiency in achieving safe drinking water.* WHO Drinking Water Quality Series, UK, 136pp.

Lindquist, A. (1999). *Emerging pathogens of concern in drinking water.* United States Environmental Protection Agency, EPA, 600/R-99/070.

McLachlan, D.R.C. (1995). Aluminum and the risk for Alzheimer's Disease. *Environmetrics*, Vol. 6, pp. 233-275.

Madrona, G.S., Serpelloni, G.B., Vieira, A.M.S., Nishi, L., Cardoso, K.C & Bergamasco, R. (2010). Study of the effect of saline solution on the extraction of the *Moringa oleifera* seed's active component for water treatment. *Water, Air, & Soil Pollution*, Vol. 211, pp. 409–415.

Madrona, G.S. (2010). Extração/purificação do composto ativo da semente da *Moringa oleifera* Lam e sua utilização no tratamento de água para consumo humano. Doctoral Thesis, Universidade Estadual de Maringá – Maringá, PR, Brasil, 176 pp. (in Portuguese).

Muyibi, S.A. & Evison, L.M. (1995). *Moringa oleifera* seeds for softening hardwater. *Water Research*, Vol. 29, No. 4, pp. 1099–1105.

Ndabigengesere A., Narasiah, K.S. & Talbot, B.G. (1995). Active agents and mechanism of coagulation of turbid waters using *Moringa oleifera. Water Research*, Vol. 29, No. 2, pp. 703-710.

Nishi, L. (2011). Estudo dos processos de coagulação/floculação seguido de filtração com membranas para remoção de protozoários parasitas e células de cianobactérias. Doctoral Thesis, Universidade Estadual de Maringá – Maringá, PR, Brasil, 203 pp. (in Portuguese).

Nkurunziza, T., Nduwayezu, J.B., Banadda, E.N. & Nhapi, I. (2009). The effect of turbidity levels and *Moringa oleifera* concentration on the effectiveness of coagulation in water treatment. *Water Science and Technology*, Vol. 59, pp. 1551–1558.

Okuda, T., Baes, A.U., Nishijima, W. & Okada, M. (1999). Improvement of extraction method of coagulation active components from *Moringa oleifera* seed. *Water Research*, Vol. 33, No. 15, pp. 3373-3378.

Okuda, T., Baes, A.U., Nishijima, W. & Okada, M. (2001). Isolation and characterization of coagulant extracted from *Moringa oleifera* seed by salt solution. *Water Research*, Vol. 35, No. 2, pp. 405-410.

Parker, D.Y., Leonard, M.J., Barber, P., Bonic, G., Jones W. & Leavell, K.L. Microfiltration treatment of filter backwash recycle water from a drinking water treatment facility. *Proceedings of American Water Works Association Water Quality Technology Conference*. Denver, CO, 1999.

Ramos, R.O. (2005). Clarification of water with low turbulence and moderate color using seeds of *Moringa oleifera*. Doctoral Thesis. State University of Campinas - Campinas,SP, Brazil. 276 pages. (in Portuguese).

Reddy, V.; Urooj, A. & Kumar, A. (2005). Evaluation of antioxidant activity of some plant extracts and their application in biscuits. *Food Chemistry*, Vol. 90, pp. 317-321.

Renault, F., Sancey, B., Badot, P.M. & Crini, G. (2009). Chitosan for coagulation/flocculation processes-an eco-friendly approach. *European Polymer Journal*, Vol. 45, pp. 1337–1348.

Rey, L. (2001). *Parasitologia* (3th edition). Guanabara Koogan, ISBN 8527706776, Rio de Janeiro, Brazil.

Rinaudo, M. (2006). Chitin and chitosan: properties and applications. *Progress in Polymer Science*, Vol. 31, pp. 603–632.

Rizzo, L., Di Gennaro, A., Gallo, M. & Belgiorno, V. (2008). Coagulation/chlorination of surface water: a comparison between chitosan and metal salts. *Separation and Purification Technology*, Vol. 62, pp. 79–85.

Roussy, J., Van Vooren, M., Dempsey, B.A. & Guibal, E. (2005). Influence of chitosan characteristics on the coagulation and the flocculation of bentonite suspensions. *Water Research*, Vol. 39, pp. 3247–3258.

Schafer, A.I., Fane, A.G. & Waite, T.D. (2001). Cost factors and chemical pretreatment effects in the membrane filtration of waters containing natural organic matter,. *Water Research*, Vol. 35, pp. 1509–1517.

Stopka, J., Bugan, S.G., Broussous, L., Schlosser, S. & Larbot, A. (2001). Microfiltration of beer yeast suspensions through stamped ceramic membranes. *Separation and Purification Technology*, Vol. 25, pp. 535-543.

United States Environmental Protection Agency [USEPA]. (1996). *National Primary Drinking Water Regulations: Monitoring Requirements for Public Drinking Water Supplies*, Final Rule, 40CFR Part 141.

Verbych, S., Bryk, M., Alpatova, A. & Chornokur, G. (2005). Ground water treatment by enhanced ultrafiltration. *Desalination*, Vol. 179, No. 1-3, pp. 237-244.

Vieira, A.M.S., Vieira, M.F., Silva, G.F., Araújo, A.A., Fagundes-Klen, M.R., Veit, M.T. & Bergamasco, R. (2010). Use of *Moringa oleifera* Seed as a Natural Adsorbent for Wastewater Treatment. *Water, Air, & Soil Pollution*, Vol. 206, pp. 273–281.

Xagoraraki, I. & Harrington, G.W. (2004). Zeta potential, dissolved organic carbon, and removal of *Cryptosporidium* oocysts by coagulation and sedimentation. *Journal of Environmental Engineering*, Vol. 130, pp. 1424-1432.

Xia, S., Li, X., Zhang, Q., Xu, B. & Li, G. (2007). Ultrafiltration of surface water with coagulation pretreatment by streaming current control. *Desalination*, Vol. 204, pp. 351–358.

4

In situ Remediation Technologies Associated with Sanitation Improvement: An Opportunity for Water Quality Recovering in Developing Countries

Davi Gasparini Fernandes Cunha[1], Maria do Carmo Calijuri[1],
Doron Grull[1], Pedro Caetano Sanches Mancuso[1] and Daniel R. Thévenot[2]
[1]Universidade de São Paulo,
[2]LEESU, Université Paris-Est,
[1]Brazil
[2]France

1. Introduction

The access to safe water is of great importance to reduce the spread of diseases caused by water-related pathogens and to assure the life quality to the human-beings. According to the World Health Organization (WHO, 2011), diarrhea, for example, is responsible for two million deaths every year, mainly among children under the age of five. The environmental effects of some pollutants (e.g. endocrine disruptors, organic compounds) remain unclear and the harmful consequences of the exposure to contaminated water are certainly an important issue for the next decades. Moreover, many research have linked water quality to health problems, such as cancer (Rodrigues et al., 2003; Han et al., 2009), insufficient uptake of nutrients and trace-metals (Lind & Glynn 1999), diabetes, cerebrovascular and kidney disease (Meliker et al., 2007).

The costs and benefits of water quality have been the topic of stimulating discussion in the scientific community (Isaac, 1998; Hajkowicz et al., 2008; Saz-Salazar et al., 2009) because water quality decrease implies not only loss of lives, but also economic damages. The costs of the anthropogenic eutrophication reach US$2.2 billion in the United States (Dodds et al., 2009) and US$187.2 million in England and Wales every year (Pretty et al. 2002). The reduction of nutrient loading to the aquatic systems worldwide is the cornerstone of artificial eutrophication control (Smith et al., 1999), with repercussions in other fields like public health and economics.

The anthropogenic impacts on the quality of urban water bodies in developing countries are frequently exacerbated by poor levels of sanitation and inadequate water and wastewater management. Pressure from urban areas on the water quality was reported in Argentina (Almeida et al., 2007), Brazil (Jordão et al., 2007), India (Suthar et al., 2010) and Mexico (Bravo-Inclan et al., 2008). Rapid shifts in the land use patterns, unplanned urbanization and inefficient resources allocation are further aggravating environmental problems in such nations. Restrictions to the water uses are increasing as the pollution of rivers and lakes is offering more risks to the human health and to the maintenance of the ecological balance.

Within this context, the water resources management plays an important role in the conciliation of the water uses and the long-term sustainability. The *in situ* remediation of rivers, lakes and reservoirs is a decentralized alternative that may be convenient in some cases in comparison to off-site solutions. The main advantages of the *in situ* approach are, besides the relative small period of time required to its implementation, the suitability of the *in situ* facilities to the regions with lack of available areas to build off-site treatment plants (e.g. highly urbanized areas) and the lower expenses with pumping structures. Although it takes more time and requires more investments, the implementation of sanitation infrastructure is also necessary.

With the increase of the negative environmental impacts induced by the anthropogenic activities, the remediation of aquatic systems became an alternative to restore the ecological functions of the ecosystems and accelerate their recovery. The first and most important step in a remediation project is to define the remedial action aims to be accomplished at the site, involving the desirable mechanisms of treatment – biological (e.g. phytoremediation), physical and/or chemical (e.g. oxidation, air stripping, ion exchange, precipitation). Most of the current technologies for aquatic systems remediation were adapted from unitary processes used for drinking water production, industrial purposes or wastewater treatment. The flotation, for example, has been used in mining activities to separate the mineral of interest from the gangue since 1893 (Hoover, 1912). The technology was then adapted to treat water and wastewater through dissolved air flotation (e.g. Heinänen et al., 1995). Ultrafiltration membranes in turn have been mainly used for drinking water production (2 million m^3/day worldwide according to Laîné et al., 2000). According to the same authors, the oldest water industry with ultrafiltration plant started to operate in 1988 in France. The membranes are becoming cheaper over the years and the technology is more attractive for remediation of surface waters at the present time.

2. Water management in developing countries: Long and short term actions

The water management in developing countries would benefit from a well-weighted balance between long and short term actions (Fig. 1). The former actions should consider sanitation planning and infrastructure, whereas the latter ones should focus on the solutions for remediating the aquatic systems or attenuating their degradation level.

The main issues involved in both long and short term actions for water resources management are:

i.   **Political commitment**. Sanitation and environmental recovery programs usually do not receive the same amount of investments in comparison to other areas (Varis et al., 2006). Local authorities willingness and specific government policies are necessary for meeting the health and environmental goals;

ii. **Institutional framework**. The effectiveness of the water management policies would be at risk with no solid institutions and skilled professionals. The technical and social challenges will just be overcome with trained planners, sector professionals and decision-makers;

iii. **Financing**. This is a complicated question because water quality recovery brings benefits to the health and to the environment, making it a public good. However, at the same time, water is also a private good (at the level of households);

iv. **Technology development and Innovation**. Sanitation and remediation of aquatic systems depend upon technology development, under a cost-benefit analysis. Moreover, the technology has to be adapted and optimized to the local peculiarities. In the case of the developing countries, technology transfer from developed countries might be necessary;

v. **Monitoring**. Monitoring programs play a pivotal role in assessing if the targets were met. Such programs must be able to provide feedback to improve the monitored system with a view to increasing efficiency and reducing costs;

vi. **Social acceptance**. It is desirable that people get involved with the decision-making process, increasing the chances of social acceptance of the water resources management policies or programs.

According to Calijuri et al. (2010), the water resources sustainability can be defined as a state of dynamic equilibrium between the disturbances imposed to the water bodies by the anthropogenic activities and the aquatic systems ability to self-regulation (i.e. their elasticity in response to a certain impact). When the impact is strong enough to prevent the self-recovery of the original condition, actions towards remediation, including palliative/temporary solutions, are required to avoid critical levels of degradation and severe impairment of different water uses.

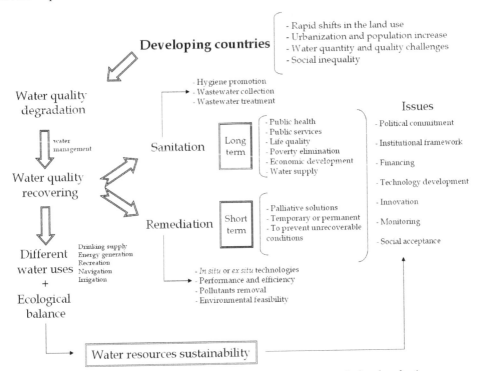

Fig. 1. Scheme of the desirable water quality management approach for developing countries, including the short term and the long term actions description.

2.1 Sanitation infrastructure

Approximately 1.1 billion people in the world do not have access to improved water supply sources and 2.4 billion people do not have access to any type of improved sanitation at all (WHO, 2011). It is clear that there is a need for additional water and sanitation services from government in partnership with other actors worldwide and especially in the megacities from the developing countries (Biswas et al., 2004). The implementation of sanitation infrastructure is a long term action that has to be continuously monitored and updated. Solid waste collection and disposal, water distribution and water and wastewater treatment are the main components of the sanitation in a country. Such items are related to life quality promotion, poverty elimination and economic development. The temporal evolution of sewage, water supply and solid waste collection services in Brazil is shown in Fig. 2.

According to the Brazilian Institute of Geography and Statistics (IBGE, 2011), there was an increase in the availability of the sewage system (i.e. wastewater collection) in the urban areas of the country from 1992 (46%) to 2009 (59%). Similar increase was observed for the availability of the solid waste collection services (62% in 1992 and 82% in 2009). The situation was worse in the rural areas, where the figures for sewage system and water supply systems reached only 5% and 33% in 2009, respectively (Fig. 2). In the year 2000, only 20% of the Brazilian municipalities treated the domestic wastewater (IBGE, 2000). The remaining loads ended up in the water bodies, contributing to water quality degradation (e.g. by increasing organic matter content and decreasing dissolved oxygen concentrations).

As shown for Brazil, the sanitation conditions in other developing countries are similar: higher levels of drinking water and sewage systems in urban areas, as compared to rural ones (Massoud et al., 2009). According to the WHO (2010), people living in low-income nations are least likely to have access to adequate sanitation infrastructure. The absence of sanitation and the access to unsafe water are therefore risk factors that are linked with increased mortality and morbidity worldwide.

2.2 *In situ* remediation technologies

Dissolved Air Flotation (DAF), Ultrafiltration Membranes (UM) and Enhanced Biological Removal (EBR) are examples of *in situ* technologies for water or wastewater treatment (Table 1). Such technologies can be used either individually or in association (Geraldes et al., 2008).

The DAF consists of the following steps:

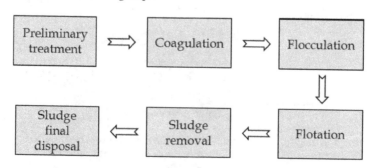

In situ Remediation Technologies Associated with Sanitation Improvement: An Opportunity for Water Quality
Recovering in Developing Countries

59

i. **Preliminary treatment** – barriers are installed in the river channel to remove coarse material;
ii. **Coagulation** – chemical compounds are added to the water to promote the coagulation (e.g. ferric chloride or aluminium hydroxide). Specific coagulation time and velocity gradients are needed;
iii. **Flocculation** – through certain flocculation time and velocity gradient, flakes are formed by aggregation of colloids;
iv. **Flotation** – small air bubbles are produced in the bottom of the river channel and their upward movement is able to bring the colloidal and particulate matter to the water surface, as a sludge;
v. **Sludge removal** – the sludge is removed from surface water through rotating blades or other mechanical device;
vi. **Sludge final disposal.**

The operation costs of a DAF system vary between 0.10 and US$0.20/m³. One of the biggest concerns in the DAF plants is the significant consumption of chemicals for aggregation and flocculation of colloids and consequently the high production of sludge. Different processes have been used for thickening and dewatering the sludge produced by DAF (i.e. increasing the solids content) and reducing its volume (Dockko et al., 2006). However, the feasibility of the alternative for final sludge disposal (e.g. application in the agriculture or disposal in a landfill) depends upon its toxicity due to the presence of metals and other persistent pollutants (Mantis et al., 2005; Luz et al., 2009).

The treatment process with UM is based in the following steps:

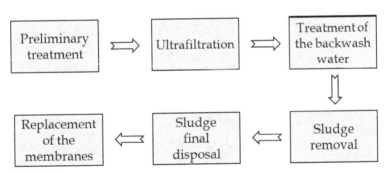

i. **Preliminary treatment** – removal of coarse material and sand;
ii. **Ultrafiltration** – the water passes through semipermeable membranes;
iii. **Treatment of the backwash water** – the contaminants rejected by the membranes accumulate on the membranes forming a fouling layer and requiring a periodical backwash to remove the debris. The water used for backwash has to be treated as it contains high concentrations of pollutants;
iv. **Sludge removal from backwash water;**
v. **Sludge final disposal;**
vi. **Replacement of the membranes** – the replacement must be achieved according to the membrane life, usually estimated as 5 to 8 years.

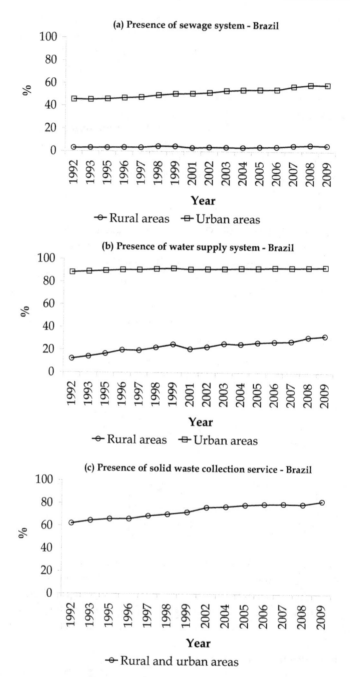

Fig. 2. Evolution (%) of the availability of the (a) sewage, (b) drinking water supply systems and (c) solid waste collection services in Brazil from 1992 to 2009. Reference: IBGE (2011).

In situ Remediation Technologies Associated with Sanitation Improvement: An Opportunity for Water Quality
Recovering in Developing Countries

61

In situ technology	Brief description	Total operating costs (US$/m³)	Benefits	Limitations
Dissolved Air Flotation (DAF)	Small air bubbles carry the impurities to the surface (as sludge) after previous coagulation and flocculation	0.10 – 0.20	- High efficiency for the removal of some nutrients (e.g. phosphorus), hydrocarbons and surfactants	- Significant amount of sludge to be managed - Chemicals consumption for coagulation and flocculation
Ultrafiltration Membranes (UM)	The water is forced against a semipermeable membrane	0.03 – 0.25	- No addition of chemicals - The membranes may be fed with the water course own pressure - Smaller footprint	- Preliminary treatment is required to remove coarse and sand material (to protect the membranes) - Treatment of the backwash water is necessary
Enhanced Biological Removal (EBR)	The growth of certain types of bacteria and the biological conversion are stimulated with the addition of specific enzymes or inoculums (commercially available: see Table 2).	0.01 – 0.15	- Minimum infrastructure is required - Some added reagents increase gas transfer rates and dissolved oxygen concentrations - Degradation of toxics - Rapid odor elimination	- Acceptability of the addition of inoculums to the aquatic systems, especially if they are exogenous and if the surface water is used for drinking water production

Table 1. Brief description of some *in situ* technologies: Dissolved Air Flotation, Ultrafiltration Membranes and Enhanced Biological Removal; comparison of their operating costs, efficiency and performance criteria.

The more stringent regulations regarding water quality associated with the reduction of costs of the ultrafiltration membranes have made this technology more attractive in the last years. UM can be used for drinking water production (Laîné et al., 2000; Xia et al., 2005) and river water treatment (Konieczny et al., 2009). The lower level of energy requirements and the lower consumption of chemicals are the main advantages of the UM systems. The need for frequent backwash preventing membrane fouling, especially when the organic matter content in the raw water is high, is an important issue to be managed.

However, the addition of a preliminary step with coagulation [e.g. with $FeCl_3$, $Fe_2(SO_4)_3$ or $Al_2(SO_4)_3$] and flocculation may increase the efficiency of the system and avoid too rapid membrane fouling (Konieczny et al., 2006; Babel & Takizawa, 2011; Bergamasco et al., 2011). The treatment of the backwash water is also an issue because the effluent from the backwash contains significant concentrations of the pollutants previously accumulated on the membranes. The backwash water is normally sent to a thickening tank, where the suspended matter settles to the bottom (as sludge). The clear water on top is then pumped to the upstream part of the treatment plant. Some recent studies consider the recycling of the backwash water as an interesting alternative, e.g. through a blending of 10% of backwash water and 90% of raw water (Gora & Walsh, 2011). The operating costs of the UM are expected to approximately range between 0.03 and US$0.25/$m^3$ and the main factors influencing the costs are the quality of the water to be treated, labor, energy and chemicals (if any) consumption as well as membrane replacement.

The EBR in turn is based in a single step, the **application of bioremediators** (Table 2) in the river or reservoir to be treated and subsequent monitoring of their efficiency in relation to the targeted pollutants.

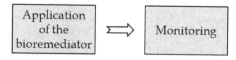

The bioremediators, which can be microorganisms (e.g. bacteria) or enzymes, are able to increase biodegradation and gas transfer rates. There is a considerable variety of products available in the market. Some examples are shown in Table 2 with the commercial name of the bioremediators, their definition, main applications and highlights according to their manufacturer. The bioremediators are expected to perform the removal, transformation or detoxification of pollutants from the aquatic environment into a less toxic form (Whiteley & Lee, 2006). The effect of the bioremediators is normally based in the combination of several processes, like solubilization (physical), oxidation (chemical) and catalysis (biological). Some recent studies have recognized the effectiveness of enzymatic processes of water remediation, mainly when associated with established technologies (Demarche, in press). However, some factors like costs and stability of the biocatalysts require further investigation.

There are two broad types of remediation through microorganisms: biostimulation (stimulation of the growth of indigenous microorganisms) and bioaugmentation (introduction of specific microorganisms to the local population). In both cases, predation, competition, adaptation to the environmental matrix (water, wastewater), possible adsorption on available solids as well as survival strategies play an important role in determining the overall efficiency of the remediation (Fantroussi & Agathos, 2005). The use of immersed biofilms (e.g. artificial plastic substrates) is also an alternative to perform water treament taking advantage of the nutrient uptake by algae (Jarvi et al. 2002) or the organic matter degradation by bacteria (Bishop, 2007). Although biofilms are usually unwanted in drinking water treatment stations due to the biofouling, they may be useful for water remediation and biodegradation of persistent compounds. Ammonia-nitrogen concentrations removal, for example, may be considerably improved with the growth of nitrifying bacteria (Jiao et al., 2011).

In situ Remediation Technologies Associated with Sanitation Improvement: An Opportunity for Water Quality Recovering in Developing Countries

63

Company / Product	Definition	Highlights
Bio-Organic Catalyst™	"Fermentation supernatant derived from plants and minerals, which is blended synergistically in combination with a non-ionic surfactant to create a broad spectrum bio-organic catalyst"	- Dissolved oxygen levels increase - Biological nitrogen removal is enhanced - Solubilization rates of insoluble fats, oils and grease increase
Advanced BioCatalytics Corporation/ Accell3 Green™	"Refined, fermentation-derived, bioactive stress proteins that are formulated with surfactants"	- Bacterial metabolism is enhanced to accelerate the breakdown of organic material by oxidation - Organic contaminants are digested more rapidly and more completely to carbon dioxide and water
Enzilimp™	"Facultative bacteria are added to boost the nitrogen cycle and accelerate the organic matter degradation"	- Odors elimination - Reduction of Total and Fecal Coliforms, Biochemical and Chemical Oxygen Demand
Bioplus Biol2000™	"Facultative bacteria that are able to promote the degradation of organic compounds from industrial effluents"	- Biochemical and Chemical Oxygen Demand reduction - Phosphorus and nitrogen concentrations decrease
BioTC™ Rinenbac/Rinenzim	"Specific bacteria to stimulate degradation of fats and organic matter"	- Odors elimination - Reduction of Coliforms and Biochemical Oxygen Demand
Realco™ Realzyme	"Enzymes that are able to transform biofilms into water-soluble organic residues"	- Avoid the contamination from a biofilm

Table 2. Examples of some bioremediators available in the market, including their definition and major benefits, according to their manufacturer.

3. Case study with DAF: The Pinheiros River (São Paulo State, Brazil)

The Pinheiros River (23°42'S; 46°40'W) is located in the Southeast Region of Brazil (Fig. 3) in an extremely urbanized area in the Metropolitan Region of São Paulo, whose population is approximately 19.7 million inhabitants. This river links the Tietê River to the Billings Reservoir (storage volume: 995 million m^3; residence time: 30 months), a multipurpose water system used for drinking supply, energy generation (through Henry Borden Power Plant), navigation and recreation.

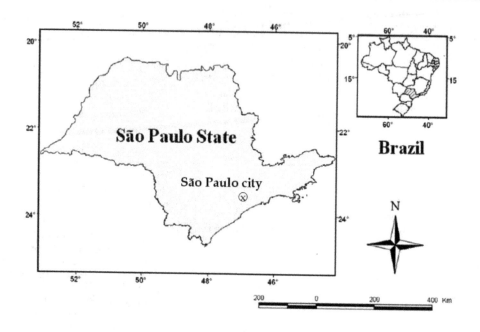

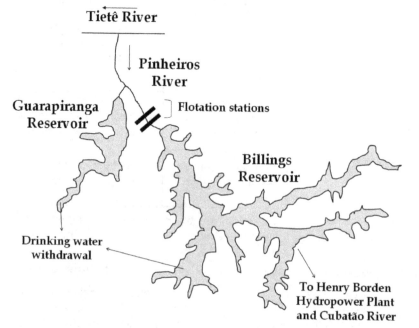

Fig. 3. Map of the São Paulo State, located in the Southeast Region of Brazil, and a scheme of the Tietê and Pinheiros Rivers and the Billings and Guarapiranga Reservoirs, two important multipurpose reservoirs in the area.

Considering the technology availability and the need for recovering the Pinheiros River water quality to assure no impacts to the Billings Reservoir, the DAF was tested. The *in situ* DAF pilot-scale system, composed by two treatment plants (Zavuvuz – 23°40'44"S; 46°41'53"W and Pedreira – 23°42'01"S; 46°40'56"W), was installed in the Pinheiros River channel to treat 10 m³/s (Table 3) with subsequent pumping of the water to the Billings Reservoir (Fig. 4). Previous coarse material removal was necessary. Coagulation with ferric chloride and flocculation were followed by the flotation step (i.e. the production of tiny air bubbles in the bottom of the river). The upward movement of such bubbles was responsible for bringing the impurities to the surface, where they were collected through rotating blades for sludge removal. The treated flow (10 m³/s) of the system installed in the Pinheiros River is significantly bigger than other similar treatment stations in Brazil: 0.05 m³/s reported by Lopes & Oliveira (1999), 0.15 m³/s (Oliveira et al., 2000) and 0.75 m³/s (Coutinho & von Sperling, 2007).

Operational parameter or variable (unit)	Value or range
Tietê River mean flow (m³/s)	120
Pinheiros River mean flow (m³/s)	5*
Zavuvuz Stream mean flow (m³/s)	0.7
Treated flow (m³/s)	10
Recycle flow (m³/s)	0.8 – 1.0
Hydraulic detention time (min)	25 – 30
Ferric chloride dosages (mg/L)	50 – 400
Rapid mixing (coagulation) time (min)	0.5
Rapid mixing velocity gradient (/s)	800
Slow mixing (flocculation) time (min)	27
Slow mixing velocity gradient (/s)	60
Sludge production in both plants (m³/day)	150
Solids content in the sludge after centrifugation (%)	20
Energy consumption in both plants (kWh/day)	42,000

Table 3. Major operational parameters or variables of the *in situ* DAF pilot-scale system (both plants) in the Pinheiros River, São Paulo, Brazil. * When the pilot-scale system was operating, there was a contribution from the Tietê River waters to the total flow of the Pinheiros River. Reference: adapted from Cunha et al. (2010).

The system was operated from August 2007 to March 2010. A comprehensive monitoring program (148 water variables and more than 200,000 laboratory analyses) was delineated to assess the efficiency and feasibility of the pilot-scale prototype. Detailed information about the monitoring results in the Tietê and Pinheiros Rivers and in the Billings Reservoir may be found in some recent papers (Cunha et al., 2010; Cunha et al., 2011a and Cunha et al., 2011b). The efficiency achieved by each flotation station and the overall effect for some variables is shown in Table 4. The global removal efficiency achieved by both DAF treatment stations was 90% for total phosphorus, 54% for apparent color, 53% for chemical oxygen demand, 48% for turbidity, 40% for total suspended solids, 31% for soluble iron and only 2% for nitrogen-ammonia. The prototype promoted an increment of about 60% in the dissolved oxygen concentrations.

Through the operation and performance assessment of the DAF system, some positive aspects have been observed:

i.   The technology was available and the local government was willing to promote the reclamation of the water quality of the Pinheiros River and to stimulate additional energy generation with the Billings Reservoir water. The conjunction of technology availability, environmental and economic issues (as shown in Fig. 1) proved to be important to boost actions towards effective water management;

ii. The treated flow that was transferred to the Billings Reservoir (10 m³/s) was convenient for favouring the different water uses in the reservoir, such as energy and drinking water production;

iii. Considering the combined effect of both treatment stations, the pilot-scale system reached a significant percentage of removal of total phosphorus, one of the targeted nutrients whose loads to the Billings Reservoir must be reduced to help preventing the artificial eutrophication.

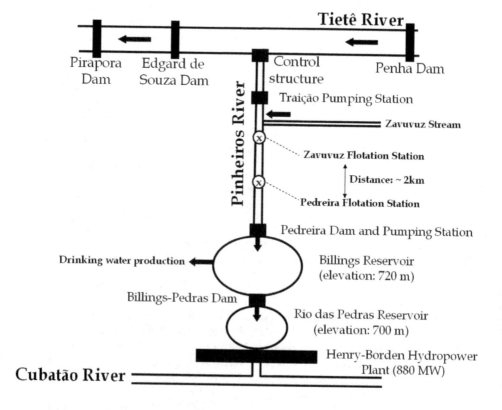

Fig. 4. Scheme of the *in situ* DAF pilot-scale systems placed in the Pinheiros River. Reference: adapted from Cunha et al. (2010).

In situ Remediation Technologies Associated with Sanitation Improvement: An Opportunity for Water Quality
Recovering in Developing Countries

67

However, our experience indicated the following negative factors and limitations:

i. The operation and maintenance of an "opened system" have to consider the potential influences of external factors, like sudden changes in the water flow and quality, both natural or human-induced;

ii. Sludge production and energy consumption were high. During the studied period, the sludge was sent to a landfill because no other alternative was considered safe due to the significant level of contamination (e.g. by heavy metals);

iii. As expected, ammonia-nitrogen was not removed by the DAF system and we assume that the high concentrations of this nutrient in the Billings waters may contribute to water quality decrease, affecting the aquatic life. A biological component with adequate residence time would probably be necessary for removing ammonia-nitrogen. Total suspended solids and metals (aluminium, chromium and iron) concentrations were also high in the treated water.

Variable	Relative efficiency (%) of removal (−) or increase (+)		
	Zavuvuz Flotation Station	Pedreira Flotation Station	Overall effect
Aluminium (soluble)	−17	null	null
Ammonia-Nitrogen	−1	−1	−2
Apparent color	−48	−27	−54
Cadmium (total)	−22	−17	−22
Chromium hexavalent	null	−33	null
Chemical oxygen demand	−41	−18	−53
Copper (total)	−13	−14	−60
Dissolved oxygen	+69	+24	+63
Ionic conductivity	null	null	null
Iron (soluble)	−19	null	−31
Lead (total)	−18	−42	−36
Total phosphorus	−48	−84	−90
Total suspended solids	−30	−10	−40
Turbidity	−37	−35	−48

Table 4. Efficiency (%) of removal (−) or increment of the *in situ* DAF pilot-system in the Pinheiros River. Reference: adapted from Cunha et al. (2010).

The DAF system in the Pinheiros River has focused on integrative approach, technology application, environmental quality and sustainability in the long-term. Nevertheless, since some inefficiencies and gaps were detected, further studies regarding sludge management and efficiency improvement for the removal of some variables (e.g. through complementary processes like those previously described in this chapter) are necessary. Urban waters in developing countries are a challenging issue and the operation of the DAF system in the Pinheiros River was an important contribution for the water resources management.

4. Conclusion

The term "developing countries" is often used to describe nations whose inhabitants have a standard of living between "low" and "medium", a growing industrial base and a rising Human Development Index. According to Kofi Annan (Secretary-General of the United Nations from 1997 to 2006), "developed country is one that allows all its citizens to enjoy a free and healthy life in a safe environment".

Therefore, environmental sustainability is an important step towards the full development of the developing countries, with a view to ensuring social equity, economic strength and environmental quality. Specifically regarding the management of water resources, two main issues should be considered. Our chapter has shown that the implementation of sanitation facilities (e.g. sewage collection and treatment systems) requires significant investments with long-term returns. On the other hand, the remediation of polluted rivers and reservoirs should be seen as a short-term palliative and emergency action to prevent these aquatic systems to reach levels of irreversible degradation, before necessary wastewater collection and treatment are available. The *in situ* approach for remediation may be desirable from the environmental point of view and also economically convenient. By analyzing the main benefits and limitations of three *in situ* technologies (Dissolved Air Flotation, Ultrafiltration Membranes and Enhanced Biological Removal), our investigation has suggested that the costs of remediation of aquatic systems ranged from US$ 0.01/m^3 (the cheapest cost in the range for treatment with bioremediators) to US$ 0.25/m^3 (the most expensive cost in the value range for ultrafiltration membranes). The technologies described in this chapter can be used simultaneously (for example, the DAF associated with the biological treatment with enzymes or with biofilms) to increase the efficiency and meet environmental standards. Although further research is required to find alternatives to solve the detected inefficiencies, our experience with the operation of a pilot-scale DAF system in the Pinheiros River (São Paulo, Brazil) was positive. It has indicated that integrated concepts of water management are necessary to explore urban waters as resources (and not risks) for human activities.

A well-balanced combination of actions, policies and programs for increasing the levels of sanitation coverage and promoting the remediation of impacted aquatic systems is a great opportunity to the developing countries. Political commitment, technology development or transfer from other country, comprehensive monitoring and involvement of the local citizens are factors that can legitimate the whole process and increase the probability of economic and environmental effectiveness and public acceptability.

5. References

Advanced BioCatalytics Corporation. (2011). Information about the bioremediator. In: *Advanced BioCatalytics Corporation*. 02 Aug 2011. Available from www.abiocat.com.

Almeida, C.A., Quintar, S., González, P. & Mallea, M.A. (2007). Influence of urbanization and tourist activities on the water quality of the Potrero de los Funes River (San Luis - Argentina). *Environmental Monitoring and Assessment*, Vol. 133, No. 1-3, (2007), pp. 459-465.

Babel, S. & Takizawa, S. (2011). Chemical pretreatment for reduction of membrane fouling caused by algae. *Desalination*, Vol. 274, No. 1-3, (2011), pp. 171-176.

In situ Remediation Technologies Associated with Sanitation Improvement: An Opportunity for Water Quality
Recovering in Developing Countries

69

Bergamasco, R., Konradt-Moraes, L.C., Vieira, M.F., Fagundes-Klen, M.R. & Vieira, A.M.S. (2001). Performance of a coagulation-ultrafiltration hybrid process for water supply treatment. *Chemical Engineering Journal*, Vol. 166, No. 2, (2001), pp. 483-489.

Bio-Organic Catalyst. (2011). Information about the bioremediator. In: *Bio-Organic Catalyst*. 02 Aug 2011. Available from <bio-organic.com>.

Bioplus. (2011). Information about the bioremediator. In: *Bioplus*. 02 Aug 2011. Available from <www.athcsm4.com.br/bioplus>.

BioTC.(2011). Information about the bioremediator. In: *BioTC*. 02 Aug 2011. Available from <biotecnicontrolambiental.yolasite.com>

Bishop, P.L. (2007). The role of biofilms in water reclamation and reuse. *Water Science and Technology*, Vol. 55, No. 1-2, (2007), pp. 19-26.

Biswas, A., Lundqvist, J., Tortajada, C. & Varis, O. (2004). Water management for megacities. *Stockholm Water Front*, Vol. 2, (2004) pp. 1–13.

Bravo-Inclan, L.A., Saldana-Fabela, M.P. & Sanchez-Chavez, J.J. (2008). Long-term eutrophication diagnosis of a high altitude body of water, Zimapan Reservoir, Mexico. *Water Science and Technology*, Vol. 57, No. 11, (2008), pp. 1843-1849.

Calijuri, M.C., Cunha, D.G.F. & Povinelli, J. (2010). *Sustentabilidade: um desafio na gestão dos recursos hídricos* (1st edition), EESC/USP, São Carlos, SP, Brazil.

Coutinho, W. & von Sperling, M. (2007). Avaliação de desempenho da estação de tratamento por flotação dos córregos afluentes à Represa da Pampulha – Belo Horizonte. *Proceedings of 24º Congresso Brasileiro de Engenharia Sanitária e Ambiental*, Belo Horizonte, MG, Brazil, 2007.

Cunha, D.G.F., Grull, D., Damato, M., Blum, J.R.C., Eiger, S., Lutti, J.E.I. & Mancuso, P.C.S. (2011a). Contiguous urban rivers should not be necessarily submitted to the same management plan: the case of Tietê and Pinheiros Rivers (São Paulo-Brazil). *Annals of the Brazilian Academy of Sciences* (in press).

Cunha, D.G.F., Grull, D., Damato, M., Blum, J.R.C., Lutti, J.E.I., Eiger, S. & Mancuso, P.C.S. (2010). On site flotation for recovering polluted aquatic systems: is it a feasible solution for a Brazilian urban river? *Water Science and Technology*, Vol. 62, No. 7, (2010), pp. 1603-1613.

Cunha, D.G.F., Grull, D., Damato, M., Blum, J.R.C., Lutti, J.E.I., Eiger, S. & Mancuso, P.C.S. (2011b). Trophic state evolution in a subtropical reservoir over years in response to different management procedures. Water Science and Technology, Vol. 64, No. 12, (2011), pp. 2338-2344.

Demarche, P., Junghanns, C., Nair, R.R. & Agathos, S.N. (in press). Harnessing the power of enzymes for environmental stewardship. *Biotechnology Advances* (in press).

Dockko, S., Park, S.C., Kwon, S.B. & Han, M.Y. (2006). Application of the flotation process to thicken the sludge from a DAF plant. *Water Science and Technology*, Vol. 53, No. 7, (2006), pp. 159-165.

Dodds, W.K., Bouska, W.W., Eitzmann, J.L., Pilger, T.J., Pitts, K.L., Riley, A.J., Schloesser, J.T. & Thornbrugh, D.J. (2009). Eutrophication of U.S. Freshwaters: Analysis of Potential Economic Damages. *Environmental Science & Technology*, Vol. 43, No. 1, (2009), pp. 12-19.

Enzilimp. (2011). Information about the bioremediator. In: *Enzilimp*. 02 Aug 2011. Available from <www.enzilimp.com.br/site/default.asp>.

Fantroussi, S.E. & Agathos, S.N. (2005). Is bioaugmentation a feasible strategy for pollutant removal and site remediation? *Current Opinion in Microbiology*, Vol. 8, No. 3, (2005), pp. 268-275.

Geraldes, V., Anil, A., Pinho, M.N. & Duarte, E. (2008). Dissolved air flotation of surface water for spiral-wound module nanofiltration pre-treatment. *Desalination*, Vol. 228, (2008), pp. 191-199

Gora, S.L. & Walsh, M.E. (2011). Recycle of waste backwash water in ultrafiltration drinking water treatment processes. *Journal of Water Supply and Technology*, Vol. 60, No. 4, (2011), pp. 185-196.

Hajkowicz, S., Spencer, R., Higgins, A. & Marinoni, O. (2008) Evaluating water quality investments using cost utility analysis. *Journal of Environmental Management*, Vol. 88, No. 4, (2008), pp. 1601-1610.

Han, Y.Y., Weissfeld, J.L., Davis, D.L. & Talbott, E.O. (2009). Arsenic levels in ground water and cancer incidence in Idaho: an ecologic study. *International Archives of Occupational and Environmental Health*, Vol. 82, No. 7, (2009), pp. 843-849.

Heinänen, J., Jokela, P. & Ala-Peijari, T. (1995). Use of dissolved air flotation in potable water treatment in Finland. *Water Science and Technology*, Vol. 31, No. 3, pp. 225-238.

Hoover, T.J. (1912). *Concentrating Ores by Flotation*. The Mining Magazine, London, 221 p.

IBGE. (2000, 2011). National Survey on Sanitation and Statistical Series on Sewage, Water Supply and Solid Waste Collection. In: *Brazilian Institute of Geography and Statistics*. 09 Jul 2011. Available from <www.ibge.gov.br/home>.

Isaac, R.A. (1998) Costs and benefits of water quality: Massachusetts as a case example. *Water Science and Technology*, Vol. 38, No. 11, (1998), pp. 15-21.

Jarvi, H.P., Neal, C., Warwick, A., White, J., Neal, M., Wickham, H.D., Hill, L.K., Andrews, M.C. (2002). Phosphorus uptake into algal biofilms in a lowland chalk river. *Science of the Total Environment*, Vol. 282-283, (2002), pp. 353-373.

Jiao, Y., Zhao, Q., Jin, W., Hao, X., You, S. (2011). Bioaugmentation of a biological contact oxidation ditch with indigenous nitrifying bacteria for in situ remediation of nitrogen-rich stream water. *Bioresource Technology*, Vol. 102, No. 2, (2011), pp. 990-995.

Jordão, C.P., Ribeiro, P.R., Matos, A.T., Bastos, R.K., Fernandes, R.B. & Fontes, R.L. (2007). Environmental assessment of water-courses of the Turvo Limpo River basin at the Minas Gerais State, Brazil. *Environmental Monitoring and Assessment*, Vol. 127, No. 1-3, (2007), pp. 315-326.

Konieczny, K., Sakol, D. & Bodzek, M. (2006). Efficiency of the hybrid coagulation–ultrafiltration water treatment process with the use of immersed hollow-fiber membranes. *Desalination*, Vol. 198, (2006), pp. 102-110.

Konieczny, K., Sakol, D., Plonka, J., Rajca, M. & Bodzek, M. (2009). Coagulation–ultrafiltration system for river water treatment. *Desalination*, Vol. 240, (2009), pp. 151-159.

Laîné, J.M., Vial, D. & Moulart, P. (2000). Status after 10 years of operation - overview of UF technology today. *Desalination*, Vol. 131, (2000), pp. 17-25.

Lind, Y. & Glynn, A.W. (1999) Intestinal absorption of copper from drinking water containing fulvic acids and an infant formula mixture studied in a suckling rat model. *Biometals*, Vol. 12, No. 2, (1999), pp. 181-187.

Lopes, M.A.B.N. & Oliveira, J.C.G. (1999). O processo de flotação em fluxo como alternativa de despoluição de lago urbano – caso Aclimação na cidade de São Paulo. *Proceedings of 20° Congresso Brasileiro de Engenharia Sanitária e Ambiental*, Rio de Janeiro, RJ, Brazil.

Luz, T.N., Tidona, S., Jesus, B., Morais, P.V. & Sousa, J.P. (2009). The use of sewage sludge as soil amendment. The need for an ecotoxicological evaluation. *Journal of Soils and Sediments*, Vol. 9, No. 3, (2009), pp. 246-260.

Mantis, I., Voutsa, D. & Samara, C. (2005). Assessment of the environmental hazard from municipal and industrial wastewater treatment sludge by employing chemical and biological methods. *Ecotoxicology and Environmental Safety*, Vol. 62, No. 3, (2005), pp. 397-407.

Massoud, M.A., Tarhini, A. & Nasr, J.A. (2009). Decentralized approaches to wastewater treatment and management: Applicability in developing countries. *Journal of Environmental Management*, Vol. 90, (2009), pp. 652-659.

Meliker, J.R., Wahl, R.L., Cameron, L.L. & Nriagu, J.O. (2007). Arsenic in drinking water and cerebrovascular disease, diabetes mellitus, and kidney disease in Michigan: a standardized mortality ratio analysis. *Environmental Health*, Vol. 6, (2007), pp. 1-11.

Oliveira, J.C.G., Netto, A.M., Angelis, J.A. & Barbosa, M.A. (2000). Estação de flotação e remoção de flutuantes do Parque do Ibirapuera – São Paulo – a aplicação do processo de tratamento por flotação em fluxo para a recuperação de lagos urbanos. *Proceedings of 21° Congresso Brasileiro de Engenharia Sanitária e Ambiental*, João Pessoa, PB, Brazil.

Varis, O., Biswas, A.K., Tortajada, C. & Lundqvist, J. (2006). Megacities and Water Management. *International Journal of Water Resources Development*, Vol. 22, No. 2, (2006), pp. 377-394

Pretty, J.N., Mason, C.F., Nedwell, D.B. & Hine, R.E. (2002). A Preliminary Assessment of the Environmental Costs of the Eutrophication of Fresh Waters in England and Wales. In: *University of Essex*. 19 Jun 2011. Available from <www.essex.ac.uk/ces/occasionalpapers/EAEutrophReport.pdf>.

Realco. (2011). Information about the bioremediator. In: *Realco*. 02 Aug 2011. Available from <www.realco.be>.

Rodriguez, M.J., Vinette, Y., Serodes, J.B. & Bouchard, C. (2003) Trihalomethanes in drinking water of greater Quebec region (Canada): Occurrence, variations and modelling. *Environmental Monitoring and Assessment*, Vol. 89, No. 1, (2003), pp. 69-93.

Saz-Salazar, S.D., Hernández-Sancho, F. & Sala-Garrido, R. (2009) The social benefits of restoring water quality in the context of the Water Framework Directive: A comparison of willingness to pay and willingness to accept. *Science of The Total Environment*, Vol. 407, No. 16, (2009), pp. 4574-4583.

Smith, V.H., Tilman, G.D. & Nekola, J.C. (1999) Eutrophication: impacts of excess nutrient inputs on freshwater, marine, and terrestrial ecosystems. *Environmental Pollution*, Vol. 100, No. 1-3, (1999), pp. 179-196.

Suthar, S., Sharma, J., Chabukdhara, M. & Nema, A.K. (2010). Water quality assessment of river Hindon at Ghaziabad, India: impact of industrial and urban wastewater. *Environmental Monitoring and Assessment*, Vol. 165, No. 1-4, (2010), pp. 103-112.

Whiteley, C.G. & Lee, D.J. (2006). Enzyme technology and biological remediation. *Enzyme and Microbial Technology*, Vol. 38, (2006), pp. 291-316.

WHO. (2010). World Health Statistics. In: *WHO Press, World Health Organization: Switzerland.* 10 Jun 2011. Available from <www.who.int/whosis/whostat/en>.

WHO. (2011). Water supply, sanitation and hygiene development. In: *Water sanitation and Health, World Health Association.* 10 Jun 2011. Available from <www.who.int/water_sanitation_health/hygiene/en>.

Xia, S., Li, X., Liu, R. & Li, G. (2005). Pilot study of drinking water production with ultrafiltration of water from the Songhuajiang River (China). *Desalination,* Vol. 179, (2005), pp. 369-374.

Water Quality Improvement Through an Integrated Approach to Point and Non-Point Sources Pollution and Management of River Floodplain Wetlands

Edyta Kiedrzyńska[1,2] and Maciej Zalewski[1,2]
[1]*International Institute of the Polish Academy of Sciences,*
European Regional Centre for Ecohydrology Under the Auspices of UNESCO, Lodz,
[2]*University of Lodz, Department of Applied Ecology, Lodz,*
Poland

1. Introduction

The world is faced with problems related to quality and quantity of water resources due to extensive industrialization, increasing population density and a highly urbanized society. Global scenarios suggest that almost two-thirds of the world's population will experience some water stress by 2025, which will accelerate the water environmental degradation to a unimaginable crisis scale (Momba, 2010).

Wetland are among the most important ecosystems on the Earth. The extent of the world's wetlands is now thought to be from 7 to 10 million km2, or about 5 to 8 % of the land surface of the Earth (Mitsch and Gosselink, 2007). Wetlands include swamps, bogs, marshes, mires, fens, and also river floodplain wetlands.

River floodplain wetlands are very important hydrosystems that retain a significant part of the global freshwater bodies, and because of their location at lower elevations in the landscape, they are also highly exposed to accumulation of large loads of nutrients and other pollutants. This results in eutrophication, which in turn leads to degradation of biological diversity and the appearance of toxic cyanobacterial blooms, which pose threats to human and animal health.

This chapter will try to answer the frequently asked question "What exactly is a wetland?" and "What is the hydrological and biological characteristics of wetlands?" and "What are point and non-point sources pollution?". A section will also be presented on the role of river floodplain wetlands as key ecosystems important for regulation of the water, sediments and nutrients retention, and as a natural buffering system that can be considered as a tool for the reduction of nutrients and other pollutants transport by a river to downstream water ecosystems, and thus contributing to freshwater quality improvement. Part of the chapter will be devoted to application of the ecohydrological sustainable management of floodplain-wetland ecosystems, which is based on the restoration of natural mechanisms determining these ecosystems and functioning of the landscape for the increasing efficiency of water purification, and reducing the negative impact of pollution on the freshwater resources. The third part of the chapter will present a general assumption of the crucial international document "The Declaration on Sustainable Floodplain Management".

2. What is a wetland?

Wetlands sometimes are described as „the kidneys of the landscape" because they function as the downstream receivers of water and waste from both national and human sources. Furthermore, wetlands stabilize water supplies and water balance of the catchment area, thus ameliorating both floods and drought, and they have been found to clean polluted waters, protect shorelines, and recharge groundwater aquifers (Mitsch et al., 2009).

These ecosystems also have been called „ecological supermarkets" due to the extensive food chain and rich biodiversity they support. They play major roles in the landscape by providing unique habitats for a wide variety of flora and fauna. Now that we have focused our attention on the health of our entire planet, wetlands are being described by some as important carbon sinks and climate stabilisation on a global scale (Mitsch and Gosselink, 2007).

Wetland definitions and terms are many and are often confusing or even contradictory. Nevertheless, definitions are important both for the scientific understanding of these systems and for their proper management (Mitsch and Gosselink, 2007), and above all for using the wetlands for water quality improvement .

The Ramsar Convention on Wetlands (signed in Ramsar, Iran 1971) defines wetlands as areas of marsh, fen, peatland or water, whether natural or artificial, permanent or temporary, with water that is static or flowing, fresh, brackish or saline, including areas of marine water, the depth of which at low tide does not exceed six meters.

According to the U.S. Environmental Protection Agency wetlands are areas where water covers the soil, or is present either at or near the surface of the soil all year or for varying periods of time during the year, including the growing season. Wetlands vary widely because of regional and local differences in soils, topography, climate, hydrology, water chemistry, vegetation and other factors, including human disturbance. Indeed, wetlands are found from the tundra to the tropics and on every continent except Antarctica.

According to the wetland definition given by Mitsch and Gosselink (2007), it should include three main components: (i) wetlands are distinguished by the presence of water, either at the surface or within the root zone; (ii) wetlands often have unique soil conditions that differ from adjacent uplands; (iii) wetlands support biota such as vegetation adapted to wet conditions (hydrophytes) and, conversely, are characterized by the absence of flooding-intolerant biota.

Floodplain wetlands are one of the types of natural wetlands and are transitional between terrestrial of the river valley and open water river ecosystems (Fig. 1). Factors such as climate and geomorphology define the degree to which wetlands can exist, however the starting point is the hydrology, which, in turn, affects the physiochemical environment, including the soils, which in turn, together with the hydrology, determines what and how much of the biota, including vegetation, is found in a wetland (Mitsch et al., 2009).

Fig. 1. The Pilica River floodplain, upstream of the Sulejów Reservoir (central Poland); A –
situation of high discharge (Q=83.2 m³ s⁻¹) in spring 2006 (Photo by Piotr Wysocki); B - low
discharge (Q= 6.7 m³ s⁻¹) in summer 2006 (Photo by Mariusz Koch).

3. Wetland hydrology

Hydrologic conditions are extremely important for the maintenance of a floodplain
wetland's structure and function, because they affect many abiotic factors, including soil
anaerobiosis, nutrient availability (Mitsch and Gosselink, 2007; Vorosmarty and Sahagian,
2000). The hydrology of a river wetland creates unique physiochemical conditions that make
such an ecosystem different from both well-drained floodplain systems and deeper old river
bed systems.

The major components of river wetland's water budget include precipitation,
evapotranspiration, surface flow, ground water fluxes, and other overbank flooding in
floodplain wetlands. Water depth, flow patterns, and duration and frequency of flooding,
sediments and nutrients transport (Kadlec and Knight, 1996; Magnuszewski et al., 2007;
Altinakar et al., 2006; Kiedrzyńska et al., 2008a; Kiedrzyńska et al., 2008b), which result from
all hydrologic inputs and outputs, influence the biochemistry of the soils and are major
factors in the ultimate selection of the biota of wetlands (Mitsch and Gosselink, 2007;
Kiedrzyńska et al., 2008a). The water status of a wetland defines its extent and determines
the species composition in a natural floodplain wetland (Mitsch and Gosselink, 1993).
However, biota components are active in altering the wetland hydrology and other
physiochemical conditions (Zalewski 2000; 2006; Mitsch and Gosselink, 2007; Kiedrzyńska et
al., 2008a).

4. Wetland biology

Hydrology affects biological processes in wetlands, such as species composition and biodiversity, efficiency of primary productivity, organic accumulation, and nutrient cycling and retention in wetlands.

Floodplain wetland environments are characterized by stresses that most organisms are ill equipped to handle. Aquatic organisms are not adapted to deal with the periodic drying that occurs in many wetlands, and terrestrial organisms are stressed by long periods of flooding. Because of the shallow water, the temperature extremes on the wetland surface are greater than would be expected in aquatic environments (Mitsch and Gosselink, 2007).

The genetic and functional responses of wetland organisms (microbial and macrophytes) are essentially limitless and result in the ability of natural systems to adapt to changing environmental conditions, such as flooding in natural wetlands or some addition of wastewaters in the treatment of wetlands (Kadlec and Knight, 1996; Kiedrzyńska et al., 2008a). This adaptation allows living organisms to use the constituents from wastewaters for their growth and biomass production. Primary productivity is the highest in wetlands with high flow of water and nutrients, but also in wetlands with pulsing hydroperiods.

When using these nutrients, wetland organisms mediate physical, chemical and biological transformations of pollutants and modify the water quality. In wetlands engineered for water treatment, design is based on the sustainable functions of organisms that provide the desired transformations (Mitsch and Gosselink, 1993; Kadlec and Knight, 1996; Mitsch and Gosselink, 2007) and in natural river floodplain wetlands, we can use autochthonic vegetation of macrophytes (Kiedrzyńska et al., 2008a; Keedy 2010).

Wetland macrophytes are the dominant structural components of most wetland treatment systems, and understanding of the growth requirements and characteristics of these wetland plants is essential for successful river floodplain and a treatment wetland design and its operation (Kadlec and Knight, 1996).

Water pollution control and water quality improvement using macrophytes has been discussed in the literature (Klopatek, 1978; Athie and Cerri, 1987; Surrency, 1993; Copper, 1994; Kadlec and Knight, 1996; Kiedrzyńska et al., 2008a). Production of macrophyte biomass differs significantly both between seasons and between particular species, and may be restricted by a range of limiting abiotic factors, such as soil quality, climate, hydrology and biotic factors, e.g. intraspecific competition and the condition of mycorrhizal symbionts (Sumorok and Kiedrzyńska, 2007).

According to Kadlec and Knight (1996) and Kiedrzyńska et al. (2008a), the biomass of *Phragmites australis*, per hectare ranges between 6,000 and 35,000 kg d.w., making this macrophyte one of the most effective ones. According to Gołdyn and Grabia (1996) and Kiedrzyńska et al. (2008a), the total harvest of wetland grasses in the summer period ranges between 4,300 and 14,000 kg d.w. ha^{-1}.

Plant productivity may be limited by the availability of phosphorus (Compton and Cole, 1998; Mainstone and Parr, 2002; Olde Venterink et al., 2002, 2003). The amount of phosphorus accumulated in the vegetation biomass depends principally on the ecology and biology of plant species and on edaphic factors (Ozimek and Renman, 1996), and usually ranges from 0.1% to 1% (Fink, 1963).

Water Quality Improvement Through an Integrated Approach to Point and Non-Point Sources Pollution and
Management of River Floodplain Wetlands

77

According to Kiedrzyńska et al. (2008a), the phosphorus content in the floodplain wetland meadow communities was maintained at a relatively constant level of 2.54–2.89 g P kg^{-1} d.w. throughout the growing season. More variation was observed in the case of *Carex* sp., which was characterized by the highest percentage of P content in spring (4.07 g P kg^{-1} d.w.) and significantly lower one for the other seasons (summer: 1.38 g P kg^{-1} d.w.; autumn: 2.17 g P kg^{-1} d.w.). The same studies have shown that the highest values of P accumulation on the floodplain were reached in spring by *P. australis* (3.75 g P kg^{-1} d.w.), which also gradually decreased towards the end of the growing season. Finally, the efficiency of phosphorus accumulation per area unit was between 0.7 and 7.3 kg P ha^{-1} for all communities except those dominated by *P. australis*, which were nearly five times higher (34.7 kg P ha^{-1}) and resulted from the very high summer biomass of this species (Kiedrzyńska et al., 2008a).

5. Wetland ecohydrology

In order to effectively improve the water quality in wetland floodplains, the knowledge of the processes taking place there is required, as well as their identification and quantification. This way of solving the environmental problems suggests the concept of Ecohydrology (Zalewski et al., 1997; Zalewski 2000; 2002; 2007).

In this context, Ecohydrology is a conceptual tool for sustainable management of water–floodplain resources and prevention of anthropogenic landscape transformation results. Therefore, introducing the ecohydrological management in a catchment area based on the restoration of natural mechanisms determining the river-floodplain ecosystems and their functioning, is very important.

Ecohydrology is a subdiscipline of hydrology focused on ecological aspects of the hydrological cycle (Zalewski et al., 1997; Zalewski 2000). It refers specifically to two phases of the hydrological cycle: terrestrial plant - water - soil interactions and aquatic biota - hydrology interactions. Ecohydrology is based on the suggestion that sustainable development of water resources depends on the ability to maintain the evolutionarily established processes of water and nutrient circulation and energy flows at the basin scale (Zalewski 2006).

Ecohydrology provides three new aspects to environmental sciences (Zalewski, 2000; 2011) that can be adopted and used for sustainable management of the river floodplain ecosystems, water quality improvement and achievement of 'good' ecological, chemical and hydrological status of water bodies (Zalewski 2011; Zalewski and Kiedrzyńska 2010):

1. Integration of the catchment, river valley, floodplain and river together with its biota into a specific superorganism (Framework aspect). This covers the following dimensions: a) the *Scale of processes* - the meso-scale cycle of water circulation within a basin (the terrestrial/aquatic ecosystem coupling) provides a template for the quantification of ecological processes; b) *Dynamics of processes* – water and temperature have been the driving forces for both terrestrial and freshwater ecosystems; c) *Hierarchy of factors* - abiotic processes are dominant (e.g. hydrological processes), biotic interactions may manifest themselves when they are stable and predictable (Zalewski and Naiman, 1985). This is based on the assumption that abiotic factors are of primary importance and once they become stable and predictable, the biotic interactions start to

manifest themselves (Zalewski and Naiman, 1985). The quantification of hydrological pulses along the river continuum (Junk et al., 1989; Vannote et al., 1980; Agostinho et al., 2004; Altinakar et al., 2006; Magnuszewski et al., 2007; Kiedrzyńska et al., 2008b) and monitoring of threats (Wagner and Zalewski, 2000; Mankiewicz-Boczek et al., 2006; Bednarek and Zalewski, 2007a, 2007b; Kiedrzyńska et al., 2008b; Urbaniak et al., in press), such as point and nonpoint source pollution (Takeda et al., 1997; Borah and Bera, 2003; Tian et al., 2010; Kiedrzyńska et al., 2010), are necessary for optimal regulation of processes towards the sustainable water and ecosystems management.

2. Increasing the carrying capacity of ecosystems that is their evolutionarily established resistance and resilience to absorb human-induced impacts (Target aspect). This aspect of ecohydrology expresses the rationale for a proactive approach to the sustainable management of freshwater resources. It assumes that it is not enough to simply protect the ecosystems, but in the face of increasing global changes, which are manifested in the growth of the population, energy consumption, material and human aspirations, it is necessary to increase the capacity of ecosystems. This can be achieved by regulation the interplay between hydrology and biota; analysis of dynamic oscillations of an ecosystem and its productivity and succession (as reflected by nutrient/pollutant absorbing capacity versus human impacts) should be the solution to process regulation (Bednarek and Zalewski, 2007a, 2007b; Kiedrzyńska et al., 2008a, Zalewski 2011).

3. Application of "dual regulation" in shaping and management of processes in river floodplain wetlands for purification and water quality improvement, biodiversity and ecosystem services for society (Methodology aspect). This means that a biotic component (macrophytes, bacteria) of a floodplain ecosystem can control and shape the chemical parameters of water and hydrological processes through effects on shaping the substrate roughness. These relationships also occur in the opposite direction - *vice versa*, what means using hydrology to regulate the biota (Zalewski, 2006, Zalewski and Kiedrzyńska, 2010). Great potential of the knowledge, which has been generated by dynamically developing ecological engineering (Mitsch 1993; Jorgensen 1996; Chicharo, 2009), should to a large extent accelerate the implementation of the above concept.

Sustainable management of the river floodplain wetlands gives a number of positive implications on the global ecosystem by improving the water quality, which depends on the development, dissemination and implementation of these principles and interdisciplinary knowledge, based on the latest achievements in environmental protection (Fig. 2).

The success of these actions depends on the profound understanding of the whole range of multi-dimensional processes involved. The first dimension is temporal: spanning a time frame from the past, paleohydrological conditions till the present, with a due consideration of future, global change scenarios. The second dimension is spatial: understanding the dynamic role of river and floodplain biota over a range of scales, from the molecular- to the valley-scale. Both dimensions should serve as a reference system for enhancing the buffering capacity of floodplain wetlands as key ecosystems important for the regulation of water, sediments and nutrients retention, and reduction of nutrients and other pollutants transport by a river to downstream water ecosystems, and thus contributing to freshwater quality improvement.

Water Quality Improvement Through an Integrated Approach to Point and Non-Point Sources Pollution and
Management of River Floodplain Wetlands

79

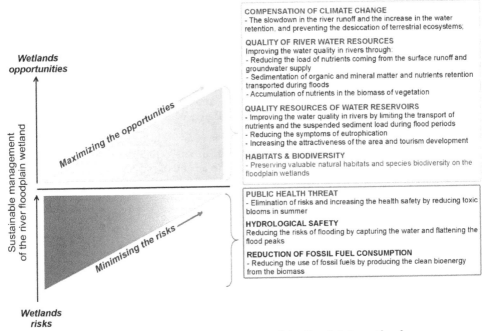

COMPENSATION OF CLIMATE CHANGE
- The slowdown in the river runoff and the increase in the water
retention, and preventing the desiccation of terrestrial ecosystems;

QUALITY OF RIVER WATER RESOURCES
Improving the water quality in rivers through:
- Reducing the load of nutrients coming from the surface runoff and
groundwater supply
- Sedimentation of organic and mineral matter and nutrients retention
transported during floods
- Accumulation of nutrients in the biomass of vegetation

QUALITY RESOURCES OF WATER RESERVOIRS
- Improving the water quality in rivers by limiting the transport of
nutrients and the suspended sediment load during flood periods
- Reducing the symptoms of eutrophication
- Increasing the attractiveness of the area and tourism development

HABITATS & BIODIVERSITY
- Preserving valuable natural habitats and species biodiversity on the
floodplain wetlands

PUBLIC HEALTH THREAT
- Elimination of risks and increasing the health safety by reducing toxic
blooms in summer

HYDROLOGICAL SAFETY
Reducing the risks of flooding by capturing the water and flattening the
flood peaks

REDUCTION OF FOSSIL FUEL CONSUMPTION
- Reducing the use of fossil fuels by producing the clean bioenergy
from the biomass

Fig. 2. Implications of the sustainable management of the floodplain wetlands.

6. Wetland water quality improvement – A new way of thinking

6.1 Water pollution – Point and non-point sources pollution

Water pollution is a crucial global problem, which requires ongoing evaluation and revision of water resource policy at all levels (from the international one down to individual aquifers and wells). It has been suggested that it is the leading worldwide cause of deaths and diseases, and that it accounts for the deaths of more than 14,000 people daily (West, 2006).

Water resources are usually referred to as polluted when they are impaired by anthropogenic contaminants and either do not support a human use, such as drinking water, and/or undergo a considerable shift in their ability to support their constituent biotic communities, such as fish. Natural phenomena, such as algae blooms, storms, and earthquakes, also cause major changes in the water quality and the ecological status of the water.

Surface water and groundwater have often been studied and managed as separate resources, although they are interrelated. Surface water seeps through the soil and becomes groundwater. Conversely, groundwater can also feed surface water sources. Sources of surface water pollution are generally grouped into two categories based on their origin (Winter, 1998).

Point source (PS) water pollution refers to contaminants that enter a waterway from a single, identifiable source, such as a pipe or ditch. Examples of sources in this category include discharges from a sewage treatment plant, a factory, or a city storm drain. Non–

point source (NPS) pollution refers to diffuse contamination that does not originate from a single discrete source. NPS pollution is often the cumulative effect of small amounts of contaminants gathered from a large area. A common example is the leaching of phosphorus and nitrogen compounds from fertilized agricultural lands. Nutrients runoff in stormwater from "sheet flow" over an agricultural field or forest are also cited as examples of NPS pollution. Excessive export of nutrients from PS and NPS pollution are the leading causes of eutrophication in lakes, reservoirs and rivers, and coastal water bodies worldwide (Alexander et al., 2008; Diaz and Rosenberg, 2008; Tian et al., 2010).

Eutrophication is a shift in the trophic status of a given water body in the direction of increasing plant biomass, by adding some artificial or natural substances, such as nitrates and phosphates, through e.g. fertilizers or sewage, to an aquatic system.

In other terms, it is a water bloom resulting from a great increase of phytoplankton in a water body. Negative environmental effects include hypoxia, the depletion of oxygen in the water, which induces reductions in specific fish and other animal populations. Thus, eutrophication of water resources leads to degradation of biological diversity and the appearance of toxic cyanobacterial blooms, which pose threats to human health and animals (Tarczyńska et al., 2001; Mankiewicz et al., 2001, 2005; Jurczak et al. 2004).

River wetlands are altered by the runoff of pollutants from point and diffuse sources of pollution flowing from the upper catchment areas and thus are purified. The effects of polluted water on wetlands have not received yet enough attention.

6.2 Wetlands as key ecosystems improving the water quality

Rivers and floodplain wetlands are the ecosystems that are particularly exposed to eutrophication and high anthropogenic stress (Meybeck 2003, Zalewski and Kiedrzyńska 2010). This is because they are situated in landscape depressions, into which the whole range of catchment anthropogenic modifications and impacts are transferred and accumulated (Altinakar et al., 2006; Zalewski, 2006; Magnuszewki et al., 2007), e.g. sediments and nutrients (Kiedrzyńska et al., 2008a; Kiedrzyńska et al., 2008b), dioxins (Urbaniak et al., 2009; Urbaniak et al., in press), microbial contamination (Gągała et al., 2009). These dramatically progressing disturbances are sometimes negatively amplified by degradation of the hydrological cycle and the loss of integrity between fluvial ecosystems and floodplains, which can result in the increased eutrophication (Tarczyńska et al., 2001; Izydorczyk et al., 2005; Izydorczyk et al., 2008) and the reduction of biodiversity and ecosystem services for societies (Zalewski 2008; Zalewski and Kiedrzyńska, 2010). However, the river valley with natural floodplain wetlands are areas that may be used in water purification.

Water quality improvement by the use of wetlands has been broadly discussed (Bastian and Hammer, 1993; Raisin and Mitchell, 1995; Nairn and Mitsch, 2000; Trepel and Kluge, 2002; Mitsch et al., 2005; Mitsch and Gosselink, 2007; Mitsch et al., 2009), especially the importance of natural floodplains for river self-purification and freshwater quality protection (Bayley, 1995; Loeb and Lamers, 2003; Zalewski 2006; Kiedrzyńska et al. 2008a; Kiedrzyńska et al. 2008b). An example can be the area of 24 km² of wetlands that collected the water from the Zala River catchment, and which has been reconstructed within the confines of a multidisciplinary research programme on the protection of the Lake Balaton (Hungary).

Water Quality Improvement Through an Integrated Approach to Point and Non-Point Sources Pollution and
Management of River Floodplain Wetlands

81

According to Pomogyi (1993), 96% of PO_4-P, 87% of NO_3-N and 58% of TP were retained in this area in 1990. Interesting studies conducted by Wassen (1995) in the Biebrza Valley in Poland reported that the floodplain vegetation is an important sink for nutrients, especially for N and P. Wetlands are also used in other European countries, e.g. in the Netherlands, Germany, Finland (Wassen et al., 2002; Olde Venterink et al., 2002) and in the United States, and around the world (Weller et al., 1996; Mitsch et al., 2005; Thullen et al., 2005; Mitsch et al. 2009).

Floodplains can optimize nutrient retention in the river ecosystem, especially in catchments with large areas of agriculture and can be considered as a tool for the reduction of nutrient transport by a river to downstream reservoirs and estuaries (Kiedrzyńska et al., 2008a; Kiedrzyńska et al., 2008b).

The highest nutrients' loads transported by rivers usually occurred during rising water stages of floods and they should be directed to floodplain areas upstream the reservoir at the very initial stages of floods, in order to diminish the load in a reservoir. The research on the Pilica River floodplain (central Poland) looked into the possibilities of enhancing this process, both through sedimentation and assimilation in the vegetation biomass. The research that was based on the DTM and hydraulic models demonstrated that sedimentation of flood sediments in the floodplain essentially reduces the transport to the reservoir. During floods, the sediment is effectively deposited and phosphorus is retained in the 30-kilometer section of the Pilica River floodplain. In the flooding area of 1007 ha, fine-grained flood sediments reached 500 t and the retention of P was 1.5 t. Furthermore, the efficiency in the assimilation of nutrients and the biomass production by autochthonous plant communities, with special emphasis on willow patches, was examined against a background of a hydroperiod. The potential of vegetation in the Pilica River floodplain (26.6 ha) for summer phosphorus accumulation was estimated at 255 kg P y^{-1}, however, a conversion of 24% or 48% of the area into fast-growing managed willow patches can increase the phosphorus retention up to 332 kg P y^{-1} or 399 kg P y^{-1}, respectively (Kiedrzyńska et al., 2008a). Theoretically, 1 kg of P can lead to some 1-2 t of algal biomass in a reservoir (Zalewski, 2005). Therefore, floodplain wetlands are mostly enriched with the riverine material and, at the same time, river water is purified by deposition of this material. Floodplains can, therefore, serve as natural, cleaning and biofiltering systems for reducing the concentrations of sediments, nutrients, micropollutants and, other pollutants coming from upper sections of the catchment area.

7. Wetland management – The Declaration on Sustainable Floodplain Management

In the 21st century, wetlands management should focus not only on the conservative protection of these valuable ecosystems, but also on the sustainable use and optimization of abiotic-biotic processes for problem solving and improving the water quality.

Floodplain wetlands are an integral part of river systems and therefore they play a fundamental role in the exchange of water masses and matter between a river and terrestrial ecosystems (Mitsch et al., 1979; Junk et al., 1989; Tockner et al., 1999; Mitsch et al., 2008; Kiedrzyńska et al., 2008b). Floodplains are "dynamic spatial mosaics", where water acts as a connector between various components (Thoms 2003; Kiedrzyńska et al., 2008a). This specific connection is crucial for maintaining the function and integrity of floodplain-river systems (Tockner et al., 1999; Amoros and Bornette, 2002; Thoms 2003). They are the hot spots of terrestrial and aquatic biodiversity in the catchment landscape due to a mosaic of plant communities and their spatio-temporal dynamics (Zalewski, 2008). Sustainable development of the river and floodplain environment needs to take into account the fact that biological structures and fundamental ecological processes, such as water and nutrients cycles, are to a large extent, suffering from deterioration (Zalewski, 2009).

The sustainable management of floodplains, which are the most diversified ecosystems and most resilient to human impact due to their hydrological pulse-driven self-regenerative capacity, is obviously very important. Therefore, there is still a further need for insights into these and other processes, whereas "engineering harmony" between river floodplain ecosystems and societies (UN MDGs) requires solutions from integrative, interdisciplinary science such as ecohydrology, a subdiscipline of sustainability science focused on ecological aspects of the hydrological cycle (Zalewski and Kiedrzyńska, 2010).

Such an integrated ecohydrological approach to sustainable management of wetlands is contained in the presented below Floodplain Declaration "Declaration on Sustainable Floodplain Management", which was elaborated based on presentations and discussions at the International Conference under the auspices of IHP of UNESCO "Ecohydrological Processes and Sustainable Floodplain Management: Opportunities and Concepts for Water Hazard Mitigation, and Ecological and Socioeconomic Sustainability in the Face of Global Changes" (19th – 23rd of May 2008, Lodz, Poland).

7.1 Declaration on Sustainable Floodplain Management

7.1.1 Recognition: Properties and values of floodplains

Floodplains are dynamic wetlands, an integral part of river basins with a high potential for biological productivity, biodiversity, flood mitigation, groundwater recharge, river purification and regulation of exchanges of nutrients between land and water, and other ecosystem services, all maintained by the pulse-regulated hydrology of running waters.

Floodplains are threatened by increasing population and improper management. Development of floodplains without consideration of the specifics of their ecological structure and dynamics thus diminishes biodiversity, reduces benefits to society related to water quality, cultural aesthetic values and – in consequence – causes economic losses.

7.1.2 Floodplains and global climate change

Floodplains are an important component of global environmental security and resilience because of their high compensatory potential to mitigate environmental change due to their capacity for water retention, food production, CO_2 sequestration, production of bio-fuels, and the diversity of habitats that they support.

7.1.3 Integrative science for problem solving

Understanding the functioning of floodplains and their potential for socio-economic benefits, requires integration of recent knowledge of:

Water Quality Improvement Through an Integrated Approach to Point and Non-Point Sources Pollution and
Management of River Floodplain Wetlands

83

- geomorphological and paleohydrological evolution of river valleys,
- hydrological processes and patterns of ecological succession,
- societal interactions and learning alliances,
- climate scenarios,
- strategic forecasts based on integrative modelling and adaptive management

In order to reverse floodplain degradation and increase ecological resilience and economic benefits, a shift in strategy from floodplain exploitation to floodplain sustainable use is necessary. Accordingly we need a change of public perception from sectoral, structural and reactive responses to an integrated, process-regulation-oriented and proactive approach.

7.1.4 Methodology for provisioning sustainable ecological services of floodplains

- *Ecohydrological management* of floodplains, will require "dual regulation" - a framework for harmonisation of biodiversity conservation with such human needs as flood mitigation, food and energy production, transport and recreation.
- *Hydrotechnical infrastructure* harmonised on the basis of integrative science and best management practices incorporating catchment scale ecosystem processes, will be a powerful tool for reversing degradation of biodiversity, and enhancing sustainable development and compensation of global changes
- *Cultural heritage* of the catchment should become an important element for spatial reconnection of floodplains to the adjacent landscape, as well as restoration of links to social, economic and cultural values.
- *People's perception* and attitudes to the changing environment can only be shaped by new solutions based on integrative science, which depend upon development of programs and methodologies for education and communication.

7.1.5 Tools for implementation

Policies by national and international institutions for water resources, energy, transportation, and environmental management must elevate the protection of pristine sections of the floodplains and promote sustainable use, and restoration of degraded floodplains on rivers, lakes and coastal zones.

Land use integrated planning, financial incentives, economic instruments, and environmental regulatory frameworks are essential tools for implementing the ecohydrological standards and criteria. In case of "novel floodplains", created by secondary succession after human impact, floodplain loss due to essential new development of e.g. transport systems should be mitigated through restoration of at least twice the area of degraded floodplain.

A network of long-term ecological processes, research sites, responsible institutions, and data bases is needed for improving progress and transfer of knowledge, and transfer and sharing of technology.

Public participation, facilitated by modern communication approaches, is fundamental to accommodating conflicting interests and uses of floodplains.

7.1.6 Recommendations for action plan

- Classification of different types of floodplains with special consideration of catchment perspective and ecosystem services;
- Development of methodology to assess rate and type of flood pulses necessary to maintain floodplain functions and structures and to reconcile protection and social needs;
- Formulation of principles for floodplain management, sustainable food and renewable energy production based on integrative science and the relevant science/policy interface.

8. Conclusion

Floodplain wetlands can purify and improve the water quality because they have a significant role in the water retention, sedimentation of mineral and organic matter, nutrients and pollutants. Furthermore, the floodplain wetland vegetation has a great biological potential for the assimilation and accumulation of nutrients in biomass and especially for the uptake of phosphorus.

Therefore well-managed river wetlands can serve as natural cleaning and biofiltering systems for reducing the concentrations of sediments, nutrients, micropollutants and, other serious pollutants.

On the one hand, in the 21st century, the floodplain wetlands management should focus on the protection of biodiversity and values of these important ecosystems, but on the other hand, also on the sustainable use and optimization of abiotic-biotic processes for problem solving and improving the water quality.

In accordance with the conclusions of the Floodplain Declaration, the successful reversal of degradation of floodplain ecosystems should become the objective for the development of a sound vision of co-evolution of Ecosphere and Anthroposphere, by engineering harmony between three dynamic and evolving components: catchment areas, water resources and a society, with an emphasis on the change from exploitative to participatory environmental consciousness. For this purpose, it is necessary to continue the integration of studies of highly specialized disciplines of environmental and social sciences into the framework of Ecohydrology - a holistic problem-solving concept. The system approach, foresight methodology and learning alliances are these important new components of the trans-disciplinary sustainability science that should be used for sustainable water management in the catchment area, and also the ecological and socio-economic potential of the basin should be used for the improvement of human health and the quality of life following the UN MDGs.

9. Acknowledgment

Part of these researches was developed within the framework of the following projects: 1) Project of the Polish Ministry of Science and Higher Education: NN 305 365738 "Analysis of point sources pollution of nutrients, dioxins and dioxin-like compounds in the Pilica River catchment area and drawing up the reclamation methods"; 2) LIFE+EKOROB project: Ecotones for reduction of diffuse pollutions (LIFE08 ENV/PL/000519); 3) The Pilica River Demonstration Project under the auspices of UNESCO and UNEP.

Water Quality Improvement Through an Integrated Approach to Point and Non-Point Sources Pollution and
Management of River Floodplain Wetlands

85

We are particularly grateful to the most active Members of the Steering Committee and the Advisory Committee of the International Conference "Ecohydrological Processes and Sustainable Floodplain Management: Opportunities and Concepts for Water Hazard Mitigation, and Ecological and Socioeconomic Sustainability in the Face of Global Changes" (19th – 23rd of May 2008, Lodz, Poland) for their methodological contribution to the Declaration on Sustainable Floodplain Management. The Floodplain Declaration is available from the following link on the II PAS ERCE under the auspice of UNESCO website at: http://www.erce.unesco.lodz.pl.

10. References

Agostinho, A.A.; Gomes, L.C. & Verissimo, S. (2004). Flood regime, dam regulation and fish in the upper Paraná river: effects on assemblage attributes, reproduction and recruitment. *Reviews of Fish Biology and Fishery*, 14(1): 11–19.

Alexander, R.B.; Smith, R.A.; Schwarz, G.E.; Boyer, E.W.; Nolan, J.V. & Brakebill, J.W. (2008). Differences in phosphorus and nitrogen Delivery to The Gulf of Mexico from the Mississippi River Basin. *Environmental Science and Technology*, 42(3): 822-830.

Altinakar, M.; Kiedrzyńska, E. & Magnuszewski, A. (2006). Modeling of inundation pattern at Pilica river floodplain, Poland. In: *Climate Variability and Change Hydrological Impacts*. Demuth, S.; Gustard, A.; Planos, E.; Scatena, F. & Servat, E. (Eds). *IAHS Publ.* 308. 579-585. *Proceedings of the Fifth FRIEND World Conference*, Havana, Cuba, November, 2006.

Amoros, C. & Bornette, G. (2002). Connectivity and biocomplexity in waterbodies of riverine floodplains. *Freshwater Biology* 47: 517-539.

Athie, D. & Cerri, C.C. (1987). The use of macrophytes in water pollution control. *Water Sci. Technol.* 19, 10.

Bastian, R.K. & Hammer, D.A. (1993). The use of constructed wetlands for wastewater treatment and recycling. In: *Constructed wetlands for water quality improvement*. G.A. Moshiri (Ed.), , CRC Press, Florida, USA, pp.59-68.

Bayley, P.B. (1995). Understanding large river-floodplain ecosystems. *BioScience* 45: 153-158.

Bednarek, A. & Zalewski, M. (2007a). Potential effects of enhancing denitrification rates in sediments of the Sulejów Reservoir. *Environment Protection Engineering*. 33(2): 35–43.

Bednarek, A. & Zalewski, M. (2007b). Management of lowland reservoir littoral zone for enhancement of nitrogen removal via denitrification. *Proceedings of International Conference. W3M Wetlands: Monitoring, Modelling and Management*. Wierzba. Warsaw Press, pp. 293–299.

Borah, D.K. & Bera, M. (2003). Watershed-scale hydrologic and nonpoint-source pollution models: Review of mathematical bases. *American Society of Agricultural Engineers*, 46(3): 1553-1566.

Chicharo, L.; Wagner, I.; Chicharo, M.; Łapinska, M. & Zalewski, M. (2009). Practical experiments guide for Ecohydrology. UNESCO, Venice, Paris. 122pp, (ISBN 978-989-20-1702-0).

Compton, J.E. & Cole, D.W. (1998). Phosphorus cycling and soil phosphorus fractions in Douglas-fir and red alder stands. *Forest Ecol. and Manag.* 110, 101-112.

Copper, J.R. (1994). Riparian wetlands and water quality. *J. of Environ. Qual.* 23, 896-900.

Diaz, R.J. & Rosenberg, R. (2008). Spreading dead zones and consequences for marine ecosystems. *Science*, 321(5891): 926-929.

Fink, J. (1963). Introduction to the plants phosphorus biochemistry. *National Publishing of Agriculture and Forestry*. Warsaw, Poland, pp. 241.

Gągała, I.; Izydorczyk, K.; Skowron, A.; Kamecka-Plaskota, D.; Stefaniak, K.; Kokociński, M. & Mankiewicz-Boczek, J. (2009). Appearance of toxigenic cyanobacteria in two Polish lakes dominated by *Microcystis aeruginosa* and *Planktothrix agardhii* and environmental factors influence. *Ecohydrology & Hydrobiology* 9 (2).

Gołdyn, R. & Grabia, J. (1996). A conception of using natural methods for additional wastewater treatment in Jarocin. In: *Constructed wetlands for wastewater treatment, II Scientific-Technical Conference*. Kraska, M. & Błażejewski, R. (Eds.), Poznań, Poland.

Izydorczyk, K.; Tarczynska, M.; Jurczak, T.; Mrowczynski, J. & Zalewski, M. (2005). Measurement of phycocyanin fluorescence as an online Early warning system for cyanobacteria in reservoir intake water. *Environmental Toxicology* 20: 425-430.

Izydorczyk, K.; Jurczak, T.; Wojtal-Frankiewicz, A.; Skowron, A.; Mankiewicz-Boczek, J. & Tarczyńska, M. (2008). Influence of abiotic and biotic factors on microcystin content in Microcystis aeruginosa cells in a eutrophic temperate reservoir. *Journal of Plankton Research* 30 (4): 393-400.

Jørgensen, S. E. (1996). The application of ecosystem theory in limnology. *Verh. Int. Verein Limnol*. Vol. 26, 181-192.

Junk, W.J.; Bayley, P.B. & Sparks, R.E. (1989). The flood pulse concept in river-floodplain systems. In: *Proceedings of the international large river symposium*. Dodge, D.P. (Ed.). *Can. Spec. Publ. Fish. Aquat. Sci*, 106, 110-127.

Jurczak, T.; Tarczynska, M.; Karlsson, K. & Meriluoto, J. (2004). Characterization and diversity of cyano- bacterial hepatotoxins (microcystins) in blooms from Polish freshwaters identified by liquid chromatography-electrospray ionisation mass spectrometry. *Chromatographia* 59, 571-578.

Kadlec, R.H. & Knight, R.L. (1996). *Treatment Wetlands*. Lewis Publishers, CPR Press. USA. 893 pp.

Keedy, P.A. (2010). *Wetland Ecology, Principles and conservation*. Second Edition. Cambridge University Press. UK. pp. 497.

Kiedrzyńska, E., Wagner-Łotkowska, I. & Zalewski, M. (2008a). Quantification of phosphorus retention efficiency by floodplain vegetation and a management strategy for a eutrophic reservoir restoration. *Ecological Engineering* 33, 15-25.

Kiedrzyńska, E.; Kiedrzyński, M. & Zalewski, M. (2008b). Flood sediment deposition and phosphorus retention in a lowland river floodplain: impact on water quality of a reservoir, Sulejów, Poland. *Ecohydrology & Hydrobiology* 8: 2-4.

Kiedrzyńska, E., Macherzyński, A., Skłodowski, M., Kiedrzyński, M., Zalewski M. (2010). Analysis of point sources of pollution of nutrients in the Pilica River catchment and use of ecohydrological approach for their reduction (in polish). In: *Hydrology in Environmental Protection and Management*. A. Magnuszewski (Ed.). Monograph of the Committee for Environmental Sciences PAS, 69, pp. 285 - 295.

Klopatek, J.M. (1978). Nutrient dynamic of freshwater riverine marshes and the role of emergent macrophytes. In: *Freshwater wetlands: Ecological processes and management potential*. Good, R.E.; Whigham, D.F. & Simpson, R.L. (Eds). Academic Press, New York, pp. 195-216.

Water Quality Improvement Through an Integrated Approach to Point and Non-Point Sources Pollution and
Management of River Floodplain Wetlands

87

Loeb, R. & Lamers, L. (2003). The effects of river water quality on the development of wet floodplain vegetations in the Netherlands. *Proceedings of the International conference "Towards natural flood reduction strategies"*. September 2003. Warsaw, Poland.

Magnuszewski, A.; Kiedrzyńska, E.; Wagner-Łotkowska, I. & Zalewski, M. (2007). Numerical modelling of material fluxes on the floodplain wetland of the Pilica River, Poland. In: *Wetlands: Monitoring, Modelling and Management*. Okruszko, T.; Szatyłowicz, J.; Mirosław – Świątek, D.; Kotowski, W. & Maltby, E. (Eds). A.A. Balkema Publishers – Taylor & Francis Group. pp. 205-210.

Mainstone, C.P. & Parr, W. (2002). Phosphorus in rivers – ecology and management. *The Science of the Total Environ.* 282/283, 25-47.

Mankiewicz, J., Walter, Z., Tarczynska, M., Fladmark, K.E., Doskeland, S.O., Zalewski, M. (2001). Apoptotic effect of cyanobacterial extract on rat hepatocytes and human lymphocytes. *Environmental Toxicology* 3 (16), 225-233.

Mankiewicz, J., Komarkova, J., Izydorczyk, K., Jurczak, T., Tarczynska, M., Zalewski, M. (2005). Hepatotoxic cyanobacterial blooms in the lakes of northern Poland. *Environmental Toxicology* 20, 499-506.

Mankiewicz-Boczek, J., Izydorczyk, K., Jurczak, T. (2006). Risk assessment of toxic *Cyanobacteria* in Polish water bodies. In: Kungolos, A.G., Brebbia, C.A., Samaras, C.P., Popov, V. (Eds) *Environmental Toxicology*. WIT press, Southampton, Boston, *WIT Transactions on Biomedicine and Health* 10, 49-58.

Meybeck, M. (2003). Global analysis of river systems: from Earth system controls to Anthropocene syndromes. Philosophical Transactions of the Royal Society of London, [B] 358 (1440), 1935-1955.

Mitsch, W. J. (1993). Ecological Engineering - a co-operative role with planetary life support system. *Environmental Science Technology* Vol. 27, 438-445.

Mitsch, W.J & Gosselink, J.G. (2007). *Wetlands*. Fourth Edition. John Wiley & Sons, Inc. USA.

Mitsch, W.J.; Dorge C.L. & Wiemhoff, J.W. (1979). Ecosystems dynamics and a phosphorus budget of an alluvial cypress swamp in southern Illinois. *Ecology* 60: 1116-1124.

Mitsch W.J. & Gosselink, J.G. (1993). *Wetlands*. Second Edition. John Wiley & Sons, Inc. USA. 722 pp.

Mitsch, W.J.; Zhang, L.; Anderson, C.J.; Altor, A.E. & Hernández, M.E. (2005). Creating riverine wetlands: Ecological succession, nutrient retention, and pulsing effects. *Ecological Engineering* 25, 510-527.

Mitsch, W.J.; Zhang, L.; Fink, D.F.; Hernandez, M.E.; Altor, A.E.; Tuttle, C.L. & Nahlik, A.M. (2008). Ecological Engineering of floodplain. *Ecohydrology & Hydrobiology* Vol.8, No. 2-4: 139-147.

Mitsch, W.J.; Gosselink, J.G.; Anderson, C.J. & Zhang, L. (2009). *Wetland Ecosystem*. John Wiley & Sons, Inc. USA. 295 pp.

Momba, MNB. 2010. Wastewater Protozoan-Driven Environmental Processes for the Protection of Water Sources. In: Momba M. and Bux F. Eds. Biomass. Croatia, downloaded from sciyo.com., pp.202.

Nairn, R.W. & Mitsch, W. J. (2000). Phosphorus removal in created wetland ponds receiving river overflow. *Ecological Engineering* 14: 107-126.

Olde Venterink, H.; Pieterse, N.M.; Belgers, J.D.M.; Wassen, M.J. & De Ruiter, P.C. (2002). N, P, and K budgets along nutrient availability and productivity gradients in wetlands. *Ecological Applications* 12, 1010-1026.

Olde Venterink, H.; Wassen, M.J.; Verkroost, A.W.M. & De Ruiter, P.C. (2003). Species richness-productivity patterns differ between N-, P-, and K- limited wetlands. *Ecology* 84, 2191-2199.

Ozimek, W. & Renman, G. (1996). The role of emergent macrophytes in the constructed wetlands for wastewater treatment. In: *Constructed wetlands for wastewater treatment. II Scientific-Technical Conference.* Kraska, M. & Błażejewski, R. (Eds). Poznań, Poland, pp. 9-17.

Pomogyi, P. (1993). Nutrient retention by the Kis-Balaton Water Protection System. *Hydrobiologia* 251, 309-320.

Raisin, G. W. & Mitchell, D. S. (1995). The use of wetlands for the control of non-point source pollution. *Water Science Technology* 32 (3): 177-186.

Sumorok, B. & Kiedrzyńska, E. (2007). Mycorrhizal status of native willow species at the Pilica River floodplain along moist gradient. In: *Wetlands: Monitoring, Modeling and Management.* Okruszko, T.; Maltby, E.; Szatyłowicz, J.; Świątek, D. & Kotowski, W. (Eds). A.A. Balkema Publishers – Taylor & Francis Group, pp. 281-286.

Surrency, D. (1993). Evaluation of aquatic plants for constructed wetlands. In: *Constructed wetlands for water quality improvement.* Moshiri, G.A. (Ed.), CRC Press. Florida, pp. 349-357.

Takeda, I.; Fukushima, A.& Tanaka, R. (1997). Non-point pollutant reduction in a paddy-field watershed using a circular irrigation system. *Water Research,* 31(11): 2685-2692.

Tarczyńska, M.; Romanowska-Duda, Z.; Jurczak, T. & Zalewski, M., (2001). Toxic cyanobacterial blooms in a drinking water reservoir-causes, consequences and management strategy. *Water Science Technology*: Water Supply 1 (2), 237–246.

Thoms, M.C. (2003). Floodplain-river ecosystems: lateral connections and the implications of human interference. *Geomorphology* 56: 335-349.

Thullen, J.S.; Sartoris, J.J. & Nelson, S.M. (2005). Managing vegetation in surface-flow wastewater-treatment wetlands for optimal treatment performance. *Ecological Engineering* 25, 583-593.

Tian, Y.W.; Huang, Z.L. & Xiao, W.F. (2010). Reductions in non-point source pollution through different management practices for an agricultural watershed in the Three Gorges Reservoir Area. *Journal of Environmental Sciences,* 22(2): 184-191.

Tockner, K.; Pennetzdorfer, D.; Reiner, N.; Schiemer F. & Ward, J.V. (1999). Hydrological connectivity, and the exchange of organic matter and nutrients in a dynamic river-floodplain system (Danube, Austria). *Freshwater Biology* 41: 521-535.

Trepel, M. & Kluge, W. (2002). Ecohydrological characterisation of a degenerated valley peatland in Northern Germany for use in restoration. *Journal for Nature Conservation* 10:155-169.

Urbaniak, M.; Zieliński, M.; Wesołowski, W. & Zalewski, M. (2009). Sources and distribution of polychlorinated dibenzo-para-dioxins and dibenzofurans in sediments of urban cascade reservoirs, Central Poland. *Environ. Protec. Engineer.,* No. 3, vol. 35, 93-103.

Urbaniak ,M.; Kiedrzyńska, E. & Zalewski, M. (in press). The role of a lowland reservoir in the transport of micropollutants, nutrients and the suspended particulate matter along the river continuum. *Hydrology Research* 00-00.

Vannote, R.L.; Minshall, G.W.; Cummings, K.W.; Sedell, J.R. & Cushing, C. E. (1980). The River Continuum Concept. *Canad. J. Fish. Aquati. Sci.,* 37, 130-137.

Water Quality Improvement Through an Integrated Approach to Point and Non-Point Sources Pollution and
Management of River Floodplain Wetlands

89

Vorosmarty, C.J. & Sahagian, D. (2000). Antropogenic disturbance of the terrestrial water cycle. *Bioscience* 50(9): 753–765.

Wagner, I. & Zalewski, M. (2000). Effect of hydrological patterns of tributaries on biotic processes in lowland reservoir – consequences for restoration. *Ecological Engineering.* Special Issue 16, 79-90.

Wassen, M.J. (1995). Hydrology, water chemistry and nutrient accumulation in the Biebrza fens and floodplains (Poland). *Wetlands Ecology and Management* 3, 125-137.

Wassen, M.J.; Peeters, W.H.M. & Olde Venterink, H. (2002). Patterns in vegetation, hydrology, and nutrient availability in an undisturbed river floodplain in Poland. *Plant Ecology* 165, 27-43.

Weller, C.M.; Watzin, M.C. & Wang, D. (1996). Role of wetlands in reducing phosphorus loading to surface water in eight watersheds in the Lake Champlain Basin. *Environmental Management* 20, 731-739.

West, L. (2006). World Water Day: A Billion People Worldwide Lack Safe Drinking Water. http://environment.about.com/od/environmentalevents/a/waterdayqa.htm. About. (March 26, 2006).

Winter, T.C.; Harvey, J.W.; Franke, O.L. & Alley, W.M. (1998). Ground Water and Surface Water: A Single Resource. *United States Geological Survey* (USGS) - Circular 1139. Denver, USA.

Zalewski, M. (2011). Ecohydrology for implementation of the EU water framework directive. Proceedings of the Institution of Civil Engineers. *Water Management* 8, 16 Issue, 375-385.

Zalewski, M. (2000). Ecohydrology – the scientific background to use ecosystem properties as management tools toward sustainability of water resources. In: Zalewski, M. (Ed.). *Ecological Engineering. Journal on Ecotechnology* 16: 1-8.

Zalewski, M., (2002). Ecohydrology – the use of ecological and hydrological processes for sustainable management of water resources. *Hydrological Sciences Journal* 47(5), 825-834.

Zalewski, M. (2005). Engineering Harmony. *Academia* 1(5), 4-7.

Zalewski, M. (2006). Flood pulses and river ecosystem robustness. In: *Frontiers in Flood Research.* Tchiguirinskaia, I.; Thein, K.N.N., K. & Hubert, P. (Eds). Kovacs Colloquium. June/July 2006. UNESCO, Paris. *IAHS Publication* 305. 212 pp.

Zalewski, M. (2007). Ecohydrology as a Concept and Management Tool. [In:] *Water and Ecosystems Managing Water in Diverse Ecosystems to Ensure Human Well-being.* King, C., Ramkinssoon, J., Clü sener-Godt, M., Adeel, Z. (Eds). UNU-INWEH UNESCO MAB, Canada: 39.53.

Zalewski, M. (2008). Rationale for the "Floodplain Declaration" from environmental conservation toward sustainability science. *Ecohydrology &Hydrobiology* Vol. 8, No. 2-4, 107-113.

Zalewski, M. (2009). Ecohydrology for engineering harmony between environment and society. *Danube News.* May 2009. No. 19. Volume 11.

Zalewski, M.; Janauer, G. S. & Jolankai, G. (1997). Ecohydrology - A new Paradigm for the Sustainable Use of Aquatic Resources. International Hydrological Program UNESCO. *Technical Document on Hydrology* No 7, Paris.

Zalewski, M. & Naiman, R.J. (1985). The regulation of riverine fish communities by a continuum of abiotic-biotic factors. In: *Habitat Modifications and Freshwater Fisheries*. Alabaster, J.S. (Ed.). FAO UN. Butterworths, London. 3-9.

Zalewski, M. & Kiedrzyńska, E. (2010). System approach to sustainable management of inland floodplains – declaration on sustainable floodplain management. *CAB Reviews: Perspectives in Agriculture, Veterinary Science, Nutrition and Natural Resources* 5, No. 056, 1-8.

Water Quality in the Agronomic Context: Flood Irrigation Impacts on Summer In-Stream Temperature Extremes in the Interior Pacific Northwest (USA)

Chad S. Boyd, Tony J. Svejcar and Jose J. Zamora

USDA-Agricultural Research Service, Eastern Oregon Agricultural Research Center,
Burns, OR
USA

1. Introduction

European arrival to the Pacific Northwest (U.S.A.) in the late 1800's signalled the beginning of an era of change for watershed use and water quality of free-flowing stream systems. Historic uses centered around trapping of beaver, utilization of livestock forage, and harvesting of anadramous fish and timber resources (Beschta, 2000). Impacts of these land uses on stream systems were severe; impaired flood plain development with beaver removal, interrupted nutrient cycling with overharvest of anadramous fish, streambank degradation due to overgrazing, and severe flooding associated with abusive logging practices (Trefethen, 1985; Meehan, 1991; Beschta, 2000). Perhaps most dramatic was the conversion of ecosystem type from stream system to impoundment associated with dam construction for hydroelectric power production (Beschta, 2000). The US Clean Water Act (CWA) of 1965 was crafted and passed into law in response to these and other issues, providing a legal and regulatory framework for managing land use practices that impact water quality (Adams 2007). Subsequent amendments of the CWA in 1972 and 1977 helped expand this framework to include impacts of non-point source pollution and mandated that states develop water protection programs (National Research Council, 1995; Adams, 2007).

While the forgoing issues continue to attract popular and political attention, legislative and regulatory mandates have dramatically reduced the acute effects of these practices and attention has now shifted to less dramatic, but nonetheless important associations between topical land use practices and water quality. There is an established and growing awareness of the impact of agriculturally-related non-point source pollution on stream systems. In this chapter we will 1) examine the scope of non-point source pollutant issues relating to the water quality/agriculture interface for streams in the Interior Pacific Northwest (PNW), 2) present a case-study examining the relationship between flood irrigation and in-stream temperature, and 3) make the case that stream temperature and non-point source water quality issues are complex problems that are best addressed by considering variation of the problem in both time and space.

1.1 Modern agriculture and water quality of free-flowing streams in the Pacific Northwest

Excessive erosion associated with row-crop agriculture has reduced water quality over much of the PNW, particularly in the region dominated by the Columbia Basin and Columbia Plateau of Eastern Washington, North-central Oregon, and Northern Idaho (Schillinger et al., 2010). Similarly, pathogens and nutrients from agricultural operations can and have disrupted PNW stream ecology. Pathogens now rank second and nutrients fourth on the list of top pollutants in U.S. water bodies and agricultural lands are a significant contributor to both pathogen and nutrient loading (Richter et al., 1997; Parajuli et al., 2008; USEPA, 2009). Nutrient by-products from agricultural operations, particularly nitrogen, can cause eutrophication of stream systems and associated reduction of dissolved oxygen and reduced biodiversity (Vitousek et al., 1997). From a management standpoint the issues of sediment, nutrient and pathogen contaminants have collectively been addressed using stream-side vegetation buffer strips in an attempt to attenuate pollutant entry into stream systems (Castelle et al., 1994; Schmitt et al., 1999; Dosskey 2002; Dorioz et al., 2006; Mayer et al., 2007). Vegetation buffers have been effective at reducing nutrient (Yates and Sheridan, 1983; Lowrance et al., 1985), pathogen (Tate et al., 2004; Knox et al., 2007) and sediment loading (Lyons et al., 2000; Lee et al., 2003) in streams. However, development of effective policy concerning the use of buffer strips has been complicated by the fact that the efficacy of buffers in reducing pollutant loading varies strongly in accordance with a number of design and environmental factors including buffer width, cover and height of plant material, slope, and soil attributes (Pearce et al., 1997, Atwill et al., 2005, George et al., 2011). For example, a review by Dorioz et al. (2006) reported variation in sediment retention by vegetation buffers of 2.5 orders of magnitude across studies. Ultimately, the use of riparian buffer strips will need to be paired with spatially explicit models (e.g., Tim and Jolly, 1994) to assist managers in integrating land-use practices with conservation measures to reduce pollutant yield at the watershed scale.

Livestock operations can have both direct and indirect effects on water quality of PNW stream systems. Direct effects can include increased nutrient and pathogen loading through fecal or urine additions (Nader et al., 1998; George et al. 2011); excessive nutrient loading can stimulate algal blooms leading to reduced dissolved oxygen concentrations (Belsky et al., 1999). Additionally, increased turbidity associated with hoof action can negatively impact aquatic organisms by decreasing primary production (Henley et al., 2000; Line, 2003). Strategic location of livestock attractants (e.g., salt, artificial water sources) can reduce livestock densities near riparian areas and associated nutrient inputs into stream systems (Tate et al., 2003; George et al. 2011). Indirect effects of livestock on water quality include impacts to streamside plant communities and physical damage to streambanks. Riparian vegetation plays a critical role in sustaining the biotic integrity of the stream ecosystem by anchoring bank material in place during high flow events (Kleinfelder et al., 1992; Clary and Leininger, 2000). Livestock grazing of this resource is generally sustainable under conditions of moderate utilization (e.g., graze to 10cm stubble height; Boyd and Svejcar, 2004; Volesky et al., 2011), but timing of grazing can also influence livestock impacts on water quality and riparian vegetation (Boyd and Svejcar, 2004). Severe utilization of streamside plants by livestock can lead to reduced plant biomass and altered stream channel conditions (e.g., wider and more shallow channel) that are associated with decreased water storage capacity,

Water Quality in the Agronomic Context: Flood Irrigation Impacts on Summer In-Stream Temperature
Extremes in the Interior Pacific Northwest (USA)

93

reduced filtration effects of vegetation buffers, and increased water temperature (Kauffman and Kruger, 1984; Winward, 1994; Toledo and Kauffman, 2001).

1.2 Agricultural impacts on stream temperature

Recently, concern has developed regarding summer in-stream temperature dynamics and agricultural practices that may be associated with elevated water temperature. Water temperature is an important attribute of the aquatic environment that has the potential to affect basic ecological processes such as nutrient cycling and can also modulate biotic ecology (Poole and Berman, 2001; Isaak and Hubert, 2001). Water temperature is a key driver of invertebrate demography and impacts fish species via temperature-dependent fluxes in dissolved oxygen content (Young and Huryn, 1998). In the Mountain and Northwest United States, concern over water temperature extremes and their impact on red band (*Oncorhynchus mykiss newberri*) and bull (*Salvelinus confluentus*) trout is an important driver of regulatory mechanisms governing agronomic and other land use practices.

Much of the interest surrounding the influence of agriculture on water temperature has focused on practices that reduce woody plants in the riparian area. A growing body of empirical evidence suggests that shade from woody plants can reduce daily maximum water temperature (Poole and Berman, 2001; Tate et al., 2005). Livestock grazing (as well as wildlife-associated herbivory) can reduce woody plant cover, particular when grazing occurs subsequent to the senescence of herbaceous vegetation (Holland et al., 2005; Clary et al., 1996; Matney et al., 2005). However, shade from woody plants is only one of a myriad of factor, including air mass characteristics, elevation, stream flow, stream gradient, adiabatic rate, channel width/depth, and groundwater inputs that can influence water temperature maxima in streams (Larson and Larson, 1996; George et al., IN PRESS). Diversion of stream water for purposes of irrigating agricultural crops can potentially impact a number of these processes including stream flow and groundwater dynamics.

2. Flood irrigation impacts on stream temperature: A case study

Flood irrigation is a common agricultural practice on interior PNW meadows used for hay and/or livestock forage. These meadows are typically low-to-mid elevation and contain, or are influenced by, a seasonal or perennial natural stream system. Water from the stream is diverted at the upstream end of the meadow into smaller irrigation ditches that parallel the stream at a distance on one or both sides. Water then disperses from the irrigation ditch across the meadow either passively through sub-surface flow, or actively via overland spillage from the irrigation ditch. Specific water management practices vary by and within drainage, however, the irrigation season generally begins in spring following the onset of mountain snowmelt. Efficiency and control of water usage is obviously low with flood irrigation, but it remains a popular form of irrigation due to low costs associated with the absence of power (i.e., electric or fossil fuel) and few major infrastructural requirements.

Excessive in-stream water temperature is the most frequent water quality impairment for streams in southeastern Oregon (Oregon Department of Environmental Quality 2002). Stream temperature regulations are associated mainly with concern over thermal requirements for cold water fish stocks (Boyd and Strudevant 1996, Gamperl. et. al. 2002).

Many streams in this region lack or have minimal amounts of woody cover (i.e. shade); under such conditions, fluctuations in water temperature are strongly influenced by parallel fluctuations in air temperature (Stefan and Preud'homme 1993, McRae and Edwards 1994). Flood irrigation can potentially increase in-stream temperature through two mechanisms: 1) decreases in in-stream discharge, and 2) overland flow of warmer flood waters re-entering the stream. Conversely, flood irrigation could act to cool in-stream water temperature if sufficient flood water returns to the stream via groundwater input.

Previous research has suggested that flood irrigation may act to moderate summer in-stream temperature extremes by elevating the groundwater table in the surrounding meadow and increasing groundwater inputs into the stream (Stringham et al. 1998). If such a moderating effect were to occur, one prediction would be that daily in-stream temperature maximums would be less sensitive to the controlling influence of air temperature. Our objective in this case-study was to determine the impact of flood irrigation on seasonal temperature dynamics of a meadow stream in southeastern Oregon. Specifically, we characterized depth to groundwater and discharge patterns in irrigated and non-irrigated years, and tested the hypothesis that air temperature would have less effect on stream temperature in an irrigated compared to non-irrigated stream reach.

2.1 Study site

Our study took place in the Lake Creek drainage in Grant Co., OR, U.S.A. (11T0370683 UTM4890874) at an elevation of approximately 1500 m. Lake Creek is a perennial Rosgen C class (Rosgen 1994) stream. Its position near the base of the Strawberry Mountains causes strong seasonal discharge fluctuations associated with snowmelt in April through early June. The local area receives approximately 330 mm of annual precipitation, most falling as snow during the winter months; maximum air temperatures (approx. 28°C) occur in late July (Oregon Climate Service 2005). Soils were sandy clay loam underlain by a clay lens at depths ranging from 75 – 90 cm. We established a 3.5 km upper study reach and a 1.0 km lower study reach approximately 1.3 km downstream from the upper reach (Figure 1). The upper reach was within the irrigated portion of the meadow and the lower reach was not. Two tributaries entered the main channel of Lake Creek within the irrigated portion of the meadow. The beginning of the upper study reach was immediately downstream from the farthest downstream tributary. No tributaries entered within the lower study reach. Shade from woody plants was non-existent in either reach but the lower reach had limited topographic shading.

The meadow surrounding Lake Creek is seasonally flooded by two irrigation ditches flowing north to south along the east and west sides of the meadow (Figure 1). The West Ditch (0.08 m³/sec⁻¹) and East Ditch (0.06 m³/sec⁻¹) diversions were opened in 2004 from April 13 to June 30 and from April 21 to June 30 in 2005. Approximately 70% of the flow in West Ditch #1 was diverted into West Ditch #2. A lateral slope downward from the irrigation ditches to streamside elevations allowed for subsurface flow from irrigation ditches to the surrounding meadow. Diversions within the West Ditch were used to augment flood irrigation from subsurface flow. Water spreading from the East Ditch relied entirely on subsurface flow from the irrigation ditch.

Water Quality in the Agronomic Context: Flood Irrigation Impacts on Summer In-Stream Temperature
Extremes in the Interior Pacific Northwest (USA)

95

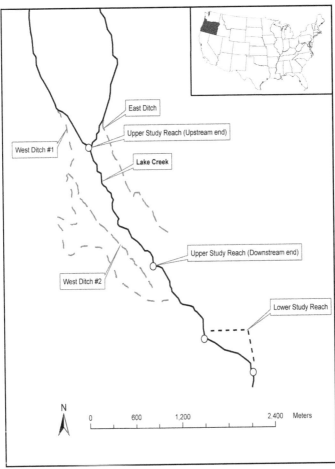

Fig. 1. Overview of study site on Lake Creek, southeast OR, U.S.A. depicting irrigation
ditches and locations for upper and lower study reaches (inset shows location of Oregon
within U.S.A.). The meadow surrounding the upper reach was flood irrigated from mid-
April to June 30 of 2004-2005.

2.2 Methods and materials

Data for a non-irrigated year were collected in 2002 and for flood irrigated years in 2004 –
2005. We measured point-in-time stream discharge at approximately 10-day intervals in
June and July using permanent main channel cross-sections located at the upstream ends of
the upper and lower study reaches. Discharge was estimated using a magnetic-head pygmy
flow meter with a top set wading rod and an Aquacalc 5000 discharge recorder (JBS Energy,
Inc; West Sacramento, CA). Measurements were made at 60% of total depth from the stream
water surface. Regression analysis was used to estimate mid-month values for June and
July. Depth to groundwater within the upper reach was measured at 10-day intervals in
June and July using shallow PVC wells spaced at 15-m intervals along 4 transects run

perpendicular to the stream channel from streamside to 135 m distance. Wells were constructed of perforated 1.9-cm diameter plastic pipe buried to 150 cm depth, or restrictive layer, whichever came first. Gravel (0.5-cm diameter) was packed around the well to within 15 cm of the soil surface and the remaining hole filled with soil. Transects were located at the beginning and end of the upper reach and at 2 intermediate points. The east or west side of the creek was randomly selected for placement of each transect; 3 transects were located on the west side and 1 transect on the east side. In 2002 we used a laser level to measure the elevation of each well relative to the deepest portion of the adjacent stream channel. Raw groundwater depth values were modified by adding the elevation of the well to generate values indicating the elevation of the meadow water table relative to channel elevation. We used regression analysis to estimate mid-month depth values, by year and distance from stream, for June and July. Values were averaged for each date across the 4 transects and within distance from stream.

In-stream, air and groundwater temperature data were collected in June and July of all years. Daily maximum stream temperature was measured using thermistors placed at the beginning, end and two intermittent positions within the upper reach and at the beginning and end of the lower reach. Thermistors were programmed for hourly readings and maximum temperatures were identified by selecting the highest hourly reading for the warming portion of the daily thermograph. Maximum in-stream temperature values were averaged within day and reach. Daily maximum air temperatures were determined as per methods for in-stream values using four stream-side thermistors interspersed throughout the upper and lower reaches. Because daily air temperatures did not differ across thermistors, we averaged values within day. Daily maximum groundwater temperature was monitored using thermistors placed in 7.62-cm diameter plastic pipe wells located 10 m from the active channel at the beginning and end of the upper study reach. Maximum daily temperatures were determined as per methods for in-stream temperatures. Values were averaged monthly, across locations and within year.

Because air temperatures vary across years, and because of the strong influence of air temperature on water temperature (Stefan and Preud'homme 1993, McRae and Edwards 1994), inter-annual variation in air temperature can obscure treatment effects on water temperature. Thus, we compared upper and lower reaches within year. A common range of maximum daily air temperatures was selected across years and within month (June or July). These data were regressed (within study reach) with the corresponding days maximum stream temperature. Within a year and month, slope and intercept values were compared between study reaches using Statistical Analysis Software (PROC SYSLIN, SAS 1999). Regression equations for maximum air and water temperatures were then used to generate predicted maximum water temperatures using air temperature values that approximated the highest observed values within month and across years (June = 26°C, July = 31°C). All mean values are reported with their associated standard error.

2.3 Results and discussion

In-stream discharge dropped sharply from the spring run-off (June) to post run-off (July) periods and was generally less in June for the upper reach as compared to the lower reach (Table 1). Discharge was similar between the upper and lower reaches in July of all years. Between-year differences in June were associated with differing run-off patterns between

Water Quality in the Agronomic Context: Flood Irrigation Impacts on Summer In-Stream Temperature
Extremes in the Interior Pacific Northwest (USA)

97

years and diversion of water in the upper reach. Data for groundwater elevation indicate that by June 15 of the non-irrigated year (2002) the elevation of the water table at 15 m distance from the stream was slightly lower than that of the stream-side well, a gradient that suggests the potential for water losses to the surrounding meadow (Figure 2); by July of the non-irrigated year this downward slope continued to 75 m distance from the stream.

	June 15		July 15	
	Upper reach	Lower reach	Upper reach	Lower reach
2002	0.75 +/-0.01	0.76 +/- 0.01	0.12 +/- 0.00	0.12 +/- 0.01
2004	0.67 +/- 0.02	0.76 +/- 0.01	0.16 +/- 0.01	0.16 +/- 0.00
2005	0.49 +/- 0.01	0.62 +/- 0.02	0.19 +/- 0.01	0.21 +/- 0.01

Table 1. Seasonal discharge (m^3/sec^{-1}) for upper and lower study reaches on Lake Creek, OR, U.S.A. Data from 2002 represent a non-irrigated year. The meadow surrounding the upper reach was flood irrigated from mid-April to June 30 of 2004-2005.

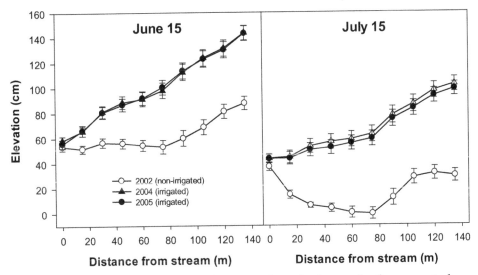

Fig. 2. Mean groundwater elevation and associated standard errors for the upper study reach as a function of distance from stream for Lake Creek, OR, U.S.A. Values represent the height for the surface of the groundwater table relative to the channel thalweg. The meadow surrounding the upper reach was flood irrigated from mid-April to June 30 of 2004-2005.

In contrast, during irrigated years the water table elevation had a positive slope with distance from stream until mid-July and this positive slope was maintained to 135 m distance from the stream. In irrigated years groundwater elevation dropped up to 50 cm (depending on distance from stream) after irrigation shut-off. A similar drop was noted in the non-irrigated year, but the absolute value of groundwater elevations after irrigation shut-off was much less as compared to the irrigated years (Figure 2). These data suggested a higher meadow water table with irrigation and a greater potential for groundwater inputs into the stream. Maximum daily groundwater temperatures increased from June to July and ranged from approximately 9 to 12°C across years and months (Figure 3). Increasing groundwater temperature in July was probably associated with increasing air temperature (Ward 1985). Groundwater temperatures were generally 8 to 14°C cooler than in-stream temperature maxima suggesting that any groundwater return flow into the main channel would have a buffering effect on maximum stream temperatures.

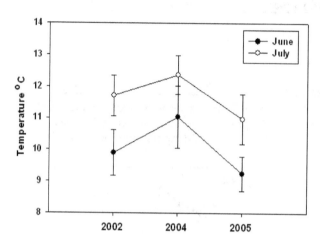

Fig. 3. Daily maximum groundwater temperature means and associated standard errors for the upper reach of Lake Creek, OR, U.S.A. The meadow surrounding the upper reach was flood irrigated from mid-April to June 30 of 2004-2005.

Maximum daily air temperature was associated positively with maximum daily water temperature in both irrigated and non-irrigated years (Figure 4). R^2 values ranged from 0.55 to 0.81, with the exception of July, 2002 which had values of 0.30 and 0.32 for the upper and lower reaches, respectively (Table 2). Explanatory power during this time period was decreased by several outlying points that were associated with cool nights followed by warm, but overcast days. If these days are excluded (n = 3), R^2 values for the upper and lower reaches increase to 0.48. In the non-irrigated year (2002) maximum daily water temperature in both study reaches responded similarly to air temperature although the intercept was slightly higher for the lower reach in June (P = 0.003, Figure 4). In the irrigated years intercepts differed (P < 0.001) between the upper and lower reaches and slopes were different (P < 0.071) for all but July of 2004 (P = 0.159, Figure 4). At a given maximum air temperature, water temperature was slightly less in the upper reach suggesting that flood irrigation helped moderate maximum daily water temperature. Although the y intercept was slightly higher for the lower reach in June of the non-irrigated year (2002), the lines of best fit for irrigated and non-irrigated reaches converge at higher air temperatures. In contrast, lines of best fit for the study reaches diverge at higher air temperatures during irrigated years (2004-2005), providing further evidence for an irrigation-related buffering effect.

Water Quality in the Agronomic Context: Flood Irrigation Impacts on Summer In-Stream Temperature
Extremes in the Interior Pacific Northwest (USA)

99

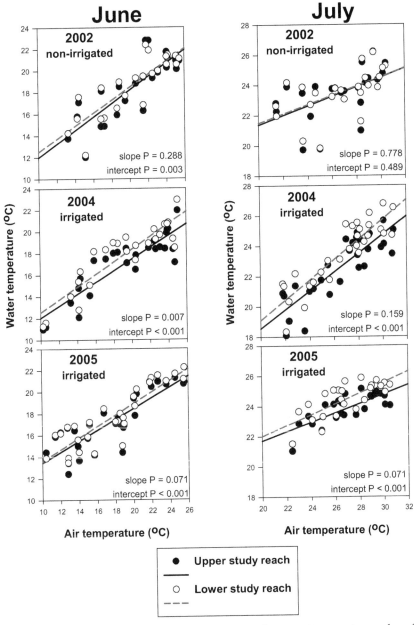

Fig. 4. The relationship between daily maximum air and stream temperature values for June and July within the upper and lower study reaches of Lake Creek, OR, U.S.A. The meadow surrounding the upper reach was flood irrigated from mid-April to June 30 of 2004-2005. Probability values for differences in slope and intercept values between irrigated and non-irrigated reaches are reported within graphs.

Year	Month	Reach	Slope	y intercept	R^2	Predicted maximum water temperature °C
2002	June	Upper	0.62	5.77	0.66	21.92
		Lower	0.60	6.51	0.71	22.03
2002	July	Upper	0.31	15.16	0.3	24.80
		Lower	0.30	15.45	0.32	24.82
2004	June	Upper	0.54	6.64	0.79	20.71
		Lower	0.58	6.71	0.81	21.84
2004	July	Upper	0.62	6.20	0.73	25.33
		Lower	0.66	5.82	0.79	26.37
2005	June	Upper	0.50	8.47	0.72	21.44
		Lower	0.53	8.31	0.78	22.11
2005	July	Upper	0.31	15.60	0.57	25.09
		Lower	0.36	14.93	0.55	25.93

Table 2. Predicted monthly maximum water temperature values for the upper and lower study reaches of Lake Creek, OR, U.S.A. The meadow surrounding the upper reach was flood irrigated from mid-April to June 30 of 2004-2005. Equations were in the form of *max water = max air(x) + b* and were generated for each year/month/location combination based on data for regression equations in Figure 4. Air temperatures of 26°C and 31°C were used to approximate observed maximum temperatures for June and July, respectively.

Without irrigation (2002) predicted maximum water temperatures in the 2 reaches were very similar. However, with irrigation (2004 and 2005) there was divergence with cooler temperatures in the upper reach. For irrigated years predicted water temperature maximums were approximately 0.9°C warmer in the lower reach as compared to the upper reach (Table 2). Temperature reduction in irrigated years was recorded under varying discharge levels including the late snow-melt run-off period in June and lower flow conditions in July (Table 1). Stringham et al. (1998) reported maximum stream temperatures of 1 to 3°C cooler for a flood irrigated vs. non-irrigated reach. The relatively smaller thermal response in our study may be associated with the presence of old irrigation ditches between the currently used ditches (Figure 1) and the active stream channel. These ditches intercepted some of the sub-surface flow and provided additional water surface area for evaporative losses. Our results are in contrast to those of Tate et al. (2005) who reported increased stream temperatures with flood irrigation. This discrepancy may be explained by the relatively low percentage of total stream discharge (about 20%) that we diverted for irrigation as opposed to that of Tate et al. (30 to 70%).

In summary, we found that air temperature exerted a measurable and strong influence on water temperature maximums, explaining up to 80% of the variation in this variable (Table 2). For the stream we studied, flood irrigation appeared to moderate the influence of air temperature on daily maximum water temperature; this effect was observed during

Water Quality in the Agronomic Context: Flood Irrigation Impacts on Summer In-Stream Temperature
Extremes in the Interior Pacific Northwest (USA)

101

irrigation and continued for at least 1 month after cessation of irrigation. Our results agree with earlier work from Stringham et al. (1998) and suggest that flood irrigation can help buffer daily maximum water temperature during summer air temperature extremes. In the present study, the magnitude of the predicted temperature reduction was generally < 1°C. However, stream size, percentage of water diverted, and proportion of surface/subsurface flows may all influence the effect of flood irrigation on water temperature.

3. Stream temperature as a complex problem

Boyd and Svejcar (2009) presented the notion that those problems facing natural resources managers today differ, in fundamental ways, when compared to those of previous generations. Specifically these authors argued that many of the issues facing managers today are "complex", in that the environmental factors associated with these issues vary in both space and time, making it difficult to generalize management prescriptions or characterize ecosystem responses. These same hindrances make it challenging to formulate biologically-realistic regulatory statues for stream temperature and other water quality parameters. In terrestrial systems, simultaneous consideration of space and time is difficult given that space has x, y and z (i.e., elevation) dimensions. Measurement of water temperature (and water quality in general) is somewhat different in that, practically speaking, space has only one dominant dimension (i.e., upstream or downstream). Thus the riparian ecosystem offers a unique opportunity for simultaneously considering environmental characteristics in space and time.

Our preceding examination of water temperature dynamics on Lake Creek is a good example of how variation in space and time can interact to in influence water quality parameters. Our approach in this case was to monitor water temperature over both space and time, in irrigated and non-irrigated years and to use this information to make inferences regarding the influence of flood irrigation on water temperature maxima. To further characterize variation in water temperature over both space and time, we plotted daily water temperature maximums for a non-irrigated year (2002) from June 1 (Julian day 152) to September 30 (Julian day 273; Figure 5). Data were interpolated using negative exponential smoothing and displayed in a contour plot (SigmaPlot 12.0; Systat Software Inc.; San Jose, CA) that allows the user to view change in temperature over time at a given point in the stream and change over space for a given day.

When viewed in this way, daily maximum water temperature is set within a background of strong intra-annual and spatial variation. From a management standpoint, these data imply that capturing variation in space and time should be an integral part of a water quality assessment strategy. For example, if we were to have measured temperature at one point within the irrigated reach (stream distance 0 – 3.5 km), and one location within the non-irrigated reach (3.5 to 5.9 km) then our conclusions regarding the influence of flood irrigation on stream temperature would have been strongly influenced by choice of sampling location, as compared to the multiple location sampling strategy that we employed. Similarly, our use of a two-month sampling window helped overcome undue influence of time at shorter sampling intervals. This same logic applies equally to determining compliance within a regulatory framework. For example, the Oregon Department of Environmental Quality stipulates that average daily maximum water temperature for cold-water fisheries should not exceed 20°C for any seven-day period, and that number drops to 12°C for streams containing bull trout spawning and juvenile rearing habitat (ODEQ, 2008). Determination of compliance with either standard for the section of Lake Creek we studied would depend strongly on when and where samples were taken (see Figure 5). The spatial and temporal variability of maximum water temperature for Lake Creek could also have strong relevance to the ecology of aquatic organisms. For example, while water temperatures at a given point/time on Lake Creek may exceed tolerance levels of cold water fishes, variability in temperature over space may act to provide thermal refugia that allow affected species to escape potentially harmful temperatures (Ebersole et al., 2001).

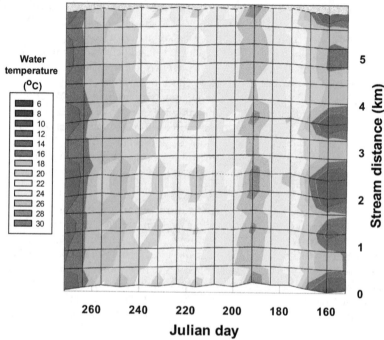

Fig. 5. Daily maximum water temperature values over space and time for Lake Creek, OR, U.S.A.

4. Conclusions

In conclusion, we suggest that complex problems such as water temperature management may not be solved with the broad, sweeping regulatory statues (e.g., the Clean Water Act and various amendments) historically used to solve more easily identifiable problems, but instead require an understanding of the inherent ecological uniqueness that defines complex problems on a case-by-case basis. We suggest that management and regulation of in-stream water temperature is a complex problem that will vary across both space and time and that natural resources professionals must recognize the dynamic nature of this relationship in designing management plans, regulatory policy, and future research. Such recognition is often at odds with contemporary policies and paradigms that focus attention on discreet temperature (or other water quality) values, thus creating tension between managers and

Water Quality in the Agronomic Context: Flood Irrigation Impacts on Summer In-Stream Temperature
Extremes in the Interior Pacific Northwest (USA)

103

regulators. Incorporating spatial and temporal variation into the concept of stream temperature will allow for a more representative characterization of water temperature regime and promote a more comprehensive understanding of the potential impacts of water temperature on the stream ecosystem and its inhabitants.

5. References

Adams, P.W. (2007). Policy and management for headwater streams in the Pacific Northwest: synthesis and reflection. *Forest Science*, Vol.53, pp. 104-118.

Atwill, E.R.; Tate, K.W.; Das Gracas, M.; Pereira, C.; Bartolome, J.W. & Nader, G.A. (2005). Efficacy of natural grass buffers for removal of *Cryptosporidium parvum* in rangeland runoff. *Journal of Food Protection*, Vol.69, pp. 177-184.

Belsky, A.J.; Matzke, A. & Uselman, S. (1999). Survey of livestock influences on stream and riparian ecosystems in the western United States. *Journal of Soil and Water Conservation*, Vol.54, pp. 419-431.

Beschta, R.L. (2000). Watershed management in the Pacific Northwest; the historical legacy. In: Land Stewardship in the 21st Century: Contributions of Watershed Management. USDA Forest Service Rocky Mountain Forest & Range Experiment Station General Technical Report.

Boyd, M. & Strudevant, D. (1996). The scientific basis for Oregon's stream temperature standard: common questions and straight answers. Oregon Department of Environmental Quality. 30 pp.

Boyd, C.S. & Svejcar, T.J. (2004). Regrowth and production of herbaceous riparian vegetation following defoliation. *Journal of Range Management*, Vol. 57, pp. 448-454.

Boyd, C.S. & Svejcar, T.J. (2009). Managing complex problems in rangeland ecosystems. *Rangeland Ecology and Management*, Vol. 62, pp. 491-499.

Castelle, A.J.; Johnson, A.W. & Conolly, C. (1994). Wetland and stream buffer requirements: a review. *Journal of Environmental Quality*, Vol.23, pp. 878-882.

Clary, W.P & Leininger, W.C. (2000). Stubble height as a tool for management of riparian areas. *Journal of Range Management*, Vol. 53, pp. 562-573.

Clary, W.P.; Shaw,N.L; Dudley, J.G.; Saab, V.A.; Kinney, J.W.& Smithman, L.C. (1996). Response of a depleted sagebrush steppe riparian system to grazing control and woody plantings. USDA, Forest Service, Intermountain Research Station Research Paper INT-RP-492.

Dorioz, J.M.; Wang, D.; Poulenard, J. & Trevisan, D. (2006). The effect of grass buffer strips on phosphorous dynamics – A critical review and synthesis as a basis for application in agricultural landscapes in France. *Agriculture, Ecosystems and Environment*. Vol.117, pp. 4-21.

Dosskey, M.G. (2002). Setting priorities for research on pollution reduction functions of agricultural buffers. *Environmental Management*, Vol.30, pp. 641-650.

Ebersole, J.L.; Liss, W.J. & Frissell, C.A. (2001). Relationship between stream temperature, thermal refugia and rainbow trout *Oncorhynchus mykiss* abundance in arid-land streams in the northwestern United States. *Ecology of freshwater fish*, Vol.10, pp. 1-10.

Gamperl, A. K., Rodnick, K. J., Faust, H. A., Venn, E. C., Byington, J. D., Bennett, M. T., Crawshaw, L. I., Kelley, E. R., Powell, M. S. & Li, H. (2002). Metabolism, swimming performance and tissue biochemistry of high desert redband trout (Oncorhynchus mykiss ssp.): Evidence of phenotypic differences in physiological function. Physiological and Biochemical Zoology. Vol.75, pp. 413-431.

George, W.F.; Jackson, R.D.; Boyd, C.S. & Tate, K.W. (2011). A scientific assessment of the effectiveness of riparian management practices. pp. 213-252. Briske, D.D. (*editor*). Conservation benefits of rangeland practices: Assessment, recommendations, and knowledge gaps. United States Department of Agricultur – Natural Resources Conservation Service.

Henley, W.F.; Patterson, M.A.; Neves, R.J. & Dennis Lemly, A. (2000). *Reviews in Fisheries Science*, Vol.8, pp. 125-139.

Holland, K.A; Leininger, W.C. & Trilica, M.J. (2005). Grazing history affects willow communities in a montane riparian ecosystem. *Rangeland Ecology and Management*, Vol.58, pp. 148-154.

Isaak, D.J.& Hubert, W.A. (2001). A hypothesis about factors that affect maximum summer stream temperatures across montane landscapes. *Journal of the American Water Resources Association*, Vol. 37, pp. 351-366.

Kauffman, J.B. & Krueger, W.C. (1984). Livestock impacts on riparian ecosystems and streamside management implications...A review. *Journal of Range Management*, Vol. 37, pp. 430-438.

Kleinfelder, D.; Swandon, S.; Norris, G. & Clary, W. (1992). Unconfined compressive strength of some streambank soils with herbaceous roots. *Journal of the Soil Science Society of America*, Vol. 56, pp. 920-924.

Knox, A.K.; Tate, K.W.; Dahlgren, R.A. & Atwill, E.R. (2007). Management reduces E. coli in irrigated pasture runoff. *California Agriculture*, Vol.61, pp. 159-165.

Larson, L.L. & Larson, S.L. (1996). Riparian shade and stream temperature: A perspective. *Rangelands*, Vol.18, pp. 149-152.

Lee, K.H.; Isenhart, T.M. & Shultz, R.C. (2003). Sediment and nutrient removal in an established multi-species riparian buffer. *Journal of Soil and Water Conservation*, Vol. 58, pp. 1-8.

Line, D.E. (2003). Changes in a stream's physical and biological conditions following livestock exclusion. *American Society of Agricultural Engineers*, Vol. 46, pp. 287-293.

Lowrance, R.R.; Leonard, R.A., Asmussen, L.E. & Todd, R.L. (1985). Nutrient budgets for agricultural watersheds in the southeastern coastal plan. *Ecology*, Vol.66, pp. 287-296.

Lyons, J.; Trimble, S.W. & Paine, L.K. (2000). Grass versus trees: Managing riparian areas to benefit streams of central North America. *Journal of the American Water Resources Association*, Vol.36, pp. 919-930.

Matney, C.A.; Boyd, C.S. & Stringham, T.K. (2005). Use of felled junipers to protect streamside willows from browsing. *Rangeland Ecology and Management*, Vol.58, pp. 652- 655.

Mayer, P.M.; Reynolds, S.K.; Steven, J., Jr.; McCutchen, M.D.; Marshall, D. & Canfield, T.J. (2007). Meta-analysis of nitrogen removal in riparian buffers. *Journal of Environmental Quality*, Vol.36, pp. 1172-1180.

McRae, G. & Edwards, C.J. (1994). Thermal characteristics of Wisconsin headwater streams occupied by beaver: Implications for brook trout habitat. *Transactions of the American Fisheries Society*, Vol.123, pp. 641-656.

Water Quality in the Agronomic Context: Flood Irrigation Impacts on Summer In-Stream Temperature
Extremes in the Interior Pacific Northwest (USA)

105

Meehan, W.R. (1991). Influences of forest and rangeland management on salmonid fishes and their habitats. American Fisheries Society, Bethesda, MD. 751p.

Nader, G.; Tate, K.W.; Atwill, R. & Bushnell, J. (1998). Water quality effect of rangeland beef cattle excrement. *Rangelands*, Vol.20, pp. 19-25.

National Research Council (1995). Wetlands: characteristics and boundaries. Committee of the Characterization of Wetlands, National Academy of Sciences, Washington, D.C. 307p.

Oregon Climate Service. (2005). Climatological data for Seneca, Oregon. Oregon State university, Corvallis, OR.

Oregon Department of Environmental Quality. (2002). Oregon's Final 2002 303(d) List. http://www.deq.state.or.us/wq/WQLData/View303dList02.asp.

Oregon Department of Environmental Quality. (2008). Temperature Water Quality Standard Implementation - A DEQ Internal Management Directive. http://www.deq.state.or.us/wq/pubs/imds/Temperature.pdf

Parajuli, P.B., Mankin, K.R. & Barnes, P.L. (2008). Applicability of targeting vegetative filter strips to abate fecal bacteria and sediment yield using SWAT. *Agricultural Water Management*, Vol.95, pp. 1189-1200.

Pearce, R.A.; Trlica, M.J.; Leininger, W.C.; Smith, J.L. & Frasier, G.W. (1997). Efficiency of grass buffer strips and vegetation height on sediment filtration in laboratory rainfall simulation. *Journal of Environmental Quality*, Vol.26, pp. 139-144.

Poole, G.C. & Berman, C.H. (2001). An ecological perspective on in-stream temperature: natural heat dynamics and mechanisms of human-caused thermal degradation. *Environmental Management*, Vol. 27, pp. 787-802.

Rosgen, D.L. (1994). A classification of natural rivers. *Catena*, Vol.22, pp. 169-199.

Richter, B.D.; Braun, D.P.; Mendelson, M.A. & Master, L.L. (1997). Threats to imperilled freshwater fauna. *Conservation Biology*, Vol.11, pp. 1081-1093.

SAS Institute Inc. (1999). SAS procedures guide, release 8.0. Cary, NC.

Schillinger, W.F.; Papendick, R.I. & McCool, D.K. (2010). Soil and Water Challenges for Pacific Northwest Agriculture. Soil and Water Conservation Advances in the United States. pp. 47-79. Zobeck, T.M. & Schillinger, W.F. (*editors*). Soil Science Society of America Special Publication 60, Madison, WI.

Schmitt, T.J.; Dosskey, M.G & Hoagland, K.D. (1999). Filter strip performance and processes for different vegetation, widths, and contaminants. *Journal of Environmental Quality*, Vol.28, pp. 1479-1489.

Stefan, H.G. & E.B. Preud'homme. (1993). Stream temperature estimation from air temperature. *Water Resources Bulletin*, Vol.29, pp. 27-45.

Stringham, T.K., J.C. Buckhouse & Krueger, W.C. (1998). Stream temperatures as related to subsurface waterflows originating from irrigation. *Journal of Range Management*, Vol.51, pp. 88-90.

Tate, K.W.; Atwill, E.R., McDougald, N.K. & George, M.R. (2003). Spatial and temporal patterns of cattle feces deposition on rangeland. *Journal of Range Management*, Vol.56, pp. 432-438.

Tate, K.W.; Das Gracas, M; Pereira, C & Atwill, E.R. (2004). Efficacy of vegetated buffer strips for retaining *Cryptosporidium parvum*. *Journal of Environmental Quality*, Vol.33, pp. 2343-2251.

Tate, K.W.; Lile, D.F.; Lancaster, D.L.; Porath, M.L.; Morrison, J.A. & Sado, U. (2005). Statistical analysis of monitoring data to evaluate stream temperature – a watershed scale case study. *California Agriculture*, Vol.59, pp. 161-167.

Tim, U.S. & Jolly, R. (1994). Evaluating agricultural nonpoint-source pollution using integrated geographic information systems and hydrologic/water quality models. *Journal of Environmental Quality*, Vol.23, pp. 25-35.

Toledo, Z.O. & Kauffman, J.B. (2001). Root biomass in relation to channel morphology of headwater streams. *Journal of the American Water Resources Association*, Vol. 37, pp. 1653-1663.

Trefethen, J.B. (1985). An American Crusade for Wildlife. Boone and Crockett Club, Alexandria, VA. 409p.

United States Environmental Protection Agency (2009). Fact Sheet: Introduction to Clean Water Act (CWA) Section 303(d) Impaired Waters Lists. U.S. Environmental Protection Agency, Washington, D.C. Available at http://www.epa.gov/owow/tmdl/results/pdf/aug_7_introduction_to_clean.pdf (accessed September 26, 2011).

Vitousek, P.M.; Aber, J.; Howarth, R.W.; Likens, G.E.; Matson, P.A.; Schindler, D.W.; Schlesinger, W.H. & Tilman, G.D. (1997). Human alteration of the global nitrogen cycle: Causes and consequences. *Issues in Ecology*, Vol.61, pp. 1225-1232.

Voleksy, J.D; Schacht, W.H.; Koehler, A.E.; Blankenship, E. & Reece, P.E. (2011). Defoliation effects on herbage production and root growth of wet meadow forage species. *Rangeland Ecology and Management*, Vol.64, pp. 506-513.

Ward, J.V. (1985). Thermal characteristics of running waters. *Hydrobiologia*, Vol.125, pp. 31-46.

Winward, A.H. (1994). Management of livestock in riparian areas, p. 49–52. In : G . A. Rasmussen and J.P. Dobrowolski (*editors*), Riparian resources: a symposium on the disturbances, management, economics, and conflicts associated with riparian ecosystems. Coll. of Natur. Resour., Utah State University, Logan, UT.

Yates, P. & Sheridan, J.M. (1983). Estimating the effectiveness of vegetated floodplains/wetlands as nitrate-nitrite and orthophosphorous filters. *Agriculture, Ecosystems and Environment*, Vol.9, pp. 303-314.

Young, R.G. & Huryn, A.D. (1998). Comment: Improvements to the diurnal upstream-downstream dissolved oxygen change technique for determining whole-stream metabolism in small streams. *Canadian Journal of Fisheries and Aquatic Sciences*, Vol.55, pp. 1784-1785.

7

The Effect of Wastes Discharge on the Quality of Samaru Stream, Zaria, Nigeria

Y.O. Yusuf and M.I. Shuaib

Department of Geography, Ahmadu Bello University, Zaria,
Nigeria

1. Introduction

Water is the most important natural resource in the world. Since without it life cannot exist and most industries would not function or operate (Tebbut, 1998). It is essential to life and in fact the basis of life, being so it was almost inevitable that the development of water resources preceded any real understanding of their origin and formation (Ward, 1975). However, water is a unique resource having no substitute with its quality and quantity varying over space and time, hence is finite. Water is equally one of the remarkable substance known which is found in vast quantities in nature; it could be in gaseous, liquid and solid state. There are nearly 14×10^8 cubic kilometers of water on the planet but 97.5% of this is salty water, fresh water account for only 2.35% of the total. Many people depend on fresh water from lakes, rivers and streams for their water supplies, and these sources contain respectively only 0.26 and 0.006% of the total volume of fresh water (UNEP, 1994)

Water plays a vital role in the development of a stable community, since human being can exist for days without food but absence of water for a few days may lead to death. Water is an essential pre-requisite for the establishment of a stable community. In the absence of which nomadic lifestyle becomes necessary and communities move from one area to another as demand for water exceed its availability. Hence, it is therefore not surprising that sources of water are often zealously guarded over the century. Many skirmishes have taken place over water right (Tebbut, 1998). Water is used for drinking, cooking, sanitation, agricultural purpose, industrial purposes and also used for generating hydro-electric power e.t.c. The amount of water used for other purposes, apart from food preparation and cooking, vary widely and are greatly influenced by the type and availability of water supply.

Furthermore, these facts show how important water is to man, so if water is contaminated it poses a great health hazard to man causing various diseases, and one of the greatest avenue for the spread of diseases is through water. Therefore the presence of water in the environment does not suffice, rather how useful it is to man is what qualifies it as a resource to man. Considering that the utility of water is limited, evaluating in terms of quality, quantity and reliability of all the possible water sources becomes expedient. World Health organization (WHO, 1976) reported that 1.5 billion people worldwide drink filthy water, and this is thought to be increasing for about 20 million each year, up to 25,000 people die each day in the world from water related diseases. Waste water is a complex mixture of natural inorganic and organic material mixed with man-made substances. It contains everything discharged to the sewer, including material washed from roads and roots, and of course where the sewer is damaged groundwater will also gain entry (Edmund *et al*, 1976 and Gray, 2000).

The contamination of water can affect aquatic animals whose survival depend on the quality of water in which they live presently, aquatic life existence is being threatened by man's activities such as industrial waste, domestic waste, waste of animals and human discharge into the stream. These activities have been affecting river quality and in turn the living organisms in the stream (Vega *et al*, 1996).

Samaru stream, which takes its source from Samaru, Zaria, Kaduna state (Nigeria) is a headwater tributary of river Kubanni. The most noticeable problem in Samaru is the inadequate or lack of pipe sewer system. This obviously creates some inefficiency in the disposal of liquid and solid wastes. Refuse disposal is by open dump system. Soak away pit are used as a means of disposal of domestic wastes. All the aforementioned constitute a major pollution in the study area.

There is a network of open gutter which convey storm water away to the major drain running parallel to Zaria-Sokoto road. This drain discharge directly into the stream, this wastes water carries with it faeces, solid materials, ashes, and several other assorted materials which if not deposited along the course of the stream end up in Kubanni dam. These problems are aggravated during the rainy season and these wastes are transported and deposited in the river channel and subsequently into the Kubanni reservoir which may increase the incidence of water borne disease.

The aim of this research is to assess the effect of wastes discharge on the quality of Samaru stream, Zaria, Kaduna state, Nigeria. In order to achieve the aim, the following objectives are imperative, to:

1.   Determine the effect of pollution on the quality of the Samaru stream.
2.   Compare upstream and downstream water quality of the Samaru stream.
3.   Give possible suggestions on measures to improve the quality of water in the Samaru stream.

2. Background to the study area

Zaria is located at approximately 11°3′ north of the equator and on longitude 7° 42′ East of the Greenwich meridian and at about 660m above sea level. It is the second largest city in Kaduna state after Kaduna town north ward along Kaduna-Kano highway (see fig.1 and 2). It has four (4) geopolitical zones (local government areas) including Zaria city local government, Sabon Gari local government area to the north, Soba local government to the east and Giwa local government area to the west.

Samaru is located in Zaria and situated in Sabon-Gari local government area of Kaduna state. It is located at latitude 11° 10′ north and longitude 7° 37′ east. The area is an extension of urban Zaria to the north along Zaria-Sokoto and Zaria Kaura Namoda railway line. It is also one of the major settlements that make up the urban Zaria. It is an educational and administrative settlement which brought about the establishment of new settlement for non-residents of Zaria City. It grows as a result of the establishment of Ahmadu Bello University, which was established in 1962 and other institutes like Federal Institute for Chemical and Leather Research, Federal College of Aviation and Industrial Development Corporation.

Fig. 1. Map of Nigeria showing Kaduna State

The Kubanni River has its source from the Kampagi hill in Shika near Zaria. It flows in a Southeast direction through the premises of Ahmadu Bello University. The Samaru stream which is one of the tributaries of Kubanni River has a stream length of 1.05km within an area of 2.28km² and has drainage density of 0.4605 km/km².

The climate of the study area is a tropical savanna climate, with distinct wet and dry seasons (Aw climate Koppens classification). Zaria experiences six (6) months of rainy season and six (6) months of dry season. The rainy season is from May to late October, while the dry season is from early November to April, this is as a result of the interplay of the two dominant air masses within the region i.e. the tropical continental air messes (cT) and Tropical maritime air masses (mT) (Iguisi and Abubakar, 1998).

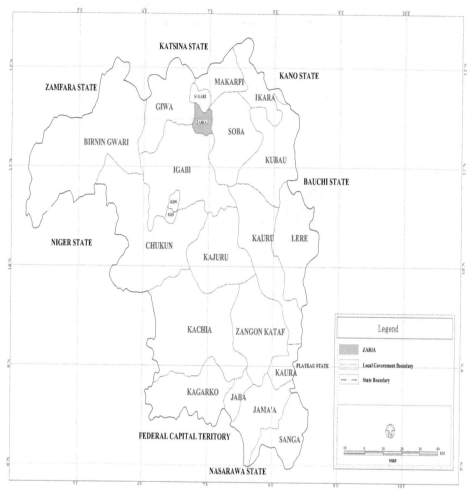

Fig. 2. Kaduna State map showing Zaria local government

The tropical continental air masses (cT) within originates from the Sahara desert and a dry cold wind comes along with it, it is dust ladden and blows from a north eastern direction while the Tropical maritime air masses (mT) originates from the oceans and blows from south western direction. It is warm and moist and therefore capable of bringing the information of rain in the Zaria region. The continental air masses are responsible for the harmattan haze in the region.

The mean daily maximum temperature is at the peak in April and about 39°C while the mean minimum temperature rises from its lowest value in December to January to its highest in July to August (Ojo, 1982).

The geology of the Zaria region is underlain by crystalline metamorphic and igneous rocks of Precambrian to lower Palaeozoic age occurring on the basement complex rock. A major part of the rock is of high grade of metamorphism mainly gneisses which suffered intense folding and granitization and has remained stable for millions of years. Others are migmatites, older granites and more recently meta- sediments (Quartz, schist, laterites and alluminium). The rate and depth of weathering is quite irregular with variabilities but thorough, ranging from as deep as 60m to as shallow as 10m (Mortimore, 1970; Wright and McCurry, 1972).

The soil type of the study area is alluvial soil, also the area constitute dark vertisol referred to as "fadama" soil (Hausa) this soil is classified as hydromorphic soil. The fadama soil is usually found in the upstream and downstream of the stream system, while the alluvial soil is predominately at the middle of the stream. The soil composed of fine grey–brown sands, clay, red sand and gravel. The upper parts of the soil are a mixture of quartz, mica and windblown particles from the savannah harmattan.

The region generally falls within the Guinea Savannah vegetation. The climax climatic vegetation of the area ought to be northern Guinea Savannah, but because nearly all vegetation within the stream system has been degraded due to man's activities such as intense cultivation, fuel wool felling, the real climax vegetation is almost absent. What is seen presently are few scattered trees interspaced with tall tree grasses about 1-15m and 2-5m respectively? Trees found here includes *Isorberlina doka*, grass type includes *Adropogonaea spp*, *Schizachirium semiberbe* and *Monocynbium ceresti* (Nyagba,1986)

The drainage system focuses on River Galma and Kubanni River. River Galma is a major tributary of River Kaduna and Kubanni River, on which Ahmadu Bello University Dam is situated, is seasonal and supply water to Ahmadu Bello University and its environs. Samaru stream flows in north south direction through the main campus of Ahmadu Bello University Zaria, situated along a valley west to Samaru village into the Kubanni River (see fig. 3).

3. Materials and methods

The primary sources include results derived from the laboratory analysis of water quality of the water samples taken from upstream and downstream of the Samaru stream. This analysis includes the physico-chemical test analysis and the bacteriological test analysis. The physicochemical test analysis includes pH, colour and total dissolved solid while the bacteriological test analysis includes the biological oxygen demand (BOD), dissolved oxygen (DO) and chemical oxygen demand (COD) by using sterilized sample bottles. The samples were collected for 30 days consecutively within the month of August and September. The samples were collected, in the morning and evening (7.00am and 6.00pm) in order to observe any variation.

The samples collected were analysed in the laboratory of the Department of Water Resources and Environmental Design, Ahmadu Bello University, Zaria. The water samples collected from two points namely, upstream and downstream were analyzed, using standard procedures, within an hour of collection in other to avoid unpredictable change in the sample (WHO, 1971; WHO, 1976).

Statistical analysis of the data collected, including average means and correlation matrices was done by using the SPSS package.

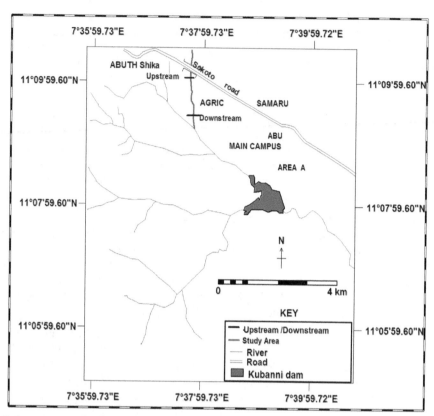

Fig. 3. Sumaru stream showing the upstream / downstream points of data collection.
Source: Adapted and modifed from Zaria SW topographivc map

4. Results and discussion

4.1 Upstream variation in water quality

Water samples were collected and tested at the upstream section of the channel using
several parameters including pH, colour, Dissolved Oxygen (DO), Biological Oxygen
Demand (BOD). These samples were collected both at the early part of the day (7am) and at
the later part of the day (6pm). The mean values for each of the parameters at the different
times period of the day over 30 days are summarized in table 1.

Parameter	Mean Upstream	
	7am	6pm
pH	7.25	7.23
Colour	15.97	22.9
DO	6.98	6.76
BOD	2.22	2.25

Table 1. Upstream mean values of parameter

4.1.1 Upstream variation in pH

From the result obtained the upstream mean pH value for the early part of the day was 7.25 while the later part of the day recorded a mean value of 7.23. This indicate that there is no much difference in upstream pH value for both morning and evening period, they are within the range of neutrality. This low variation in both time periods is further exemplified by the trend line graph shown in figure 4a.

4.1.2 Upstream variation in colour

From the derived result the upstream mean colour value for the early part of the day was 15.97, while the later part of the day recorded the mean colour value of 22.9, giving a range of 6.95, which indicate a wide variation for both time periods (morning and evening). This wide variation is further revealed by the trend line graph in figure 4b, which shows a wide and haphazard distribution over the time period.

4.1.3 Upstream variation in dissolved oxygen (DO)

The DO has a mean upstream value of 6.98(mg/l) recorded in the morning while 6.76(mg/l) was the mean value recorded for the evening period, this gave a low range of 0.22(mg/l), this means that there is no significant variation in DO upstream for both morning and evening. These trends are confirmed in figure 4c.

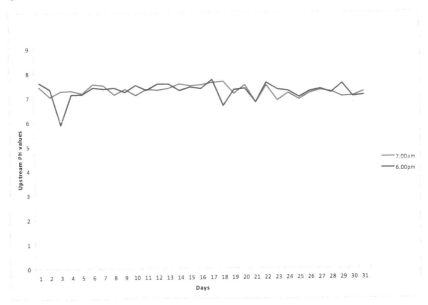

Fig. 4a. Upstream trend in pH for 30 days

4.1.4 Upstream variation in BOD

The result obtained reveals that upstream mean value for Biological oxygen demand is lower in the morning with 2.25(mg/l) than that recorded for the evening period with

2.5(mg/l), giving a range of 0.03mg/l indicating that there is low variation in BOD in both morning and evening. This low variation is depicted in the trend line for both morning and evening; this is shown in figure 4d.

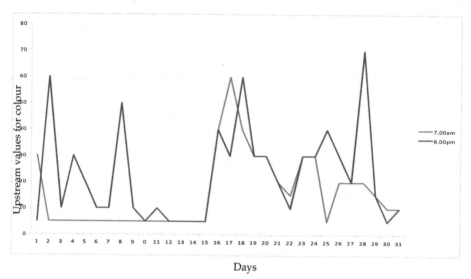

Fig. 4b. Upstream trend in colour for 30 days

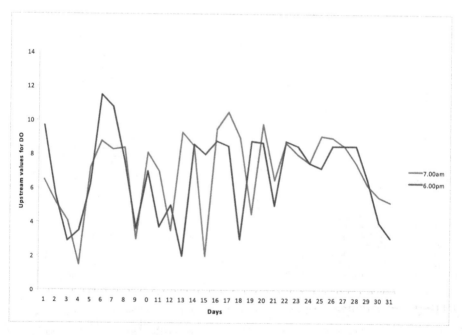

Fig. 4c. Upstream trend in Dissolved Oxygen for 30 days

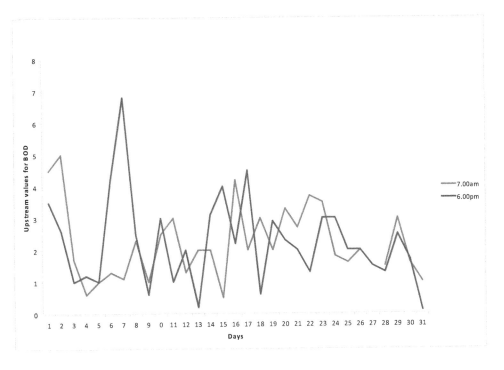

Fig. 4d. Upstream trend in BOD for 30 days

4.2 Downstream variation in water quality

Water samples were collected and tested at the downstream section of the channel using same parameters. These samples were collected both at the early part of the day and at the late part of the day. The mean values for each of the parameter at different time period of the day over 30 days are summarized in table 2.

Parameter	Downstream Mean	
	7am	6pm
pH	7.20	7.28
Colour	16.61	18.06
TDS	162.42	154.09
DO	6.71	6.83
BOD	2.15	2.69
COD	568.45	567.74

Table 2. Downstream mean values of parameters

4.2.1 Downstream variation in pH

From the result obtained, the downstream mean pH value for the early part of the day was 7.20 while in the later part of the day it was recorded as 7.20; this shows that there is no significance difference between the downstream (morning and evening), they fall within the neutrality level, giving a range of 0.08 which is relatively low. This low variation is further revealed by the trend line graph in fig 5a.

4.2.2 Downstream variation in colour

The downstream mean value of colour in the early part of the day was 16.61 while in the later part of the day it was recorded as 18.06, this shows that it was higher in the later part of the day, giving a range of 1.45 which is relatively high. This variation is depicted in the close trend line graph for both morning and evening as shown in fig 5b.

4.2.3 Downstream variation in dissolved oxygen (DO) mg/l)

From the result obtained it was observed that the mean DO was higher in the later part of the day i.e. 6.83 mg/l, while in the early part of the day it was 6.71 mg/l. Given a range of 0.12 which is relatively low, means there is no significant variation in dissolved oxygen downstream for both morning and evening. This low variation is further revealed in the trend line graph shown in fig 5c.

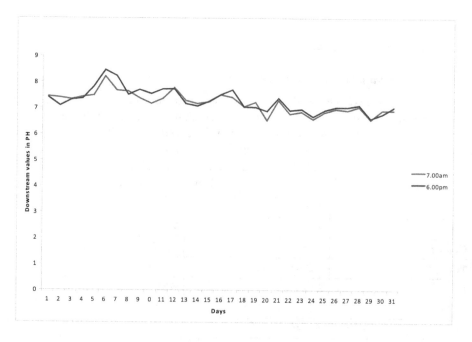

Fig. 5a. Downstream trend in pH for 30 days

4.2.4 Downstream variation in biological oxygen demand (BOD) (mg/l)

Mean BOD downstream value was 2.15 in the morning while 2.69 was the mean value recorded for the evening. This gave a range value of 0.54 which is relatively low. The trend line graph is presented in fig 5d.

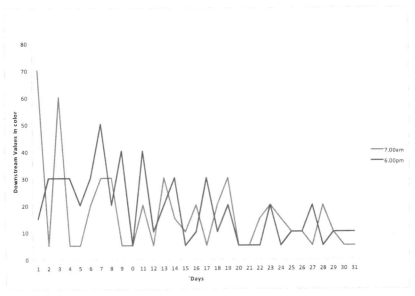

Fig. 5b. Downstream trend in color for 30 days

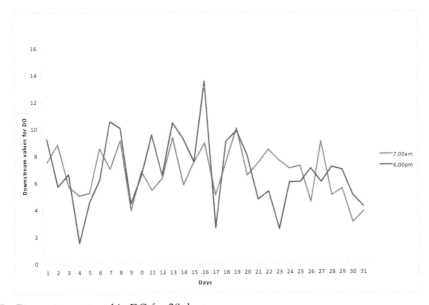

Fig. 5c. Downstream trend in DO for 30 days

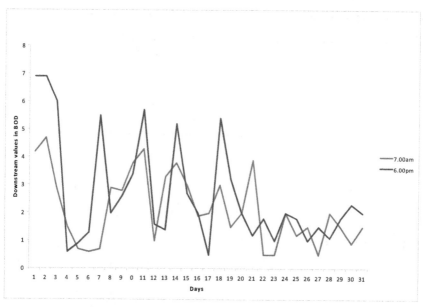

Fig. 5d. Downstream trend in BOD for 30 days

4.3 Upstream-downstream trend for the morning and evening

The upstream and downstream values for the early part of the day were correlated for the various parameters to examine the relationship between the upstream and downstream trend for each parameters at the early part of the day. From the correlation carried out, it was observed that the upstream and downstream trend for all the parameters were all positively correlated but weak with the exception of total dissolved solid (TDS) and BOD which have correlation co-efficient r values of 0.7219 and 0.5915. This shows that for an upstream increase in any of the parameters there is also a downstream increase for such parameters and vice-versa (see table 3).

Parameters	r^2	R
pH	0.0827	0.2876
Colour	0.0041	0.0640
TDS	0.5211	0.7219
DO	0.0736	0.2713
BOD	0.3499	0.05915
COD	0.1212	0.3481

Table 3. r^2 and r value for upstream – downstream relationship in the morning.

The upstream and downstream values for the later part of the day were correlated for the various parameters to examine the relationship between the upstream and downstream trend for each of the parameters at the later part of the day. From the correlation carried out, it was observed that the upstream and downstream trend for all the parameters were positively correlated but weak with the exception of TDS which has a correlation coefficient (r) value of 0.7198 (Table 4).

Parameter	r^2	R
pH	0.0149	0.1221
COLOUR	0.0261	0.1616
DO	0.0447	0.2114
BOD	0.0283	0.1682

Table 4. r^2 and r values for upstream – downstream relationship in the evening

5. Conclusion and recommendation

The result obtained from this study infers that the Samaru stream is still well oxygenated and can be said to be safe. However, the safety of this stream is being threatened by the continuous deposition of wastes into it from Samaru. This is why it can be classified as Class 2 as presented by Audu (2002) which can be said to be of doubtful water quality and needing improvement especially as it is found to be of low aesthetic quality.

Similarly the pH values observed from the study falls within the maximum permitted limit of 6.5 – 8.5 as specified by the Nigeria standard for drinking water quality. On the other hand the colour levels observed in the study is above the maximum permitted level of 15 as specified by the Nigerian standard for drinking water (NIS, 2007).

Parameters used in this study which were found to be positively and strongly correlated upstream and downstream suggest that there is no significant difference in the upstream and downstream water quality of the Samaru stream.

Based on the results obtained from the study the following recommendations became imperative.

1. Other water quality parameters, in addition to those observed in this study should be tested during the dry season to check the pollution, such as nitrate, ammonia, calcium and sulphate because of human habitation due to the discharge of waste water and agricultural practises.
2. It will be of significance to carry out another study collecting water sample both in dry season and rainy season's so as to critically examine the variation between the two seasons.
3. Location of domestic, industrial and agricultural wastes should be made far away from water bodies, this will greatly reduce the amount of wastes reaching the water bodies.
4. There is need for public enlightenment on the need to protect our environment and the benefit to be derived therefrom through the mass media such as radio, television and newspaper.
5. Control of the farming activities within the catchment should be fully implemented to avoid concentration in flow of organic and inorganic materials into the rivers.

6. References

Audu, C. (2002): The Effect of Sabon-Gari Discharge on the Quality of a Portion of River Galma, Zaria. *Unpublished M.Sc theses*, Department of Water Resources and Environmental Design, A.B.U Zaria.

Ayoade, J.O. (1988): *Tropical Hydrology and Water Resources*: Macmillan Publishers Ltd, London Pp.275

Edmund, B., Besselievre, P.E. and Schwartz, M. (1976) *The Treatment of Industrial Waste*. Second Edition, Mcgraw Hill Inc. New York.

Gray, N.F. (2000): *Water Technology*. First Edition Vinod Vasistha Publishers London.

Iguisi, E.O. and Abubakar, S.M. (1998). "The Effect of Land use on Dam Siltation" Paper presented at 41st Annual Conference of Nigeria Geographical Association, University of Uyo, Akwa Ibom.

Mortimore, M.J. (ed.) (1970): *Zaria and Its Region*. No. 4 Department of Geography, A.B.U. Zaria.

NIS (2007) *Nigerian Standard for Drinking Water*. Standards Organization of Nigeria, Lagos.

Nyagba, L. (1986) *Soils, Climate and Vegetation*. Institute of Education Press, A.B.U. Zaria

Ojo, O. (1982) *The Climate of West Africa*. HEB, London.

Tebbut, T.H.Y. (1998) *The Principle of Water Quality Control*. Fifth Edition. Butter Worth, Heineman, Linacre House Jordon Hills Oxford.

Vega, M.M., Fernández, C., Blázquez, T., Tarazona, J.V. and Castaño, A. (1996) *Biological and Chemical Tools in the Toxicological Risk*. Assessment of Jarama River Spain Environmental Pollution, 135 – 139 pp.

Ward, R.C. (1975) *Principles of Hydrology*. McGraw-Hill, London.

WHO (1971): *International Standard for Drinking Water* 3rd Edition Geneva, www.rivernetwork.org.

WHO (1976): *International Standard for Drinking Water* 3rd Edition Geneva, www.rivernetwork.org.

Wright, J. B. and McCury, P (1970) *Land Forms in Zaria and the Surrounding Regions*. Third World Planning Review, Liverpool University Press.

8

Elimination of Phenols on a Porous Material

Bachir Meghzili[1,2], Medjram Mohamed Salah[1],
Boussaa Zehou El-Fala Mohamed[1] and Michel Soulard[3]
[1]Laboratoire LARMACS, Université de Skikda Algérie,
[2]Université de Biskra,
[3]Equipe Matériaux à Porosité Contrôlée, IS2M, LRC CNRS 7228, UHA, ENSCMu,
France

1. Introduction

The surface water which feeds the majority of the stations of treatment of drinking water is charged by organic composed, including one great part makes up of humic substances. It is thus important to eliminate them to avoid the formation of generally toxic organohalogen compounds (Dore, 1989; Meier, 1988; Boudhar, 1999). The coagulation and the flocculation followed by a clarification remain the most frequent treatment to withdraw colloids present in water, which they are of organic or mineral origin (Bersillon, 1983; Lefebvre and Legube, 1990).

Conventional treatment of the clarification which could indeed eliminate these macromolecules from the humic type is not always sufficient; it often requires optimal conditions not very compatible with the practical conditions of operation and of the treatment as with the criteria of potability. Processes based on adsorption often constitute a technique of choice, complementary to treatment basic. The adsorption is one of the processes of the separation which finds its application in several fields, such as extraction, purification and depollution.

Among the most recent progress in the water treatment, the advanced processes of oxidation (advanced oxidation process AOP) considered to be effective, allow mineralization in aqueous medium of the toxic organic molecules with respect to the man and of the environment. The advanced processes of oxidation are based on the generation and the use of a very strong oxidant which is the radical hydroxyl. This last can be produced by various processes chemical, photochemical biological, electrochemical (Andreozzi and Al, 1999; Chiron and Al, 2000; Galze and Al, 1992; Safarzadeh-Amiri and Al, 1996; Dussert, 1997). These methods rest on the formation of very reactive chemical entities which will biologically break up the most recalcitrant molecules into molecules degradable or in mineral compounds (Golich and Bahnemann, 1997). The reactions generally studied on this level can be classified in three categories (Hoang, 2009):

- Reactions with the reagents électrophiles (O_3, $HOCl$, ClO_2 and NH_2Cl),
- Reactions with the radicalizing species and initiating reactions of production of radicals (HO and inorganic radicals: CO_3^-, SO_4^-, Cl_2^-, catalyse homogeneous by Fe^{II} and Fe^{III}, radiolysis of water, photocatalysis, catalytic ozonization etc…)
- Reactions of phototransformation (UV and solar) with or without catalyst.

In Algeria, organic material can represent, with it only, a big part of the organic load of surface waters, in particular in the case of water of stopping. The presence of natural organic matter in a surface water east at the origin of many problems encountered during the various stages of treatment of potabilisation. Initially, the natural organic matter is undesirable because it reacts with chlorine during disinfection to form volatile organic compounds (trihalométhanes [THM], acid dichloroacetic [DCAA], etc), produced potentially carcinogenic (Lefebvre and Legube 1990; Hooper and Al, 1996; Stevens and Al, 1976; Najm and Al, 1993).

The natural organic matter is also known for its role in the transport and the trapping of organic and/or inorganic pollutants (Bartschat and Al, 1992; Tippinge, 1993). It represents also a potential substrate for the biological growth in the distribution network of drinking water. The weak dehydration of muds resulting from the treatment of drinking water and the filling of the membranes of filtration are also related to the presence of the natural organic matter in water (Dulin and Knocke 1989; Wiesner and Al, 1989; Bersillon and Al, 1999). Among the organic compounds, the phenols are regarded as harmful pollutants even with weak concentrations because of the potential dangers on health and environment (Dutta and Al, 1992, 1998).

The choice of a adsorbent material depends inter alia its type of porosity, its specific surface and nature of the element to be trapped. Our choice was made on a bentonite of M'zila (Mostaganem), rich in montmorillonite, because of its properties particular to fix many substances, of its availability in Algeria and its low costs (Essington, 1994; Amar and Gaid, 1987; Boufatit and Al, 2007). Indeed, some phyllosilicates have the property to easily adsorb water molecules or organics in interfoliar space. This phenomenon called swelling depends on the load of the layer, the localization of this one and the nature of the cations of compensation (Cailliere and Al, 1982). The bentonite is a material which contains approximately montmorillonite 75% and whose size of the particles is lower than 2 µm (Bergaya and Al, 2006). The argillaceous mineral term or phyllosilicate corresponds to hydrated aluminium silicates, of lamellate structures. These clays generally constitute a considerable fraction of grounds (Auerbach and Al, 2004). The layers of the phyllosilicates are consisted a stacking of octahedral layers (O) and tetrahedral (T). The tetrahedral layer generally consists of atoms of silicon surrounded by four oxygen atoms and bound between them by covalent bonds Si-O. The octahedral layer is formed by hexagonal units, composed of atoms of coordinate magnesium or aluminium with six oxygen atoms or with functions hydroxyls. The layers T and O are bound by covalent bonds and imply apical oxygens. The space located between the two layers is called interfoliar space. A layer and an interfoliar space form a structural unit. The phyllosilicates have the possibility of easily adsorbing water molecules in this interfoliar space.

During a isomorphic substitution of an element by another of lower oxidation step, in tetrahedral or octahedral layer, the deficit of load "+" of the layer is made up by interfoliar cations known as of compensation, exchangeable by mineral or organic cations (Cailliere and Al, 1982; Decarreau, 1990; Bouras, 2003). The most frequent substitution for a montmorillonite is that of Al^{3+} by Mg^{2+} in the octahedral layer. For this clay, the distance between the negative sites located at the level of the octahedral layer and the exchangeable cation located at the surface of the layer are such as the forces d' attraction are weak (Cailliere and Al, 1982). Substitutions of So by Al in the tetrahedral layer are also possible.

The interfoliar cations are in general exchangeable by organic and mineral cations being in solutions put in contact with the phyllosilicate. One then characterizes each phyllosilicate by his Capacity of Cation Exchange (CEC). In the case of montmorillonites, the values of CEC lie between 75 and 160 milliéquivalents for 100 grams d' clay (Viallis-Terrisse, 2000). In tables 1 and 2 we show the characteristics physicochemical of bentonite of M' Zila, Algeria (ENOF 1997).

Surface Spécifique (m²/g)	Masse spécifique (g/cm³)	pH	Capacité d'échange (meq/100g)	Cations échangeables (meq/100g)			Na/Ca
				Ca²⁺	Na²⁺	Mg²⁺	
65,00	2,71	9,00	75,8	43,6	25,2	4,8	0,58

Table 1. Physico-chemical characteristics of bentonite

Montmorillonite	Quartz	Carbonates	Feldspaths	Biotites
45 à 60%	15 à 20%	8 à 10%	3 à 5%	8 à 10%

Table 2. Mineralogical characteristics of bentonite

2. Material and method

2.1 Procedure

For these tests, distilled water used has a pH ranging between 6 and 6,3. The initial solution of phenol equal to 100 mg/L is prepared starting from the dissolution of phenol crystallized in distilled water. The solutions are prepared by dilution in the water distilled according to the desired concentrations. For the tests of adsorption, we maintained the concentration of the constant aqueous solution (5mg/L) and varies the mass of the adsorbent m= 5,10,15,20,30,40 and 50 mg. After agitation during 5 hours with room temperature with magnetic stirrers, these solutions are centrifuged with 2000 revolutions per minute during 45 minutes for the analysis of phenol by spectrophotometer UV with a wavelength of 270nm by means of a spectrophotometer of the type SHIMADZU UV-1605.

For the kinetics of adsorption one proceeds in the same way, in beakers of 500mL containing distilled water, one adds the optimal bentonite amount (30 mg/L) as a optimal given by the jar-test. Sampling carried out during time make it possible to follow the evolution of the concentrations of phenol remaining in solution. Balance is reached after one 5 hours duration. For the analysis of the solids after adsorption of phenol sampling of bentonite (1 gram) are put in contact with phenol solutions (V=1L, C= 5 and 100 mg/L). After agitation throughout 6 hour to room temperature, the solids are separated from the liquids by filtration. After drying in a drying oven with 80°C the solids are collected for analysis by diffraction of x-rays, thermogravimetry and infra-red spectroscopy.

2.2.1 Diffraction of x-rays (DRX)

This technique allows inter alia, obtaining information on the interfoliar distance from material before and after adsorption of phenol. The apparatus X' PERT Pro of PANALYTICAL uses an assembly in reflexion θ-θ equipped with a tube with copper anode and with a detector RTMS (Real Time Multiple Strip) of type X' celerator. The recording is done uninterrupted during 90 minutes between 3 and 70 ° into 2 θ with the wavelength CuKα under a tension of 50 Kv and an intensity of 40 my.

2.2.2 Thermogravimetric analyses

The thermogravimetric analyses (TG) give indications on the variation of mass of a sample subjected to a linear rise in temperature. In this present study, we used an thermo-analyzer TG-DSC Sensys evo of SETARAM working between 25 and 750°C, under reconstituted air (mixture O_2 /N_2) and with a speed of rise of 5°C/minute. The analyses related to samples of clays saturated and unsaturated with phenol (C=100mg/L and 5mg) and of mass m=40mg. These thermogravimetric analyses under oxidizing atmosphere make it possible, indeed, to differentiate the organic part of the inorganic part, in particular water.

2.2.3 Analyzes by infra-red spectrophotometer

The Infra-red Spectroscopy with Transform of Fourier (IRTF) is based on l' absorption an infra-red radiation by analyzed material. It allows via the detection of the vibrations characteristic of the chemical bonds, to carry out l'analyzes chemical functions present in material. The bentonite samples, initially out of powder, are mixed in a small proportion with KBr (2.3 mg in 100 mg of KBr), then transformed in the form of a pastille. For the analyses, we used an apparatus BRUKER Equinox 55 in the area 400-4000 cm-1, with the following conditions of recording: detector MCT, number of scans 8, resolution 4 cm-1.

3. Results and discussion

3.1 Analyses of the solids

Diffractogram X of bentonite in a rough state (figure 1), indicates a basal distance (d001) from 12,50 Å. Moreover, it highlights the presence of several crystalline phases (quartz, feldspar). After contact with the phenol solution, the diffractogram shows that the basal distance increased, because (d001) is of 14,94 Å. Since d001 represents the thickness of the layer plus the interfoliar spacing, the spacing of the layers of initial bentonite would be equal to: 12,50 Å - 9,60 Å= 2,90 Å

The spacing of the layers of bentonite with exchange out of phenol would become equal to: 14,94 Å - 9,60 Å = 5,34 Å. A difference of 2,44 Å is thus measured between the spacing of the layers of initial bentonite and those of modified bentonite. It is thus possible that phenol molecules can be easily adsorbed in this interfoliar space.

The results obtained by thermal analysis are illustrated on figure 2 where three curves are presented: that of mass (TG), its derivative (dTG) and heat flow (HF). Curve TG presents several losses of mass according to the temperature. For temperatures going of 25 with 200°C, one observes a first loss from approximately 7%, which corresponds at the water beginning. It is followed of a weak loss (~ 0,9%) between 200 and 450°C, allotted to the oxidation of the organic matter imprisoned between the layers of clay and a third loss of mass beginning towards 450°C corresponding to the deshydroxylation from material. Dehydration as well as the deshydroxylation is accompanied by endothermic effects on the curve HF, while the oxidation of phenol results in two weak exothermic peaks. For a stronger phenol concentration (curve not represented), the organic matter loss is of approximately 1,8%. She thus increases with the concentration of phenol of departure.

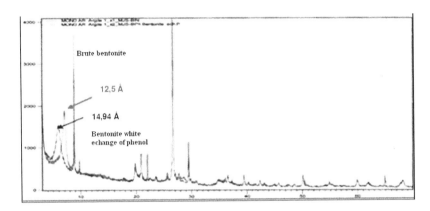

Fig. 1. Powder X-ray Diffractograms of initial bentonite and after phenol exchange

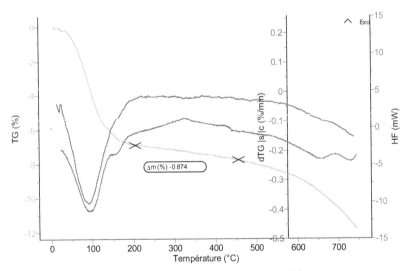

Fig. 2. TG Bentonite containing phenol (concentration at 5 mg/L)

On figure 3 are presented spectra FTIR of the various studied samples: bentonite brut, phenol and bentonite having adsorbed phenol. The presence of phenol should be evaluated by the presence of the characteristic bands, such as the connection OH (δO-H towards

$1360cm^{-1}$) and δGo around 1223 cm^{-1}). The absence of these bands on the spectra could be explained by a too weak phenol concentration adsorbed in clay, the technique not being sufficiently sensitive in this case.

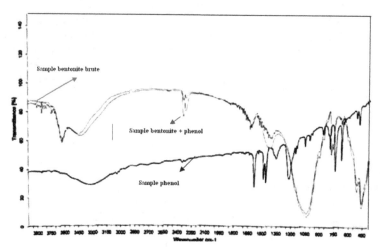

Fig. 3. Spectra FTIR of the studied samples

3.2 Analyzes phenol solutions

The residual concentrations are obtained starting from the absorbance in UV for a wavelength λ max=270nm. Figure 4 shows the whole of spectrum UV for an initial solution of phenol 20mg/L. The results relating to the study of the solutions are gathered in tables 3,4 and are illustrated by figures 4,5 and 6. Figure 5 shows the retention of phenol according to the mass of bentonite and the optimum is for a mass of 30 mg.

Co (mg/L)	Masse bentonite (mg)	Concentration d'équilibre Ce (mg/L)	Rendement % 100 (Co−Ce)/Co
5	0	5.00	0
5	5	3,98	20.4
5	10	3,40	32
5	15	2,78	44,4
5	20	2,28	54.4
5	30	1,80	64
5	40	1,80	64

Table 3. Determination of the concentrations of balance for the bentonite

Co (mg/L)	Masse bentonite (mg)	Ce (mg/L)	X=Co-Ce(mg/L)	x/m (mg/g)	Log(Ce)	Log(x/m)	m/x (g/mg)	1/Ce (L/mg)
5	5	3,98	1,02	204	0,5998	2,309	0,00490	0,251
5	10	3,40	1,60	160	0,5314	2,204	0,00625	0,294
5	15	2,78	2,22	148	0,444	2,170	0,00675	0,359
5	20	2,28	2,72	136	0,3579	2,135	0,00735	0,438
5	30	1,80	3,2	106	0,2552	2,025	0,00943	0,555
5	40	1,80	3,2	80	0,2552	1,903	0,0125	0,555
5	50	1,80	3,2	64	0,2552	1,806	0,0156	0,555

Co: concentration initiale
Ce: concentration d'équilibre

Table 4. Determination of the isotherms of Freundlich and Langmuir

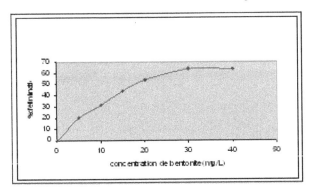

Fig. 4. Yield of elimination of phenol according to the bentonite concentration

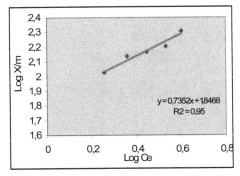
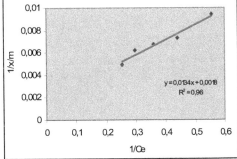

Fig. 5. Freundlich isotherm Figure 6. Langmuir isotherm

3.3 Isotherms of adsorption of phenol

Several models were quoted in the literature to describe the isotherms adsorption. The models of Langmuir and Freundlich are most often used. Balance is described by curves with the room temperature, expressing the quantity of aqueous solution adsorbed per unit of mass of adsorbent according to the concentration of the aqueous solution to the state of balance.

It is a law in the form

$$\frac{x}{m} = f(ce)$$

X (mg): express the quantity of adsorbed aqueous solution
m: mass adsorbent (g)
Ce: concentration of the aqueous solution to the state of balance (mg/L)

From table 4 we determined the equations relating to the isotherms of adsorption (figures 6 and 7) as well as the parameters relating to the law of Freundlich (N and K) and to that of Langmuir (B and qm).

3.3.1 The equation of Freundlich

The equation of Freundlich is form:

$$\frac{x}{m} = KCe^{\frac{1}{n}}$$

The linearization of this expression in form logarithmic curve gives the following function.

$$Log\ (x/m) = 1/n.logCe + logK$$

K and N are constants of balance

For our test:

$$Y = 0{,}7352x + 1{,}8468$$

$$\text{Thus } logK = 1{,}8468 \rightarrow K = 70{,}27$$

$$1/n = 0{,}7352 \rightarrow n = 1{,}36$$

$$R2 = 0{,}95$$

3.3.2 Equation to langmuir

$$q = \frac{x}{m} = qmX\frac{bCe}{1+bCe}$$

x/m: Quantity of adsorbed in mg/g to balance
qm = maximum capacity (ultimate) of adsorption
b= constant related to the energy of adsorption
Ce: Concentration with the balance (mg/l) of the organic compound

After passage to the opposite of function

$$\frac{x}{m},$$

the linearized form is the following one:

$$\frac{1}{q} = \frac{1}{qm} + \frac{1}{qmb}\left(\frac{1}{Ce}\right)$$

Figure 7 shows the variation

$$\frac{1}{x/m}$$

according to

$$\frac{1}{Ce}$$

$$Y = 0,0134\ X + 0,0018$$

$$R^2 = 0,96$$

$$\frac{1}{x/m} = 0,0134 \text{ thus } \frac{x}{m} = 74,62 \text{ mg/g}$$

$$b = 7,44$$

The calculation

$$\frac{x}{m}$$

of watch which the bentonite could adsorb ~75 mg of phenol substances per gram of bentonite. We also note that the isotherms obtained follow best the law of Langmuir (R^2 =0,96). Table 5 presents the variation of the outputs of elimination of phenol according to time and figure 8, representing the abatement of phenol according to time, shows that the kinetics of adsorption slows down according to the reaction time. One can separate the phenomenon in two distinct stages:

-   The first stage shows a fast increase in the yields of elimination during the first two hours of contact, which can be explained by the fixing of phenol on the surface of the adsorbent
-   The second phase corresponds to the external mass transfer which is fast. shows a slow increase in the yield of elimination until the time of balance, which means that there is an internal mass transfer of the absorbable which generally corresponds to a phenomenon of diffusion in the internal porosity of the adsorbent.

Temps (min)	0	5	15	30	45	60	120	180	240	300	360
Co (mg/l)	5	5	5	5	5	5	5	5	5	5	5
Ct (mg/l)	5	4,14	3,98	3,70	3,5	2,96	2,59	2,10	1,95	1,8	1,8
R (%)	0	17,12	20,4	26	30	40,8	48,2	58	61	64	64
Ct/Co		0,828	0,796	0,74	0,70	0,592	0,518	0,42	0,39	0,36	0,36
$T^{1/2}$ (min$^{1/2}$)		2,236	3,872	5,477	6,708	7,745	10,954	13,416	15,491	17,320	18,973
1-(Ct/Co)		0,172	0,203	0,260	0,300	0,408	0,482	0,580	0,610	0,640	0,640

Table 5. Variation yield of elimination of phenol according to time

A few hours of contact with the adsorbent are enough to trap phenol by adsorption effectively. Indeed, the speed of adsorption is one kinetics of first order, function of the surface of the particles but inversely proportional to the diameter of these. The right obtained on figure 9, according to the square root of time, explains the diffusion (coefficient of correlation R^2 = 0,98).

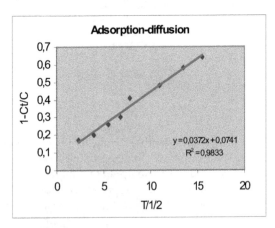

Fig. 8. Adsorption–diffusion on distilled water

Results can be compared with those of published diverse works. For example, Dali-Youcef and al., (2006) measured the adsorption of phenols on a local (bentonite) clay and a mud of dam. The results indeed confirm the capacities of the bentonite retaining more phenol with regard to the mud of dam and his capacity of adsorption is 32 mg /g. Banat and al., (2000) determined a weather of balance for the adsorption of the phenol about 6 hours and showed that the retention rate depended on the initial concentration and on the mass of bentonite used.

Al-Asheh and Al, (2003) examined the possibilities of using a bentonite for the retention of phenols and noticed that the increase in the mass of the adsorbent reduces the residual quantities of phenol in the final solution. Tests of adsorption of phenol on peat ashes and a bentonite were carried out by Viraraghavan and Alfaro, (1997). Their results indicate that the bentonite could retain phenol 46%. YU and Al, (2004) also showed that the adsorption of the phenol compounds by a montmorillonite increases with the initial phenol concentrations. In addition, more recently, of the tests of adsorption carried out by Nayak and Al, (2007) related to the retention of phenol on natural clay and modified clays. They concluded that it is possible to improve the capacity of adsorption of clays by specific treatments. Lastly, Boufatit and Al, (2007), which studied l' adsorption of organic compounds by Algerian clays rich in montmorillonite, show that the chlorinated phenols resulting from the degradation of the aromatic compounds (pesticides), are easily adsorbed on these clays.

Other materials were used for the retention of phenols in water. In particular, Guesbaya, (2005) used coagulation-flocculation with aluminium sulphate for water containing of phenol, but the retention remains very weak because the simple organic compounds tested are slightly eliminated whatever the amount from coagulant (sulphate of alumina) or the initial concentration of the compound. According to Al-Asheh and Al, (2003), in comparison with bentonite, the activated carbons have a higher adsorption capacity for the phenolic compounds. However, because of the relatively high cost of the activated carbon, the natural adsorbents, in spite of a lower performance, remain a solution interesting for the elimination of the contaminants of worn water.

4. Conclusion

The goal of this work was to study, experimentally, potential of bentonite gross of M' Zila in the adsorption of phenol pollutants. The study of the isotherms of adsorption enabled us to evaluate characteristics the characteristics adsorption of this bentonite. The results of the tests carried out in laboratory show the elimination of 64% of the concentrations (5 mg/L) of phenol and the bentonite can adsorb 75 mg/L phenol substances per gram and could be interesting in the elimination of the organic matter present in water. However, the optimum conditions for its use remain to be determined. The analysis of the isotherms makes it possible to determine if purification by adsorption can be carried out or not according to the initial phenol rate in water as well as an estimate of the mass of adsorbent making it possible to reduce in an important way the concentrations in pollutants. The preliminary reduction in the organic matter rate should limit the formation of potentially carcinogenic organochlorinated compounds likely to occur during a secondary treatment to chlorine.

5. References

Amar H. & A. Gaid (1987). Fixation on bentoniques clays of metallic ions in residual waters, *Rev. Sci. de l'Eau*, 3, 33-40.

Al-Ashehs., F. Banat & L. Abu-Aitah (2003). Adsorption of phenol using different types of activated bentonites. *Separation and Purification Technology*, 33, 1-10.

Andreozzi R., V. Caprio., A. Insola & R. Marotta (1999). Advanced oxidation process (AOP) for water purification and recovery. *Catal. Today*, 53, 51-59.

Auerbach S.M., K.A. Carrado & P.K. Sutta (2004). *Handbook of Layered Materials*, Marcel Dekker, New York, USA, 650 p.

Banat F. A., B. Al-Bashir., S. Al-Asheh & O. Hayajneh (2000). Adsorption of phenol by bentonite. *Environmental Pollution*, 107, 391-398.

Bartschat B.M., S.E. Cabanis & F.M. Morel (1992). Oligoelectrolyte model for cation binding by humic substances. *Environ. Sci. Technol.*, 26, 284-294.

Bergaya F., B.K.G. Theng &LAGALY G. (2006). *Handbook of clay science*, Developpement in clay science, series volume 1, 124 p.

Bersillon J.L. (1983). The mechanism of organic removal during coagulation. Thesis, Mc Master University, Hamilton Ontario, USA, 188 p.

Bersillon J.L., B. Lartiges, F. Thomas & L. Michot (1999). The transformation of water quality: from contribution to water treatment. Effect of mineral-organic-microorganism interactions on soil and Freshwater Environments. Kluwer Academic, Plenum Publisher, pp. 367-373.

Boudhar H. (1999). Improvement of the functioning of the small units of treatment of drinking water to eliminate néoformés elements, THM et turbidité. *Office International de l'Eau*.

Boufatit M., H. Ait-Amar & W.R. Mcwhinnie (2007). Development of Algerian material montmorillonite clay. Adsorption of phenol, 2-dichlorophenol and 2,4,6-trichlorophenol from aqueous solutions onto montmorillonite exchanged with transition metal complexes. *Desalination*, 206, 394-406.

Bouras O. (2003). Properties adsorbantes of pontées organophiles clays: synthesis and characterization, Doctoral thesis, Univ. Limoges, France.

Cailliere S., S. Henin & M. Rantureau (1982). *Minéralogie des argiles*. Masson, tome 1 et 2, pp. 184-189.

Chiron S., A. Fernandez-Alba, A. Rodriguez & E. Garcia-Calvo (2000). Pesticide chemical oxidation: State of the art. *Water Research*, 34, 366-377.

Dali-Youcef Z., H. Bouabdasselem & N. Bettahar (2006). Elimination of organic compounds by local clays. *Comptes Rendus de Chimie*, 9 (10), 1295-1300.

Decarreau A. (1990). Clayey materials: structure, properties and applications (collective work),Ed. Société Française de Minéralogie et de Cristallographie, pp. 8-21.

Dore, M. (1989). *Chimie des oxydants et traitement des eaux*, Ed.Tech. & Doc ; Lavoisier, Paris.

Dulin B.E. & W.R. Knocke (1989). The impact of incorporated organic matter on the dewatering characteristics of aluminium hydroxide Sludge. *Journal AWWA*, 81, 74-79.

Dussert B.W (1997). Advanced oxidation. UK: *Industrial wastewater*, pp.29-34.

Dutta n.N., G.S. Patil et S. Brothakur (1992). Phase transfer catalyzed extraction of phenolic substances from aqueous alkaline stream. *Separation Science and Technology*, 27, 1435.

Dutta N.N., S. Brothakur et R. Baruah (1998). A novel process for recovery of phenol from alkaline wastewater: laboratory study and predesign cost estimate. *Water Environ. Res.*, 70, 4-9.

ENOF (1997). Rapport d'analyse de la bentonite de M'Zila, Mostaghanem, Laboratoire Enof, Algérie.

Essington M.E. (1994). Adsorption of aniline and toluidines on montmorillonite, *Soil Science*, 158(3), 180-188.

Galze H., F. Beltran, T. Tuhkanen & J.W. Kang (1992). Chemical models of advanced oxidation processes. *Water Poll. Res. J. Canada*, 27, 23-42.

Guesbaya A. N. (2005). Coagulation-floculation by the sulfate of aluminum of phenolic organic compounds and humiques substances. *Larhyss Journal*, 4, 153-168.

Golich r. & D. Bahnemann (1997). Solar Water Treatment: Principles and Reactors, Intercalibration of Radical Sources and Water Quality Parameters, Swiss Federal Institute for *Environmental Science and Technology EAWAG*, Switzerland, pp. 137-148.

Hoang L. (2009). *Comparison of the energy returns on degradation of organic three by several processes of oxidation advanced(moved) in aqueous environment(middle)*, Doctoral thesis, Univ. Poitiers, France.

Hooper S.M., R.S. Summers, S. Gabriele & D.M. OWEN (1996). Improving GAC performance by optimized coagulation. *Journal AWWA*, 88, 107-120.

Lefebvre e. & B. Legube (1990). Iron III Coagulation of humic substances extracted from surface waters: effect of pH and humic substances concentration. *Water Research*, 24, 591-606.

Meier, J.R. (1988). Genatoxic, Activity of organic chemicals in drinking water, *Mutation Res.*, 196, 211- 245.

Najm i.n., V.L. Snoeyink & Y. Richard (1993). Removal of 2, 4, 6-trichlorophénol and natural organic matter from water supplies using PAC in floc- blanket reactors. *Water Research*, 27, 551-560.

Nayak P. S. & B. K. Singh (2007). Removal of phenol from aqueous solutions by sorption on low cost clay. *Desalination*, 207, 71-79.

Safarzadeh-Amiri A., J.R. Bolton & S.R. Caster (1996). The use of iron in advanced oxidation processes. *J. Adv. Oxid. Technol.*, 1, 18-26.

Terrisse V. (2000). Interaction of constituent hydrated, main Silicates of Calcium of the cement, with the chlorides of alkaline. Analogy with clays, Doctoral thesis, Univ. de Bourgogne, Dijon, France.

Tippinge. (1993). Modelling the competition alkaline earth cations and trace species for binding by humic substances. *Environ.Sci.Technol.*, 27, 520-529.

Stevens A.A., C.J. Slocum, D. SeegeR & G.G. Robeck (1976). Chlorination of organics in drinking water. *Journal AWWA*, 68, 615-620.

Viraraghavan T. & F. Alfaro (1997). Adsorption of phenol from wastewater by peat, fly ash and bentonite. *Journal of Hazardous Materials*, 57, 59-70.

Wiesner M.R., M.M. Clark & J. Malleviale (1989). Membrane filtration of coagulated suspensions. *J. Environ. Engin.*, 115, 20-40. .

YU J., M. Shin, J. Noh & J. Seo (2004). Adsorption of phenol and chlorophenols on Ca-montmorillonite in aqueous solutions. *Geosciences Journal*, 8(2), 185-189.

9

Impact of Agricultural Contaminants in Surface Water Quality: A Case Study from SW China

Binghui He and Tian Guo
Southwest University
China

1. Introduction

Water is a very precious natural resource and the matter foundation that the whole lives rely on in the earth, and it is also the matter that cannot be replaced in the natural resources on the earth again. With increasingly intensive population, various community economy activity's demand for water increases, and the space-time distribution of water resource become uneven, and the supply of water resource in the whole world become tight. Soil erosion has taken away the mass of rich topsoil and makes soil become more and more thin, reduces land productive, and poses serious threat to water environment. In addition, water body pollution has negative effect on effective use of water resource and aggravates the contradiction. The management of point source pollution and circularly making use of in the developed countries currently up make more ideally, but a lot of still directly exhaust in various water body in the extensive developing countries.

Soil erosion is the reason that results in surface source pollution, and runoff is the medium of surface source pollution, while the amount of runoff and water and soil conservation are closed tightly, therefore, the basic way to resolving surface source pollution is reducing soil erosion by water and soil conservation measures. Most of analysis of runoff pollution mechanism and runoff creation, space-time distribution of runoff, and the effect analysis of different measures are all carried on runoff plots.

With the improvement of the controlling of point source pollution, water environment issues caused by the non-point source pollution are increasingly conspicuous. Especially in the developed countries, such as the America, the non-point source pollution has been major factor of the water environment pollution (Fu et al., 2011). The agriculture runoff pollution is the main fraction of surface source pollution, and agriculture surface source pollution and soil erosion is a pair of symbiosis (Zhang & Huang, 2011). Thus the most effective and the most practical approach to solve agriculture pollution problem is to implement water and soil conservation measure. With the development of small watershed comprehensive management and surface source pollution in China, the way to resolve surface source pollution by the effective small watershed comprehensive management has been a hot point in the environment science and water and soil conservation science field nowadays.

The environmental benefit of small watershed comprehensive management includes natural environment, eco-environment and social environment. The water and soil conservation works since 1980s indicate that the small watershed comprehensive management could produce obvious environmental benefit (David et al., 1998).

Water eutrophication is one of the water pollution issues that perplexed developed countries nowadays and is also the realistic issue that the developing countries facing (Leeds-Harrison et al., 1999; Heitz et al., 2000). After the water eutrophication, the excessive reproduction of algae had leaded to water hypoxia, clarity step-down, smelly, even some toxins, thus had damaged the normal function of water body. According to calculating, the eutrophication degree of most lack in china will turn worse further with the development of the industrial and agricultural. The pollutants resulting in Water eutrophication were mainly plant nutritional substances, such as nitrogen and phosphorus etc. The search indicate that it mainly exist in closed water area, such as lake, reservoir etc.

The control of runoff pollution is an important aspect in the whole water pollution controlling progress. The search and control for non-point source pollution start relatively late in china, and the search for the control way about non-point source pollution has been valued since the last few years, But the abroad has made a great deal of works and has accumulated prolific experience in this field (Chartes & Roben, 2000; Prato & Shi, 1990). For example, In 70's, the United states proposed "BMP-Best Management Practices" in the control and management fulfillment of non-point source pollution. What it points is the combination of several measures after analysis and compassion by the government or designation which is the most useful practical measure for making the identified water by perverting and cutting non-point source pollution burden. Usually it can be divided into the control of source and the runoff pollution.

To know what effects does the watershed comprehensive management had on the runoff pollution, this search adapt the combined method of runoff plot and small watershed. The small watershed model is the main demonstration for water and soil conservation - the reservoir peripheral of Gezi channel, Lianglu town, Yubei plot, Chongqing city (managed watershed) and Changxi small basin(not managed watershed), the soil in experimental runoff plot is the same as the adapted soil in small watershed. The reservoir in Gezi has played a key role in the development of agriculture economy, and the water of reservoir is the source of peripheral farmland irrigation and the headwaters that the farm tourism keeps fish again. To know the influences of different measures on runoff pollution, the search of the distribution regulation in time and space about the reservoir agriculture runoff, the estimating of amount runoff pollutant burden of reservoir stores in warehouse by the methods of synchronous monitor and field investigation to the runoff water and sediment respectively in runoff plot and basin, and calculated the decreasing runoff pollution burden after management by the compassion of the density of runoff pollutant in managed plot and not managed plot's, and then to analyze the dimension of synthesize management and measure the effect after administering.

Therefore, there are three purposes to establish this search, the first one is evaluating the control action of small watershed comprehensive management to runoff pollution; The second one is directing the measure of water and soil conservation; The third one is defining the best sloping plant model from ecology direction. Thus it can offer a certain theory basis for predicting the control action of small watershed comprehensive management to the surface source pollution in this region.

2. Research content

The soil erosion is the main reason of slope runoff pollution, the different land use and different cropping-plants influence the occurrence and development of runoff. According to the spot investigation, it indicated that the local resident plant some farm crops, such as bean, maize, sweet potato and fruit tree under the 25° sloping fields, these farm plant are been planted alone or been intercropped, and the different intercropped area proportion could had different influence on runoff. This search adapt the method of monitoring surface runoff in plot in consideration of that the natural water is the main recharge source of surface runoff, the amount of burden of different using type of land's surface runoff is relevant with this area, which also has something to do with the pollutant density of surface runoff.

This search mainly investigates the relevant intensity of surface runoff pollutant of different type of using land, thus we can use surface runoff pollutant total nitrogen and total phosphorus, Pb and Cd, chemical oxygen demand and the average density of the amount of sediment to express the discussion of different plant model's impacts on the runoff pollution. According to it, we can compare the differences of the degree of influence of different plant model's impact on runoff pollution. The study contents were as follows: 1) the influence of different plant models on runoff pollution; 2) the timely distribution of runoff pollution; 3) the spatial distribution of runoff pollutant.

2.1 Basic situation of experimental plot and research methods

The experimental plot located in Lushan village, Lianglu town, Yubei district, Chongqing city and was shallow landform, soil type is purple soil which belongs to the Jurassic formation, the soil texture was formed by the physical chemistry elegance of sandstone and shale, soil bulk density was 1.42-1.68g/cm³, total porosity was 40.28%. And perennial mean temperature was 18.2 °C, annual yearly rainfall was 1110mm, the rainfall in May to October was about 80.62% of the whole year's, and the amount of rainfall in August was the most largest. Before the planting tillage in plot, to know what influence does different land use and slope will have on runoff pollution, we adapt the method of runoff plot to research, and the soil texture fertilizer applications was the same in each plot. The basic situation of experimental plots was as follows:

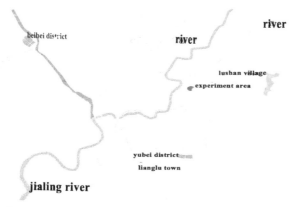

Fig. 1. Experiment Area

Plot number	Slope(°)	Aspect	Specification	Crop and planting pattern	Coverage (%)
1	25	SE	5m×20 m	bean (50%)+ sweet potato(50%)(intercropping along the slope)	20
2	25	SE	5m×20 m	seedlings of economic fruit forest(along the slope)(20cm*20cm)	5
3	25	SE	5m×20 m	seedlings of economic fruit forest (transforming slope into terrace)+ (20cm*20cm)	5
4	15	SE	5m×20 m	bean + sweet potato(25% intercropping)+ sweet potato(75%)(along the slope)	30
5	15	SE	5m×20 m	seedlings of economic fruit forest (transforming slope into terrace) (20cm*20cm)	5
6	15	SE	5m×20 m	sweet potato +maize(intercropping along the slope)	30

Table 1. Basic situation of experimental plots

Index	Plot number					
	1	2	3	4	5	6
Minimum permeability (mm/h)	3.64	3.61	3.72	3.66	3.70	3.67
Bulk density (g/cm³)	1.47	1.48	1.45	1.34	1.44	1.30
Water content(100%)	17.5	16.9	19.5	19.8	19.5	21.2

Table 2. The physical characteristic of soil in plots Basic situation of experimental plots

Soil samples were collected from five sampling spots ("S" type) of two replicates in each plot, then combined within plots before analysis. Once after a runoff, swept the sludge of gutter into runoff pool mixed it with water, and collected three columnar samples from pool (gross 1000-3000ml). Then put mixing sample (500-2500ml) into closed container, 4ml concentrated H_2SO_4 to terminate microbial activity, and then put it in the refrigerator. The surplus water sample (500ml) was precipitated, filtered and dried, to determine sediment content.

Precipitation capacity was determined by hydrocone type journal rain gauge; Runoff capacity was calculated by SWZ type journal water column and 45° triangular weir; Sediment content was measured by oven drying method, and measure method of water sample: total nitrogen(semimicro Kjeldahl), total phosphorus(molybdo-antimony anti-colorimetry), Cd (atomic absorption spectrometry), chemical oxygen demand (potassium dichromate method).

2.2 Variation and analysis of runoff pollution

On the basis of the size of runoff pollution index content , the order of effect of runoff pollution among six cropping patterns was plot 5 > plot 3 > plot 6 > plot 4 > plot 1 > plot 2. To reduce the slope runoff pollution, the slope farmland under 25 degree should be transformed into terrace, and at the seedlings of economic fruit forest, it didn't show a strong controlling effect on runoff pollution, so it should intercrop other plants according to fruit tree's biological habit and ecological characteristics.

The species of crop and the reasonable density and its structure of ecological function should be considered when adapted to the intercropping pattern.The management effect of plot 1 and plot 2 was the most obvious in the total ratio of rainfall collection, but the management effect of plot 5 and plot 6 was the worst, the Cd content of plot 4 was the highest, and Pb content of plot 5 and plot 6 was the highest.

To decrease the slope runoff pollution, the slope farmland under 25 degree should be transformed into terrace, and at the seedlings of economic fruit forest, it didn't show a strong controlling effect on runoff pollution, so it should intercrop other plants according to fruit tree's biological habit and ecological characteristics. The species of crop and the reasonable density and its structure of ecological function should be considered when adapted to the intercropping pattern. Analyzing the each index and getting its average by the runoff samples of three times, the results were as follows:

Date	Plot number	Total nitrogen (mg/l)	Total phosphorus (mg/l)	Pb (mg/l)	Cd (mg/l)	Chemical oxygen demand (mg/l)	Sediment amount (mg/l)
8.16	1	5.9	0.68	0.1064	0.0109	14.3	6.3
	2	6.29	0.71	0.1086	0.0111	14.38	6.7
	3	4.45	0.54	0.0919	0.0099	10.32	4.7
	4	5.81	0.64	0.1092	0.0107	13.94	5.9
	5	3.95	0.49	0.0897	0.0095	9.87	4.2
	6	5.47	0.62	0.106	0.0104	13.02	5.6
8.20	1	5.82	0.61	0.1057	0.0102	12.7	5.6
	2	6.22	0.64	0.1079	0.0104	13.92	6
	3	4.4	0.47	0.0912	0.0092	9.58	4
	4	5.74	0.57	0.1087	0.01	11.72	5.2
	5	3.92	0.43	0.089	0.0088	8.88	3.5
	6	5.4	0.55	0.1055	0.0097	11.81	4.9
8.23	1	5.83	0.6	0.1056	0.0101	11.78	5.5
	2	6.23	0.63	0.1078	0.0103	12.64	5.9
	3	4.39	0.46	0.0911	0.0091	9.32	3.9
	4	5.73	0.56	0.1085	0.0099	11.08	5.1
	5	3.91	0.42	0.0889	0.0087	7.76	3.4
	6	5.39	0.54	0.1053	0.0096	10.05	4.8
the average of index of three runoff pollution	1	5.85	0.63	0.1059	0.0104	12.93	5.8
	2	6.24	0.66	0.1081	0.0106	13.65	6.2
	3	4.42	0.49	0.0914	0.0094	9.74	4.3
	4	5.76	0.59	0.1088	0.0102	12.25	5.4
	5	3.94	0.45	0.0892	0.009	8.84	3.7
	6	5.42	0.57	0.1056	0.0099	11.63	5.1

Table 3. The index value of three runoff

The general order of change rate is sediment>total nitrogen> chemical oxygen demand > total phosphorus > Pb > Cd except of a little deviation of the change of pollution index. Evaluation of different planting patterns on the environmental quality pollution is as follows:

Index	Evaluation standard				
	I	II	III	IV	V
Total nitrogen	0.5	0.5	1	2	2
Total phosphorus	0.02	0.025	0.05	0.2	0.2
Pb	0.01	0.05	0.05	0.05	0.1
Cd	0.001	0.005	0.005	0.005	0.01
Chemical oxygen demand	lower than 15	lower than 15	15	20	25

Table 4. The environmental standard of surface water (GB3838-88) (mg/L)

Index	Plot number					
	1	2	3	4	5	6
Total nitrogen	0.185	0.197	0.140	0.182	0.125	0.171
Total phosphorus	0.186	0.195	0.145	0.174	0.133	0.168
Pb	0.174	0.178	0.150	0.179	0.146	0.173
Cd	0.175	0.178	0.158	0.171	0.151	0.166
Chemical oxygen demand	0.187	0.198	0.141	0.177	0.128	0.168

Table 5. Index weight under different land utilization patterns

Index	Plot number					
	1	2	3	4	5	6
Total nitrogen	4.875	5.200	3.683	4.800	3.283	4.517
Total phosphorus	6.364	6.667	4.949	5.960	4.545	5.758
Pb	2.037	2.079	1.758	2.092	1.715	2.031
Cd	2.000	2.038	1.808	1.962	1.731	1.904
Chemical oxygen demand	0.718	0.758	0.541	0.681	0.491	0.646
Total	2.92	3.21	1.86	2.74	1.59	2.52

Table 6. Composite index of runoff pollution in plots

The total tendency of pollutant content by observing the experimental data of each plot's visually was plot 5 < plot 3 < plot 6 < plot 4 < plot 1 < plot 2, and by the calculation of pollution integrated index in each plot and the average of three runoff pollutant, runoff pollutant had a same tendency in the six cropping patterns, so it was reasonable and feasible to conduct environmental quality assessment by the organic combination of the evaluation method of environmental pollution index and the one-level fuzzy evaluation method. Compared to the expert evaluation, this method could reduce workload greatly.

There were economic fruit forest in plot 2, 3 and 5, and there were not obvious differences of the runoff pollutant content between plot 3 and plot 5 , but the runoff pollutant content of plot 3 and plot 5 was smaller than the runoff pollutant content of plot 2, which indicated that slope's effects on runoff pollution is not obvious, while it made a remarkable effect on the control action of runoff pollution by transforming slope into terrace; the effect of the economic fruit forest in plot 2 was the worst because of its stage of seedlings and the small converge that couldn't control soil erosion well.

The pollutant loss in plot 6 was smaller than the effect in plot 1, which indicated that the model of maize intercropping sweet potato was better than the model of bean intercropping sweet potato, because the underground root system and big leaf of maize could make a stronger effect on loosing soil and the rainfall interception than the root and leaf of bean during the experimental period, we can also explained this phenomenon according to soil bulk density, water content and permeability; the effect of controlling pollutant loss in plot 4 was better than that in plot 1, that was to say, planting the single sweet potato on slope land had a stronger controlling effect on runoff pollutant than the model of bean intercropping sweet potato, but which didn't match the former regulation that intercropping pattern was better than single planting pattern, that because the bean was too young in plot 1, so the rainfall interception was not strong, and for planting bean, the row spacing of sweet potato extended, the total coverage of plot 1 changed small.

2.3 The spatial distribution characteristics of runoff pollution

The runoff pollution degree of the downhill of plot 2 by the filtration of buffer plot was the smallest, and the runoff pollution degree of new citrus located in the downhill of plot 4 was the maximum. Total nitrogen content and Cd content in runoff had significant correlations with sediment, and total phosphorus content and Pb content had extremely significant correlations with sediment. And chemical oxygen demand content had the extremely significant correlations with total nitrogen, Pb and Cd, chemical oxygen demand content had significant correlations with total phosphorus and sediment, and total nitrogen content was the main affecting factor of chemical oxygen demand content. During the management of basin, we could adopt intercropping pattern to increase the vegetation coverage of initial harnessing area to come to the target of controlling runoff pollutant. Moreover, it also needed to take measures to control the quality of runoff from upslope.

The total tendency in runoff was that total nitrogen, total phosphorus, Pb and chemical oxygen demand content increased at first then decreased. Total nitrogen content came to peak during one hour to two hours, and the last content had a tendency to stability and was lower than the initial value, the time of runoff pollutant content coming into peak varied with the difference of pollutant and plot, and during the time of peak appearing, there was a big difference between each plot, the total nitrogen content of each plot was generally same at the last period of the occurrence of runoff; Total phosphorus content came to peak during 0.8 hour to two hours, the total tendency of total phosphorus content at the same time was 4>5>6>7>3>1>2, during the period of peak's appearance, there were big differences between each plot, total phosphorus content had a tendency to stability at the last time of the occurrence of runoff; Pb content came to peak during one hour to two hours, the order of Pb content of each plot after the former one hour of occurrence of runoff was 1>3>7>2>6>5>4, during the period of peak's appearance, there were big differences between each plot; Total nitrogen content of each plot was generally same at the last time of the occurrence of runoff; The order of Pb content after the one hour to three hours and a half of the occurrence of runoff was 5>6>4>7, there were no differences of Pb content between 1 plot 1, plot 2 and plot 3. And Pb content of each plot had a tendency to same at the last period runoff. Cd content of each plot came to peak during 1.8 hours to two hours, the order of Cd content after the occurrences of runoff but the former two hours before the occurrence of runoff peak was 1>3>7>2>6>5>4, Cd content of each plot was generally same after the appearance of runoff peak; Cd content of plot 4, 5, 6 and 7 changed sharply, Cd content of plot 1,2 and 3 changed gently; Chemical oxygen demand content of each plot came to peak during 1.8 hours to two hours, the order of chemical oxygen demand content in the total runoff progress was 5>6>3>7; Chemical oxygen demand content of plot 1, 2, 4 and 7 changed irregularly. Generally speaking, the change range of chemical oxygen demand content of plot 4, 5, 6 and 7 was larger than plot 1,2 and 3's. The effect of runoff management of plot 1, 2 and 3 was stronger, secondly for plot 4 and 7, the effect of runoff management of plot 5 and 6 was the worst; the time of the appearance of runoff pollution peak was fixed.

Plot number	pH	Organic matter (g/kg)	Total nitrogen (g/kg)	Total phosphorus (g/kg	Total potassium (g/kg)	Cd (mg/kg)	Pb (mg/kg)
1	7.96	22.54	1.431	0.657	10.2	0.498	27.58
2	8.19	4.69	0.786	0.716	14.6	0.4	29.64
3	6.64	22.31	1.447	0.904	12.9	0.243	24.69
4	8.17	4.87	1.426	0.992	14.8	0.552	25.01
5	8.11	22.21	1.431	0.667	12.8	0.346	29.12
6	8.14	22.19	1.429	0.662	12.7	0.345	29.12
7	6.94	4.91	0.786	0.772	15	0.377	24.68
8	8.18	4.69	1.084	0.828	13.8	0.483	26.66

Table 7. The physicochemical characteristic of soil

Plot number	Slope position	Aspect	Total nitrogen (g/kg)	Total phosphorus (g/kg)	Pb (mg/kg)	Cd (mg/kg)	Chemical oxygen demand (mg/kg)	Sediment Amount (mg/kg)
1	incomingwater	NE	3.525	0.136	0.0628	0.0086	19.567	2.9
	upslope	NE	1.872	0.088	0.0431	0.0058	10.269	1.4
	midslope	NE	1.47	0.191	0.0322	0.0027	9.157	2.6
	downslope	NE	1.311	0.152	0.0257	0.001	7.683	2.2
2	incoming water	NE	2.521	0.137	0.0555	0.0063	20.233	3.3
	upslope	NE	1.966	0.112	0.0501	0.0047	12.565	2
	midslope	NE	1.659	0.159	0.0554	0.0055	8.461	3.1
	downslope	NE	1.085	0.126	0.0252	0.0008	8.282	0.8
3	incoming water	E	2.349	0.161	0.0489	0.0057	16.243	3.1
	upslope	E	1.693	0.142	0.0432	0.0044	10.856	2.2
	midslope	E	2.344	0.199	0.0456	0.0048	15.381	2.6
	downslope	E	1.192	0.161	0.0451	0.0026	7.416	2.1
4	incoming water	SE	2.657	0.136	0.0550	0.0068	16.835	2.9
	upslope	SE	2.615	0.102	0.0441	0.006	11.235	3.8
	midslope	SE	2.057	0.208	0.0487	0.0098	26.261	3
	downslope	SE	1.878	0.341	0.0537	0.0011	12.541	2.2
5	incoming water	S	2.101	0.161	0.0612	0.0052	14.235	2.6
	upslope	S	1.898	0.159	0.0551	0.0052	8.561	2.2
	midslope	S	2.551	0.188	0.0594	0.0076	15.623	2.7
	downslope	S	3.002	0.234	0.0664	0.009	24.652	3.8
6	incoming water	SW	1.921	0.167	0.0619	0.0051	14.237	2.4
	upslope	SW	1.797	0.134	0.0431	0.0049	8.141	2
	midslope	SW	2.349	0.09	0.069	0.008	12.563	2.6
	downslope	SW	3.015	0.216	0.0584	0.0092	21.695	3.7
7	incoming water	NW	2.43	0.145	0.0432	0.0051	15.623	3.3
	upslope	NW	2.234	0.099	0.0392	0.0042	12.666	2.4
	midslope	NW	2.001	0.252	0.0414	0.0071	9.283	3.2
	downslope	NW	1.647	0.206	0.0554	0.0083	8.28	2.8
8	incoming water	NW	2.53	0.147	0.0445	0.0062	15.741	3.6
	upslope	NW	2.66	0.245	0.0524	0.0071	18.564	4.1
	midslope	NW	4.36	0.612	0.0926	0.0113	51.48	5.8
	downslope	NW	5.21	0.633	0.0942	0.0113	51.88	6.1

Table 8. The spatial distribution characteristics of runoff pollution

The management effect of plot 1 and plot 2 was the most obvious in the total ratio of rainfall collection, but the management effect of plot 5 and plot 6 was the worst, and the Cd content of plot 4 was the highest, Pb content of plot 5 and 6 plot was the highest. the runoff pollution degree of the downhill of plot 2 by the filtration of buffer plot was the smallest, and the runoff pollution degree of new citrus located in the downhill of plot 4 was the biggest. Total nitrogen content and Cd content in runoff had significant correlations with sediment, and total phosphorus content and Pb content had extremely significant correlations with sediment. And chemical oxygen demand content had the extremely significant correlations with total nitrogen, Pb and Cd, chemical oxygen demand content had significant correlations with total phosphorus and sediment, and total nitrogen content was the main affecting factor of chemical oxygen demand content. During the management of basin, we could adopt intercropping pattern to increase the vegetation coverage of initial harnessing area to come to the target of controlling runoff pollutant. Moreover, it also needed to take measures to control the quality of runoff from upslope.

2.4 The evaluation of water quality of space runoff in watershed

According to the comments on runoff quality, it indicated the runoff pollution degree of the downhill of plot 2 by the filtration of buffer plot was the smallest and the runoff pollution degree of new citrus located in the downhill of plot 4 was the biggest. The not managed slope pollution index was higher than the managed slope's, which indicated that the regulation effect was significant in Gezi channel. To analyze the sample clustering of interval runoff quality by conducting the spatial runoff pollution synthetic index using SPSS and analyze the variable cluster of runoff quality in all position of the whole reservoir, the cluster result were as Fig.2 and Fig.3.

Slope position	Plot number							
	1	2	3	4	5	6	7	8
upslope	1.09 (1-1)	1.14 (2-1)	1.13 (3-1)	1.39 (4-1)	1.29 (5-1)	1.15 (6-1)	1.20 (7-1)	1.593 (8-1)
midslope	1.25 (1-2)	1.24 (2-2)	1.54 (3-2)	1.88 (4-2)	1.64 (5-2)	1.41 (6-2)	1.70 (7-2)	6.558 (8-2)
downslope	1.05 (1-3)	0.88 (2-3)	1.09 (3-3)	2.22 (4-3)	2.00 (5-3)	1.95 (6-3)	1.56 (7-3)	7.333 (8-3)

Table 9. Average of Composite index of runoff pollution in plots on Aug 16th and Aug 20th

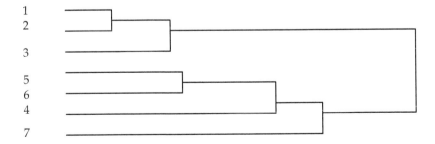

Fig. 2. Runoff water quality clustering in plots

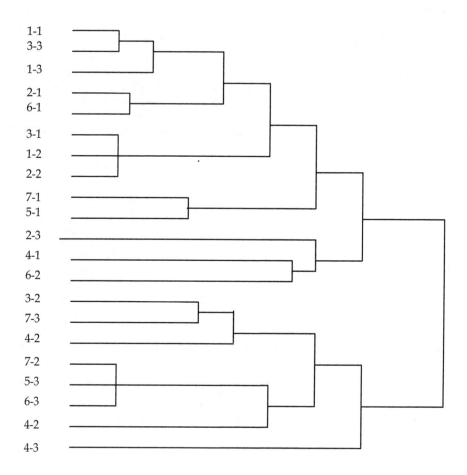

Fig. 3. Runoff water quality clustering of different position in plots

2.5 The space-time distribution characteristics of runoff pollution

The degree of runoff pollution in each plot first increased to peak then decreased gradually, and the degree of runoff pollution at the beginning was larger than in the end; the runoff pollution synthetic index of plot 1, 2 and 3 changed slowly, it appeared peak at the period 5. While the runoff pollution synthetic index of plot 5 and 6 changed sharply, the runoff pollution synthetic index of plot 4 and 7 also changed sharply.

The latter four plot appeared peak much earlier, it indicated that the runoff regulation effect of plot 1, 2 and 3 was stronger, secondly for plot 4 and 7, the runoff regulation effect of plot 5 and 6 was the worst.

The evaluation method of runoff quality is as mentioned above, the evaluation results are as follows:

Plot number	Date								
	16.50-17.20	17.20-17.40	17.40-18.00	18.00-18.30	18.30-18.50	18.50-19.20	19.20-19.50	19.50-20.20	20.20-20.50
1	0.841	0.934	0.963	1.053	1.107	0.827	0.651	0.537	0.536
2	0.786	0.856	0.911	0.973	1.027	0.803	0.657	0.603	0.551
3	0.952	1.012	1.150	1.225	1.309	0.975	0.818	0.742	0.669
4	1.444	1.738	2.125	2.154	1.959	1.730	1.332	1.074	0.976
5	1.389	1.563	1.940	2.243	2.111	1.574	1.165	0.838	1.013
6	1.293	1.422	1.727	2.115	2.019	1.312	1.041	0.903	0.887
7	1.199	1.164	1.407	1.607	1.416	1.077	0.912	0.821	0.780

Table 10. The space-time distribution characteristics of runoff pollution composite index

2.6 The timely clustering of runoff pollution composite index

The period cluster of each section: the nine periods mentioned above could be expressed by TIME1-TIME9, to cluster the runoff pollution synthetic index in each period by the Hierarchical Cluster to Q cluster, the cluster method was Euclidean distance, the cluster results were as follows:

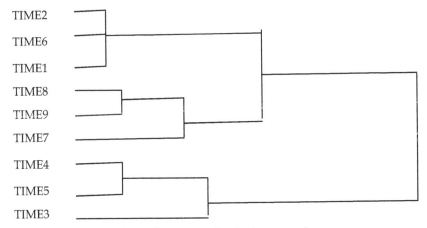

Fig. 4. The timely clustering of runoff water quality in the same plot

In the term of cluster graph, the temporal and spatial distribution characteristics of the degree of runoff pollution were as follows: the degree of runoff pollution of TIME1, TIME2 and TIME6 were the same, the degree of runoff pollution of TIME4 and TIME5 were the same, and the degree of runoff pollution of TIME8 and TIME9 were the same, the runoff pollution synthetic index of TIME4 and TIME5 were the highest, that was to say, the time of coming into peak of runoff pollution is fixed. The interval cluster results of each period were as follows:

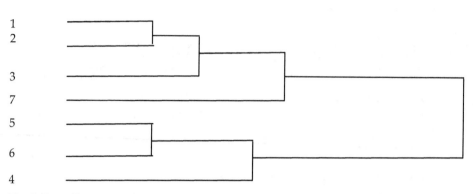

Fig. 5. Runoff water quality clustering during the same time in different plots

The cluster results were that the degree of runoff pollution of plot 1 was similar to the plot 2's and the degree of runoff pollution of plot 4 was similar to the plot 5's, the degree of runoff pollution of plot 7 was different with the other plots, it also showed the land use method and vegetation conditions influenced the development and progression of runoff pollution, which matched the chapter about the analysis of the spatial distribution characteristics of runoff pollution.

3. Conclusion

The effect of planting pattern on the runoff contamination: Adopt total nitrogen, total phosphrous, Pb, Cd, chemical oxygen demand, sedimentation and heavy metal content in surface runoff to evaluate the comprehensive effect. The order is : terrace of 15°+ economic fruit tree seedling>terrace of 25° + economic fruit tree seedling > sweet potato +maize(intercrop and along with the slope)> bean + sweet potato(25% intercropping)+ sweet potato(75%)(along with the slope)> bean(50%)+ sweet potato(50%)(intercrop and along with the slope)> economic fruit tree seedling(along with the slope).

The spatial distribution characteristic of runoff contaminate: the Gezigou reservoir small watershed is divided into 7 sections based on the land-use ways and the planting canopy percent and the origin of runoff around the watershed. In a whole, the first and the second sections are harnessed effectively, the 5th and 6th are mostly worst. The content of Cd in the 4th and the content of Pb in 5th and 6th are the highest.

The contamination degree in the second section is the least in which the lower slope is filtrated trough buffer zone, but the most contamination degree appealed in 4th in which the lower slope planted newly orange. There are remarkably correlation between total nitrogen, total phosphrous, Pb, Cd, chemical oxygen demand and sedimentation in surface runoff; there are also remarkably correlation between the content of chemical oxygen demand and total phosphrous sedimentation, moreover the content of total nitrogen is the principal factor affected chemical oxygen demand in surface runoff.

The timely distribution characteristic of runoff contaminate: In the runoff, the general trend of the content of total nitrogen, total phosphrous, Pb, Cd, chemical oxygen demand increase at first and decrease at end. The time of the contents reach the peak value different from the kinds of the contamination and the sections. At the latest time, the content of the contamination is steady to every section. The content of total nitrogen will reach the peak value within the first to second hour, at end which become steady and less than the initial value. The content of total phosphrous reach the peak valve within 0.8-2 hour; At the same time, the rank of the content of total phosphrous in every section is 4>5>6>7>3>1>2.

The content of Pb will reach the peak value within the first to second hour and at the first hour after the runoff appealed the rank of the content in every section is 1>3>7>2>6>5>4, after 1-3.5 hour is 5>6>4>7 and the difference of the content in 1, 2, 3 section is tiny; At the latest time, they are consistent. The content of Cd reach the peak value within 1.8-2 hour and at the former two hours between the runoff and peak valve appealed the rank of the content in every section is 1>3>7>2>6>5>4, the content in 4, 5, 6, 7 sections changed rapidly, but smoothly in 1, 2, 3 sections. The content of chemical oxygen demand reach the peak value within 1.8-2 hour, the content rank is 5>6>3>7 trough the whole runoff time, but the content change have not regulation in 1, 2, 4, 7 sections. In general the time of peak value of contamination is within 1.5-3.5 hours after runoff happened.

The estimation of total contamination entered into reservoir: To linearity regression between runoff and contamination of total nitrogen, total phosphrous, Pb, Cd, chemical oxygen demand according with SPSS and build a model. Based the model the contamination load of total nitrogen, total phosphrous, Pb, Cd, chemical oxygen demand in runoff in Gezigou reservoir small watershed on Aug 16th respectively 8411.69g, 1014.43g, 402.65g, 25.90g, 59829.82g, the total runoff is 6181.598m³.

Effect analysis of comprehensive Harness: Apply with the runoff load model in Zhuchangxi small watershed to calculate the runoff load of un-harnssed Gezigou reservoir small watershed and compare with the load based on the build model on Aug 16th, results show out: the total of total nitrogen, total phosphrous and Pb are less 7.87%, 44.66%, 47.09% than un-harnessed respectively.

4. Acknowledgment

The authors are thankful to Key Laboratory of Eco-environments in Three Gorges Reservoir Region, Ministry of Education, College of Resources and Environment, Southwest University for providing the laboratory facilities to do our research work.

5. References

Chartes, A.C. & Roben, P.B. (2000).A Comparision of the Hydrologic Characteristic of Natural and Created Maimstream Flood Plain Wetlands in Pennsyvania. *Ecological Engineering*, Vol.14, pp. 221-231. ISSN 0925-8574

David, J.W. (1998). Modeling Residence Time Instreamwater Ponds. *Ecological Engineering*, Vol.10, pp. 247-262. ISSN 0925-8574

Fu, L.L.; Shuai, J.Bing.; Wang,Y.b.; Ma, H.J.& Li, J.R.(2011). Temporal Genetic Variability and Host Sources of Escherichia Coli Associated with Fecal Pollution from Domesticated Animals in the Shellfish Culture Environment of Xiangshan Bay, East China Sea. *Environmental Pollution*, Vol.159, No.10, (October 2011), pp. 2808-2814, ISSN 0269-7491

Heitz, F.L. ; Khostowpanah, S. & Nelson, J. (2000). Sizing of Surface Runoff Detention Ponds for Water Quality Improvement · *Journal of the American Water Resources Association,* Vol. 36, No. 3, pp. 541-548. ISSN1093-474X

Leeds-Harrison, P.B. ; Quinton, J.N.&Walker, M.J. (1999) Grassed Buffer Strips for the Control of Nitrate Leaching to Surface Waaters in Headwater Catchments. *Ecological Engineering,* Vol.12, pp. 299-313. ISSN 0925-8574

Prato, T. & Shi H. (1990).A Comparison of Erosion And Water Pollution Control Strategies for an Agricultural Watershed. *Water Resources Research,* Vol.26, No.2, pp. 111-114. ISSN 0043-1397

Zhang, H. &Huang, G.H.(2011). Assessment of Non-point Source Pollution Using a Spatial Multicriteria Analysis Approach. *Ecological Modelling,* Vol.222, No.2, (January 2011), pp. 313-321, ISSN 0304-3800

10

Effects of Discharge Characteristics on Aqueous Pollutant Concentration at Jebel Ali Harbor, Dubai-UAE

Munjed A. Maraqa[1], Ayub Ali[2], Hassan D. Imran[1],
Waleed Hamza[1] and Saed Al Awadi[3]
[1]*United Arab Emirates University*
[2]*Griffith University*
[3]*Ports, Customs and Free Zone Corporation*
[1,3]*United Arab Emirates*
[2]*Australia*

1. Introduction

The Arabian Gulf is an important geographical location. The Gulf has been extensively used for transport purposes. Meanwhile, countries in the region benefit from the Gulf's diverse marine habitats and utilize its water for desalination or some industrial needs. Several pollutants are induced into the Gulf including those resulting from oil spill accidents, offshore exploration processes, ballast water discharge, reject brine discharge, dredging activities, and coastal construction projects. Meanwhile, some of the Gulf countries are developing new coastal industrial facilities or expanding existing ones. These facilities are not without an adverse impact on the marine environment.

Most of the work related to the quality of Arabian Gulf water has focused on understanding the flow dynamics and the impact of oil spills. A number of researchers, for example, used numerical modeling to investigate residual circulation and flow pattern of the Arabian Gulf (Hughes and Hunter, 1979; Lardner et al. 1987, 1993; Chao et al. 1992; Horton et al., 1994; Elshorbagy et al., 2006; Azam et al., 2006a, 2006b; Thoppil & Hogan, 2010). Other researchers assessed the Gulf water quality as affected by oil spills (El Samra et al., 1986; Lardner et al., 1988; Al-Rabeh et al., 1992; Spaulding et al., 1993).

Little attention, however, has been directed to investigate the impact of discharges from coastal industrial facilities on water quality in the Arabian Gulf. In this study, we will consider the case of Jebel Ali Harbor to numerically assess the harbor's water quality as affected by discharges from industrial facilities located at Jebel Ali Free Zone (JAFZ) area in Dubai, United Arab Emirates (UAE). The harbor at JAFZ area (Fig. 1) is one of the largest man-made ports in the world. The harbor receives discharge consisting of several treated industrial effluents. Water discharged into the harbor must adhere to the effluent quality criteria set forth in the Environmental Requirements established by the Ports, Customs and Free Zone Corporation (PCFC, 2003) at JAFZ area. PCFC has also established harbor water quality objective limits (PCFC, 2003) in order to protect marine life and to minimize the impact of industrial activities on the surrounding ecosystem. Regular monitoring of discharged treated wastewater as well as harbor water and sediments is conducted by the PCFC to assure adherence to effluent standard and quality objective limits.

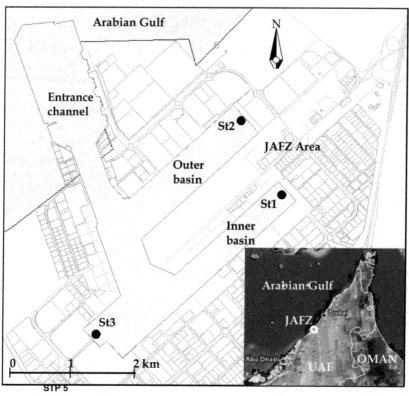

Fig. 1. Map of Jebel Ali Harbor. The circle in the bottom map is the location of JAFZ area.

Future expansion of industrial activities at JAFZ area, in addition to port activities and on-going as well as planned coastal construction projects in the vicinity of the harbor, may increase pollutant loading to the receiving water body. Limited work, however, has been conducted to assess the water quality of Jebel Ali Harbor. Maraqa et al. (2007) studied the fate of selected pollutants in the harbor and concluded that induced pollutants tend to accumulate in the harbor due to its limited flushing capacity. Furthermore, Maraqa et al. (2008) found that the main flow regime in the harbor follows alternate paths during flooding and ebbing, which creates eddy-like circulations in net flow distribution (see Fig. 2). Maraqa et al. (2008) also showed that dead-end locations at Jebel Ali Harbor have low water circulation and that flushing of a conservative pollutant discharged into the harbor takes a few months to several years depending on the discharge location.

This study expands on the work of Maraqa et al. (2008) to investigate variations in the concentration of pollutants induced into the harbor due to variations in the loading rate, discharge location and discharge concentration. A similar approach was used by Kashefipour et al. (2002; 2006) to assess the impact of various bacterial input rates on the receiving water in coastal basins in the UK. This study further explores the relationship between a continuous pollutant loading rate and average pollutant concentration in Jebel Ali Harbor water. As such, the study is of importance to modeling experts and managers interested in the hydrodynamic and transport properties of this harbor. The outcome of this study could further assist managers of the harbor decide on proper input rates and discharge locations so that water quality objective limits are not exceeded. Meanwhile, the general approach presented here may be of value in application to other systems.

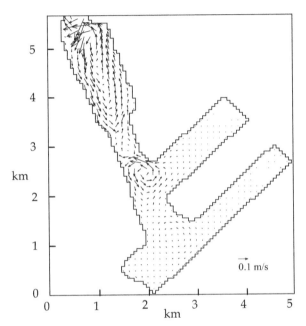

Fig. 2. Net flow over a tidal cycle of Jebel Ali Harbor (adopted from Maraqa et al., 2008)

2. Governing equations

Numerical modeling can be used to help achieve discharge conditions that meet pre-set environmental limits. Numerical modeling has been extensively applied to simulate water circulation and contaminant transport in harbors and semi-enclosed coastal areas. For example, efforts were made to better understand the hydrodynamic regimes (Estacio et al., 1997; Vethamnoy et al., 2005; Azam et al., 2006a; Dias and Lopes, 2006a,b; Maraqa et al., 2008; Montano-Ley et al., 2008), to investigate pollutant dispersion (Gesteira-Gomez et al., 1999; Das et al. 2000), and to assess water quality (Tao et al., 2001; Copeland et al., 2003; Fiandrino et al., 2003; Lopes et al., 2005; Cerejo and Dias, 2007). Other efforts were made to establish surveillance procedures (Lopes et al., 2005) and to quantify the impact of effluent discharge (Ganoulis, 1991; Kashefipour et al., 2002; Gupta et al., 2004, Kashefipour et al., 2006; Rucinski et al., 2007; Brewer et al., 2008).

Modeling of fluid flow is based on the principles of continuity of mass and conservation of momentum. For flows, which show little variation in the vertical dimension, it is acceptable to integrate these equations over the depth of water, resulting in two-dimensional (2D) equations of motion. In a 2D hydrodynamic (HD) model, the continuity equation is:

$$\frac{\partial \zeta}{\partial t} + \frac{\partial p}{\partial x} + \frac{\partial q}{\partial y} = 0 \tag{1}$$

where, ζ is the water level (m); p and q are flux densities in x and y directions (m³/s/m); t is time (s); x and y are space coordinates (m). The x-and y-momentum are given by Eq. (2) and (3), respectively:

$$\frac{\partial p}{\partial t} + \frac{\partial}{\partial x}\left(\frac{p^2}{h}\right) + \frac{\partial}{\partial y}\left(\frac{pq}{h}\right) + gh\frac{\partial \zeta}{\partial x} + $$
$$\frac{gp\sqrt{p^2+q^2}}{C^2 h^2} - \frac{1}{\rho_w}\left[\frac{\partial}{\partial x}\left(h\tau_{xx}\right) + \frac{\partial}{\partial y}\left(h\tau_{xy}\right)\right] - \tag{2}$$
$$\Omega q - fVV_x \frac{h}{\rho_w}\frac{\partial}{\partial x}\left(p_a\right) = 0$$

$$\frac{\partial q}{\partial t} + \frac{\partial}{\partial y}\left(\frac{q^2}{h}\right) + \frac{\partial}{\partial x}\left(\frac{pq}{h}\right) + gh\frac{\partial \zeta}{\partial y} + $$
$$\frac{gq\sqrt{p^2+q^2}}{C^2 h^2} - \frac{1}{\rho_w}\left[\frac{\partial}{\partial x}\left(h\tau_{yy}\right) + \frac{\partial}{\partial x}\left(h\tau_{xy}\right)\right] - \tag{3}$$
$$\Omega p - fVV_y \frac{h}{\rho_w}\frac{\partial}{\partial y}\left(p_a\right) = 0$$

where, h is water depth (m); C is Chezy resistance (m¹/²/s); f is the wind friction factor (dimensionless), V, V_x, and V_y are wind speed and components in x and y directions (m/s), respectively; Ω is Coriolis parameter (s⁻¹); p_a is atmospheric pressure (kg/m/s²); ρ_w is the density of water (kg/m³); and τ_{xx}, τ_{xy} and τ_{yy} are components of effective shear stress (N/m²).

The advection-dispersion (AD) model simulates the spreading of a substance in an aquatic environment under the influence of fluid transport and dispersion processes. The substance may be treated conservatively or with decay. The governing equation for a 2D AD model is given as (Adams and Baptista, 1986):

$$\frac{\partial}{\partial t}(hc) + \frac{\partial}{\partial x}(uhc) + \frac{\partial}{\partial y}(vhc) = \frac{\partial}{\partial x}\left(hD_x\frac{\partial c}{\partial x}\right)$$
$$+ \frac{\partial}{\partial y}\left(hD_y\frac{\partial c}{\partial y}\right) - khC + Q_s(c_s - c) \tag{4}$$

where, c is substance concentration (mg/l); u and v are horizontal velocity components in x and y directions (m/s), respectively; D_x and D_y are dispersion coefficients in x and y directions (m²/s), respectively; k is the linear decay rate coefficient (s⁻¹); Q_s is the source/sink discharge per unit horizontal area (m³/s/m²); c_s is substance concentration in the source/sink discharge (mg/l).

3. Methodology

3.1 Model description

Jebel Ali Harbor has an approach channel that starts 15 km offshore. The approach channel has a depth of 14-15 m and a width of 280 m reducing to 235 m. It bends after 10 km and becomes the entrance channel. It widens to 300 m at the bend and to 340 m at the entrance channel. There are two basins within the port. The outer 14-m deep basin is 2.3 km long and 600 m wide. The inner basin is 3.7 km long and 425 m wide, with a depth of 11.5 m. All channel and basin bottoms are sandstone. The surface area of the harbor is about 5.3 million m^2 and the total water volume is about 75 million m^3.

Maraqa et al. (2008) developed a 2D model using the MIKE21 modeling system of the Danish Hydraulic Institute (DHI, 2003a; 2003b) to simulate the HD and the AD processes within the Jebel Ali Harbor. Justification of the use of a depth-integrated 2D model was based on the nearly uniform temperature and salinity profiles found at different locations in the harbor (Maraqa et al., 2008). Since the AD model developed by Maraqa et al. (2008) was used in this study, a brief description of the model setup is presented below. For more details about the model setup the readers are referred to Maraqa et al. (2008).

The HD model of Jebel Ali Harbor is the basis for the AD model. The model was constructed with a rectangular grid system of 60×60 m^2. The dimensions of the grid were selected as a compromise between resolution and computational time. The origin of the model was 24°58'03" latitude and 55°01'28" longitude, taking east-west and north-south directions as the x and y directions, respectively. The entrance to the harbor was selected as the open boundary and the flow direction was considered perpendicular to the boundary. The closed side and bottom boundaries were considered as no flow boundaries. A constant water level and zero velocities were used as initial conditions at all grid points. Tide level was used as the boundary condition and the flow direction was considered perpendicular to the boundary. Latest topographical description of the harbor area was incorporated in the model (Jan de Nul Dredging Ltd., 2004). Although, the hydraulic regime of Jebel Ali Harbor is mainly dependent on the tide (Maraqa et al., 2008), meteorological forces were incorporated to improve the accuracy of the model. Observed meteorological conditions during January to December 2004 at the site were applied to the model in the first simulation year and similar meteorological conditions were used for a simulation period of 12 successive years.

The predicted tide level at the entrance of the harbor was used as the boundary condition for the HD model. The prediction was carried out using the Admiralty method (DHI, 2003a) facilitated in MIKE21 tools using major tidal constituents (see Table 1) with necessary seasonal corrections of -0.1 during February, March and April and +0.1 during July and August (ATT, 2003). Predicted tide levels were referenced to mean sea level datum and converted to local chart datum (CD) adding 1.02 m (ATT, 2003). The HD model was calibrated against tide level and flow data measured in December 2004. Through a rigorous calibration process, constant Chezy number and eddy viscosity were selected as 40 and 1.0 m^2/s, respectively. Simulated tide levels compared quite well with measured levels at three locations within the harbor. Also, simulated flow values through the entrance channel matched quite well with the measured ones at the same location (Maraqa et al., 2008).

In the development of the AD model of Jebel Ali Harbor, Maraqa et al. (2008) used a spatially varied dispersion coefficient determined by a formula suggested by Fischer et al. (1979):

$$D = 0.011 \frac{\bar{u}^2 W^2}{du^*} \tag{5}$$

where, D is the dispersion coefficient (m²/s); $\bar{u}$ is the average velocity (m/s); W is the width of the channel (m); d is the depth of the channel (m); u^* is the shear velocity (m/s) which is expressed as $(ghS)^{0.5}$; g is the gravitational acceleration = 9.81 (m/s²); h is the hydraulic radius ≈ depth of the channel (m); and S is the water surface slope.

Constituent name	Amplitude (m)	Phase (°)
Principal lunar semidiurnal (M₂)	0.43	359
Principal solar semidiurnal (S₂)	0.17	49
Luni-solar declinational diurnal (K₁)	0.25	155
Lunar declinational diurnal (O₁)	0.17	100
First overtide of M₂ (F₄)	0.0	0
Second overtide of M₂ (F₆)	0.0	0

Table 1. Tidal constituents at Jebel Ali Harbor (ATT, 2003).

Values of the dispersion coefficient in the x and y directions were calculated using Eq. (5) based on a channel width of 300 m, an estimated average water surface slope of 2×10⁻⁵ m/m, and a hydraulic radius (considered as the average depth of flow) of 12.0 m (Maraqa et al., 2008). Furthermore, the mean velocity at each grid point during an ebb tide in spring was calculated from the HD model and was used as an average velocity for calculating the dispersion coefficient. Estimated dispersion coefficients at the dead-end locations within the inner and outer basins were found to be much lower (<0.0001 m²/s) than those in the main channel (>0.01 m²/s).

3.2 Model applications

In this study, the effect of variation in the loading rate on the average pollutant concentration in the harbor (C_{avg}) was numerically investigated by simulating pollutant concentration in harbor water subject to different continuous loading rates at specified locations. Three discharge points were selected (Fig. 1); one in the corner of the inner basin (St1), another in the corner of the outer basin (St2), and a third point at the west corner of the inner basin (St3). These locations are currently used to discharge treated industrial wastewater into the harbor (Maraqa et al., 2007). Conservative and degradable pollutants were considered with a loading rate (LR) that varied from 0.01-1.0 g/s. The lower limit of loading rates nearly corresponds to the current total rate of discharge of phosphate, while the upper limit is close to the rate of discharge of nitrate or BOD₅ (Maraqa et al., 2007). It should be noted that conservative pollutants discharged into Jebel Ali Harbor could include reject brine from desalination plants or any pollutant that does not undergo transfer and transform reactions. On the other hand, degradable pollutants could include BOD, coliform bacteria, or any other pollutant that undergoes transformation, not transfer, reactions.

Three different sets of simulations were conducted in this study. In the first set, the loading rate of a conservative pollutant varied while fixing the discharge concentration at 20 mg/l. This discharge concentration was chosen based on the current discharge concentration of some contaminants (Maraqa et al., 2007). Since the concentration at the source was fixed in this set of simulations, variations in the loading rates are due to variations in the discharge flow rates. The second set of simulations was carried out to investigate the impact of changing the discharge concentration, with fixed loading rates, on the value of C_{avg}. Discharge concentrations of 10, 20, and 40 mg/l at St1 were simulated for discharge rates of 0.01, 0.1 and 1.0 g/s of a conservative pollutant. The third set of simulations was conducted to study the impact of pollutant degradation on C_{avg}. Simulations of the latter cases were accomplished using the ECO Lab module of MIKE21 (DHI, 2003c) along with the AD model. A decay rate constant of 0.1, 0.2 and 0.5/yr were used with pollutant input rates of 0.01, 0.1 and 1.0 g/s at St1.

All simulated cases in this study are summarized in Table 2. In all simulated cases, a pollutant concentration background value of zero in harbor water was used as the initial and boundary conditions. For each simulated case, the concentration level at different locations and the total mass of the pollutant within the harbor were numerically estimated using the developed model by Maraqa et al. (2008) to find out C_{avg}. For a continuous and constant loading rate, steady-state conditions were assumed to be reached when the mean pollutant concentration over a tidal cycle at any point in the harbor did not change over time. This definition of steady-state concentration is similar to the definition of stationary-state concentration used by Edinger et al. (1998).

The value of C_{avg} at steady-state conditions was calculated by averaging the spatial concentration values over the entire harbor-modeling area when steady-state conditions prevailed. As indicated by the California Regional Water Quality Control Board (CRWQCB, 2007), an average concentration value of the main water mass is typically used when comparison with a water quality objective limit is intended. The CRWQCB (2007) also indicated that objective limits cannot be applied at or immediately adjacent to zones of initial dilution within which higher concentration can be tolerated.

4. Results and discussion

4.1 Time of steady-state conditions

The time to reach steady-state conditions due to a continuous discharge was almost the same for a certain discharge location independent of the loading rate. However, the time to reach steady-state was dependent on the discharge location with a value of about 12 yrs for discharge at St1, 7 yrs for discharge at St2, and 4 yrs for discharge at St3 (Table 2). It should be noted that the time to reach steady-state is not directly comparable to flushing time or residence time. Generally, flushing time is defined as "the ratio of the scalar in a reservoir to the rate of renewal of the scalar" (Geyer et al., 2000). Flushing time describes the exchange characteristics of a waterbody without identifying the underlying physical processes or their spatial distribution (Monsen et al., 2002). Residence time, on the other hand, is the time it takes a waterparcel to leave a semi-enclosed waterbody through its outlet (Monsen et al., 2002). Residence time is measured from an arbitrary start location within the waterbody, whereas the time to reach steady-state concentration used in this study depends primarily on the mixing characteristics of the entire waterbody. However, steady-state times were found quite similar to residence times simulated by Maraqa et al. (2008) for these stations since both of these time scales depend on the same physical processes.

Case	Location	Concentration (mg/l)	Discharge flow (m³/s)	Loading rate (g/s)	Degradation rate constant (yr⁻¹)	Time to reach steady-state (yr)
1	St1	20	0.0005	0.01	0	12
2	St1	20	0.0025	0.05	0	12
3	St1	20	0.0050	0.10	0	12
4	St1	20	0.0250	0.50	0	12
5	St1	20	0.0500	1.00	0	12
6	St2	20	0.0005	0.01	0	7
7	St2	20	0.0050	0.10	0	7
8	St2	20	0.0500	1.00	0	7
9	St3	20	0.0005	0.01	0	4
10	St3	20	0.0050	0.10	0	4
11	St3	20	0.0500	1.00	0	4
12	St1	10	0.0010	0.01	0	12
13	St1	10	0.0100	0.10	0	12
14	St1	10	0.1000	1.00	0	12
15	St1	40	0.00025	0.01	0	12
16	St1	40	0.0025	0.10	0	12
17	St1	40	0.0250	1.00	0	12
18	St1	20	0.0005	0.01	0.1	12
19	St1	20	0.0050	0.10	0.2	12
20	St1	20	0.0500	1.00	0.5	12
21	St1	20	0.0005	0.01	0.1	12
22	St1	20	0.0050	0.10	0.2	12
23	St1	20	0.0500	1.00	0.5	12
24	St1	20	0.0005	0.01	0.1	12
25	St1	20	0.0050	0.10	0.2	12
26	St1	20	0.0500	1.00	0.5	12

Table 2. Description of the simulated cases.

4.2 Spatial variations

Spatial variations of pollutant concentration in harbor water due to a loading rate of 1.0 g/s at St1 is shown in Fig. 3. Simulation shows that the concentration distribution changes significantly from year 2 to year 8, whereas it increases slightly after year 8 until it reaches steady-state conditions. From the circulation pattern of the harbor, as presented by Maraqa et al. (2008), pollutant distribution in the inner and outer basin is dominated by diffusion while that in the entrance channel is greatly affected by advection.

To examine the effect of pollutant source location, the simulations were repeated relocating the point source at St2 (Fig. 1). It was found (Fig. 4) that the concentration distributions in the inner and outer basins are different than the distribution with the source location at St1. But, the concentration distributions in the entrance channel were almost similar for the two cases (see Fig. 3 and Fig. 4). The concentration in the entrance channel is mostly influenced by the Gulf water rather than the inside basins because of the dominant advection processes. In any case, the highest concentration occurs at the source location, while the lowest concentration generally occurs at the west side of the entrance channel due to the inward net flow conditions (Maraqa et al., 2008). For a loading rate of 1.0 g/s, the average pollutant concentration in the harbor water reached 0.35 mg/l with the discharge point located at St2 and 1.3 mg/l with the discharge point located at St1. The reasons behind reaching steady-state conditions with lower concentration at St2 are faster transport due to advection and more dilution with Gulf water. Thus, it is necessary to examine the impact of pollutant loading using modeling technique before selecting the discharge location.

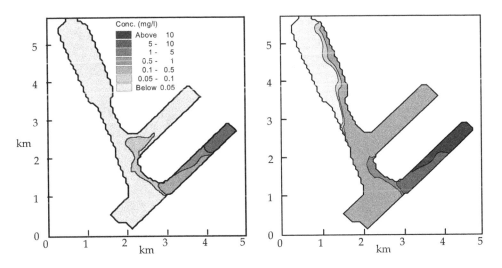

Fig. 3. Concentration map after 4 yrs (left) and 12 yrs (right) from start of simulation with a continuous loading rate of a conservative pollutant of 1.0 g/s at St1.

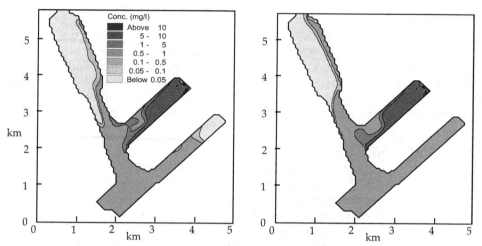

Fig. 4. Concentration map after 2 yrs (left), and 7 yrs (right) from start of simulation with a continuous loading rate of 1.0 g/s at St2.

4.3 Temporal variations

Temporal variation of pollutant concentration at St3 due to pollutant loading of 1.0 and 0.1 g/s at St1 is shown in Fig. 5. The figure shows that the concentration at St3 increases sharply at initial times and levels of at later times until it reaches a plateau value. For a conservative pollutant, the mechanisms of solute transport within the harbor are associated with the advection-dispersion processes. Advection driven by the tide was the principal transporting process within Jebel Ali Harbor. Winds and waves play a minor role in mixing and transport of pollutants because of the bottle-neck shape of the harbor.

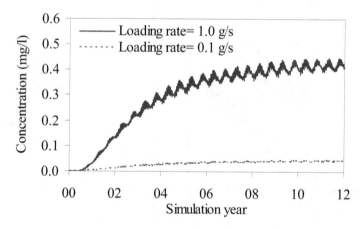

Fig. 5. Long-term variation of concentration at St3 subject to a loading rate of a conservative pollutant of 0.1 and 1.0 g/s at St1.

Further inspection of Fig. 5 shows that there are seasonal fluctuations in the concentration level. Such fluctuation occurs at hourly levels due to variations of tidal levels as demonstrated in Fig. 6. In general, the average concentration reduces during summer season when the tide levels are high and the concentration increases during winter season when the tide levels are low. Also, the concentration reduces during high water and increases during low water because of hourly tide level changes. This indicates that short-term monitoring of water quality in the harbor may not reflect on the long-term changes. For example, the concentration of a conservative pollutant at St3 reaches 0.46 mg/l in April of year 11 as a result of a loading rate of 1.0 g/s at St1. The concentration drops in July (of that year) to about 0.41 mg/l for the same loading rate at St1.

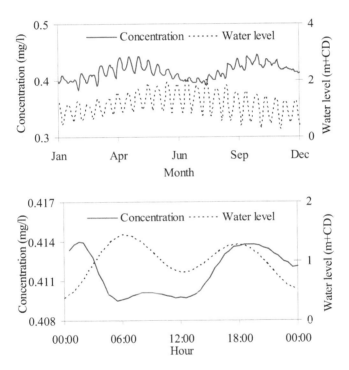

Fig. 6. Concentration at St3 due to a loading rate of 1.0 g/s at St1 showing seasonal variations (top) and daily variations (bottom) at the end of the 10th yr along with the tide levels.

4.4 Steady-state concentration of conservative pollutants

For conservative pollutants, the values of C_{avg} resulting from different loading rates at St1, St2, and St3 are presented in Fig. 7. As the figure shows, C_{avg} varies with both the loading rate and the discharge location. Higher concentration was observed when the source was located at St1. On the other hand, relatively lower concentrations were observed when the source was located at St2 or St3. Discharging at St2 and St3 produces almost the same average concentration in harbor water for similar loading rates. Similar concentration distributions were also observed in the entrance channel whether the source was located at St2 or St3, but different concentration distributions were observed in the inner and outer basins.

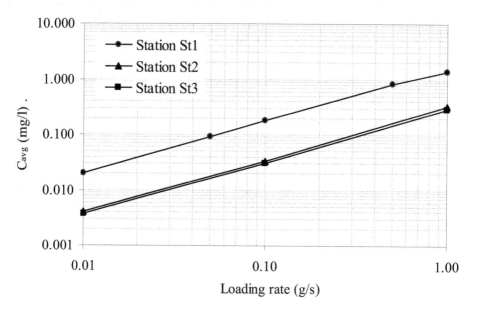

Fig. 7. Average concentration in harbor water for different discharge locations and loading rates of a conservative pollutant.

Based on the data presented in Fig. 7, the following best fit equations were produced relating the average concentration (mg/l) in harbor water to the loading rate (g/s):

$$\text{Discharge at St1: } C_{avg} = 1.468 \, (LR)^{0.926} \tag{6}$$

$$\text{Discharge at St2: } C_{avg} = 0.322 \, (LR)^{0.956} \tag{7}$$

$$\text{Discharge at St3: } C_{avg} = 0.286 \, (LR)^{0.953} \tag{8}$$

Equations 6-8 have a coefficient of determination (r^2) of 0.999. From Eqs. (6)-(8), C_{avg} correlates almost linearly with the pollutant loading rate. At a given loading rate, discharge at St1 results in values of C_{avg} that are 4-5 times higher that those due to discharge at either St2 or St3. However, the maximum concentration in the harbor always occurred close to the discharge location. This is consistent with the findings of Kumar et al. (2000) who reported higher bacterial pollution near diffuser locations (discharge points). Also, the maximum concentration in this work was almost an order of magnitude higher than the average concentration in the harbor. For example, the maximum concentration due to a discharge of 1.0 g/s at St1 was 13.26 mg/l compared to an average concentration of 1.31 mg/l for this case.

The effect of changes in the discharge concentration on the value of C_{avg} is shown in Fig. 8. The figure shows that C_{avg} is more dependent on the loading rate and less dependent on the discharge concentration. In other words, it is the mass input rate, rather than the discharge concentration itself, that influences the concentration of the pollutant in the harbor. Thus, Fig. 7 (or Eqs. 6-8) can be used to determine the allowable discharge rate of a pollutant such that C_{avg} does not exceed a pre-set harbor objective limit. Further inspection of Fig. 8 shows that C_{avg} increases with the increase in the input concentration at a loading rate of 1 g/s, while it maintains nearly the same value regardless of the input concentration at lower loading rates. Such observation could be due to the high volume of discharge water associated with a high loading rate and a relatively low input concentration.

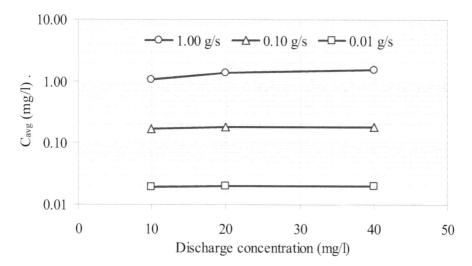

Fig. 8. Average concentration in harbor water for different discharge concentrations and loading rates of a conservative pollutant at St1.

4.5 Steady-state concentration of degradable pollutants

The simulations carried out under the previous cases were limited to pollutants that are conservative. The effect of degradation on the average concentration in harbor water is presented in Fig. 9 for degradable pollutants discharged at St1. As expected, an increase in the decay rate constant (k) results in a reduction in C_{avg}. This reduction is almost independent of the loading rate and averages 18%, 36% and 62% with a decay rate of 0.1, 0.2 and 0.5/yr, respectively. Meanwhile, the average concentration of a degrading pollutant introduced at St1 could be well predicted from that of a conservative pollutant using a decreasing exponential function with an average time (t) of 2 yrs as shown in Eq. (9):

$$(C_{avg})_k = (C_{avg})_{k=0}\, e^{-2k} \tag{9}$$

where, $(C_{avg})_k$ is the average concentration of a degrading pollutant in harbor water and $(C_{avg})_{k=0}$ is the average concentration of a conservative pollutant and k is in units of yr^{-1}.

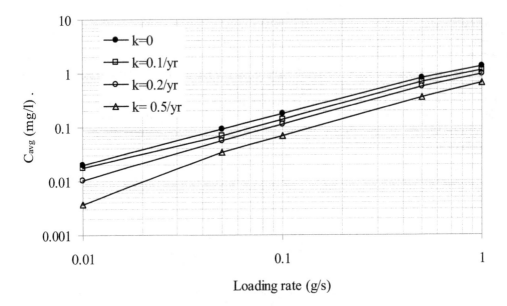

Fig. 9. Average concentration of pollutants with different degradation rates as a result of discharge at St1.

5. Conclusion

With a continuous pollutant discharge, the aqueous concentration at any point in Jebel Ali Harbor reaches steady-state conditions with duration that depends on the discharge location. The longest duration (about 12 yrs) occurs with discharge in the east corner of the inner basin. Results of this study show that the average steady-state pollutant concentration in harbor water varies with both the loading rate and the discharge location, but is independent of the discharge concentration. For a conservative pollutant discharged in Jebel Ali Harbor, developed relationships of average pollutant concentration in harbor water were found to correlate almost linearly with the discharge loading rate. It was also observed that discharging in the east corner of the inner basin results in an average steady-state concentration of about 4-5 times higher than values associated with discharge at the east corner of the outer basin or the west corner of the inner basin. For a degrading pollutant, the reduction in the steady-state average concentration is almost independent of the loading rate, but could be adequately predicted from that of a conservative pollutant using a decreasing exponential time function. Derived relationships of average aqueous pollutant concentration in the harbor versus the discharge loading rate will be useful for better management of harbor water quality.

6. Acknowledgement

This project was funded by the Ports, Customs and Free Zone Corporation at Jebel Ali, Dubai and the UAE University.

7. References

Adams, E. E. & Baptista, A.M. (1986). Ocean Dispersion Modeling. In: *Encyclopedia of Fluid Mechanics, Vol. 6, Complex Flow Phenomena and Modeling,* N.P. Cheremisinoff (ed.), pp. 865–895, Gulf Publishing, ISBN: 978-0872015180, Houston, Texas.

Al-Rabeh, A. H.; Cekirge, H. M. & Gunay, N. (1992). Modeling the fate and transport of Al-Ahmadi oil spill, *Water, Air, and Soil Pollution,* Vol. 65, No. 3-4, pp. 257-279, ISSN: 0049-6979.

ATT (2003). Admiralty Tide Tables. Admiralty Charts and Publications, the United Kingdom Hydrographic Office, Vol.3, NP 203-01.

Azam, M. H.; Elshorbagy W. & Nakata, K. (2006a). Three-dimensional modeling of the Ruwais Coastal Area of United Arab Emirates, *J. Waterway, Port, Coastal and Ocean Engineering,* Vol. 132, No. 6, pp. 487-495, ISSN: 0733-950X.

Azam, M. H.; Elshorbagy, W.; Ichikawa, T.; Terasawa, T. & Taguchi, K. (2006b). 3D model application to study residual flow in the Arabian Gulf, *J. Waterway, Port, Coastal and Ocean Engineering,* Vol. 132, No. 5, pp. 388-400, ISSN: 0733-950X.

Brewer, H. M.; Godin, G.; Cunderlik, J. M.; Houghton, E.; Firman, M. & Green, D. (2008). Assimilative capacity modelling of Collingwood Harbour: A case study on the selection of effluent limits for municipal wastewater treatment plant expansion, Proceedings of the Water Environment Federation, ISSN: 1938-6478, pp. 2172-2185(14).

Cerejo, M. & Dias, J. M. (2007). Tidal transport and dispersal of marine toxic microalgae in a shallow, temperate coastal lagoon, *Marine Environmental Research,* Vol. 63, No. 4, pp. 313–340, ISSN: 0141-1136.

Chao, S.-Y.; Kao, T. W. & Al-Hajri, K. R. (1992). A numerical investigation of circulation in the Arabian Gulf, *J. Geophys. Res.,* Vol. 97, No. C7, pp. 11219–11236, ISSN 0148-0227.

Copeland, G.; Monteiro, T.; Couch, S. & Borthwick, A. (2003).Water quality in Sepetiba Bay, Brazil, *Marine Environmental Research,* Vol. 55, No. 5, pp. 385–408, ISSN: 0141-1136.

CRWQCB (2007). San Francisco Bay Basin (Region 2) Water Quality Control Plan. California Regional Water Quality Control Board, Oakland, CA.

Das, P.; Marchesiello, P. & Middleton, J. H. (2000). Numerical modelling of tide-induced residual circulation in Sydney Harbour, *Marine and Freshwater Research,* Vol. 51, No. 2, pp. 97-112, ISSN 1323-1650.

Dias, J. M. & Lopes, J. F. (2006a). Implementation and assessment of hydrodynamic, salt and heat transport models: the case of Ria de Aveiro lagoon (Portugal), *Environmental Modelling and Software,* Vol. 21, No. 1, pp. 1–15, ISSN 1364-8152.

Dias, J. M. & Lopes, J. F. (2006b). Calibration and validation of hydrodynamic, salt and heat transport models for Ria de Aveiro lagoon (Portugal), *J. Coastal Research,* Vol. SI 39, pp. 1680–1684, ISSN 0749-0208.

DHI (2003a). Scientific Documentation, MIKE 21 Hydrodynamic Module, DHI Water and Environment, DHI Software, Denmark.

DHI (2003b). Scientific Documentation, MIKE 21 Advection-Dispersion Module, DHI Water and Environment, DHI Software, Denmark.

DHI (2003c). Scientific Documentation, MIKE 21 ECO Lab Module, DHI Water and Environment, DHI Software, Denmark.

Edinger, J. E.; Buchak, E. M. & Kollubru, V. S. (1998). Modeling flushing and mixing in a deep estuary, *Water, Air, and Soil Pollution*, Vol. 102, No. 3-4, pp. 345-353, ISSN: 0049-6979.

El Samra, M.I.; Emara, H.I. & Shunbo, F. (1986). Dissolved petroleum hydrocarbon in the northwestern Arabian Gulf, *Marine Pollution Bulletin*, Vol. 17, No. 2, pp. 65-68, ISSN: 0025-326X.

Elshorbagy, W.; Azam M. H. & Taguchi, K. (2006). Hydrodynamic characterization and modeling of the Arabian Gulf. *J. Waterway, Port, Coastal and Ocean Engineering*, Vol. 132, No. 1, pp. 47-56, ISSN: 0733-950X.

Estacio, F. J.; Garcia-adiego, E. M.; Fa, D. A.; Garcia-Gomez, J. C.; Daza, J. L.; Hortas F. & Gomez-Ariza, J. L. (1997). Ecological analysis in a polluted area of Algeciras Bay (southern Spain): External 'versus' internal outfalls and environmental implications, *Marine Pollution Bulletin*, Vol. 34, No. 10, pp. 780-793, ISSN: 0025-326X.

Fiandrino, A.; Martin, Y.; Got, P.; Bonnefont, J. L. & Troussellier, M. (2003). Bacterial contamination of Mediterranean coastal seawater as affected by riverine inputs: simulation approach applied to a shellfish breeding area (Thaulagoon, France), *Water Research*, Vol. 37, No. 8, pp. 1711-1722, ISSN: 0043-1354.

Fischer, H. B.; List, E. J.; Koh, R. C. Y.; Imberger, J. & Brooks, N. H. (1979). *Mixing in Inland and Coastal Waters*, Academic Press Inc. ISBN: 9780122581502, London, UK.

Ganoulis, J. G. (1991). Water quality assessment and protection measures of a semi-enclosed coastal area: The Bay of Thermaikos (NE Mediterranean sea), *Marine Pollution Bulletin*, Vol. 23, pp. 83-87, ISSN: 0025-326X.

Gesteira-Gomez, M.; Montero, P.; Prego, R.; Taboada, J. J.; Leitao, P.; Ruiz-Villarreal, M.; Neves, R. & Perez-Villar, V. (1999). A two-dimensional particle tracking model for pollution dispersion in A Coruna and Vigo Rias (NW Spain), *Oceanologica Acta*, Vol. 22, No. 2, pp. 167-177, ISSN: 0399-1784.

Geyer, W. R.; Morris, J. T.; Prahl, F. G. & Jay, D. A. (2000). Interaction between physical processes and ecosystem structure: A comparative approach. In: *Estuarine Science: A synthetic approach to research and practice*, J. E. Hobbie (ed.), pp. 177-210, Island Press, ISBN: 9781559637008, Washington, D.C.

Gupta, I.; Dhage, S.; Chandorkar, A. A. & Srivastav, A. (2004). Numerical modeling for Thane creek, *Environmental Modelling and Software*, Vol. 19, No. 6, pp. 571-579, ISSN 1364-8152.

Horton, C.; Clifford, M.; Schmitz, J. & Hester, B. (1994). SWAFS: Shallow water analysis and forecast system. Overview and Status Rep., Naval Oceanographic Office, Stennis Space Center, Miss.

Hughes, P. & Hunter, J. R. (1979). A Proposal for a physical oceanography program and numerical modeling of the KAP Region. UNESCO, Div. Mar. Sci., Paris, MARINF/27, 16 Oct. 1979. 102 pp.

Jan de Nul Dredging Ltd. (2004). Dredging and Land Reclamation, A company commissioned for dredging Jebel Ali Harbor, Personal Communication, June 2004.

Kashefipour, S. M.; Lin, B.; Harris, E. & Falconer, R. A. (2002). Hydro-environmental modelling for bathing water compliance of an estuarine basin, *Water Research*, Vol. 36, No. 7, pp. 1854-1868, ISSN: 0043-1354.

Kashefipour, S. M.; Lin, B. & Falconer, R. A. (2006). Modelling the fate of faecal indicators in a coastal basin, *Water Research*, Vol. 40, No. 7, pp. 1413-1425, ISSN: 0043-1354.

Kumar, R.; Subramanium, J. & Patil, D. (2000). Water quality modeling of municipal discharge from sea outfalls, Mumbai, *Environmental Monitoring and Assessment*, Vol. 62, No. 2, pp. 119–132, ISSN: 0167-6369.

Lardner, R. W.; Lehr, W. J.; Fraga, R. J. & Sarhan, M. A. (1987). Residual circulation in the Arabian Gulf. I: Density-driven flow. *Arabian J. Sci. Eng.*, Vol. 12, No. 3, pp. 341–354, ISSN: 1319-8025.

Lardner, R.W.; Cekirge, H.M.; Al-Rabeh, A.H. & Gunay, N. (1988). Passive pollutant transport in the Arabian Gulf, *Advances in Water Resources*, Vol. 11, No. 4, pp. 158-161, ISSN: 0309-1708.

Lardner, R. W.; Lehr, W. J.; Fraga, R. J. & Sarhan, M. A. (1989). A model of residual currents and pollutant transport in the Arabian Gulf, *Mathematical and Computer Modelling*, Vol. 12, No. 9, pp. 1185-1186, ISSN: 0895-7177.

Lardner, R. W.; Al-Rabeh, W.; Gunay, N.; Hossain, M.; Reynolds, R. M. & Lehr, W. J. (1993). Computation of residual flow in the Gulf using the Mt. Mitchell data and the KFUPM/RI hydrodynamic model. *Mar. Pollution Bull.*, Vol. 27, pp. 61–70, ISSN: 0025-326X.

Lopes, J. F.; Dias, J. M.; Cardoso, A. C. & Silva, C. I. V. (2005). The water quality of the Ria de Aveiro lagoon, Portugal: From the observations to the implementation of a numerical model, *Marine Environmental Research*, Vol. 60, No. 5, pp. 594–628, ISSN: 0141-1136.

Maraqa, M. A.; Ali, A. & Khan, N. (2007). Modeling selected water quality parameters at Jebel Ali Harbor, Dubai- UAE, *J. Coastal Research*, SI 50, pp. 794-799, ISSN 0749-0208.

Maraqa, M. A.; Ali, A.; Imran, H. D.; Hamza, W. & Al Awadi, S. (2008). Simulation of the hydrodynamic regime of Jebel Ali Harbor, Dubai-UAE, *Aquatic Ecosystem Health and Management*, Vol. 11, No. 1, pp. 105-115, ISSN: 1463-4988.

Monsen, N. E.; Cloern, J. E.; Lucas, L. V. & Monismith, S. G. (2002). A comment on the use of flushing time, residence time and age as transport time scales, *Limnology and Oceanography*, Vol. 47, No. 5, pp. 1545-1553, ISSN 0024-3590.

Montano-Ley, Y.; Peraza-Vizcarra, R. & Paez-Osuna, F. (2008). Tidal hydrodynamics and their implications for the dispersion of effluents in Mazatlan Harbor: An urbanized shallow coastal lagoon, *Water, Air, and Soil Pollution*, Vol. 194, No. 1-4, pp. 343-357, ISSN: 0049-6979.

PCFC (2003). Environmental Control Rules and Requirements (3rd edition), Environment, Health and Safety Department, Ports, Customs and Free Zone Corporation, Dubai, UAE.

Rucinski, D.K.; Auer, M.T.; Watkins, D.W.; Effler, S.W.; Doerr, S.M. & Gelda R.K. (2007). Accessing assimilative capacity through a dual discharge approach, *Water Resources Planning and Management*, Vol. 133, No. 6, pp. 474-485, ISSN: 0733-9496.

Spaulding, M.L.; Anderson, E. L.; Isaji, T. & Howlett, E. (1993). Simulation of the oil trajectory and fate in the Arabian Gulf from the Mina Al Ahmadi spill, *Marine Environmental Research*, Vol. 36, No. 2, pp. 79-115, ISSN: 0141-1136.

Tao, J.; Li, Q.; Falconer, R.A. & Lin, B. (2001). Modelling and assessment of water quality indicators in a semi-enclosed shallow bay, *Hydraulic Research*, Vol. 39, No. 6, pp. 611-617, ISSN: 0022-1686.

Thoppil, P.G. & Hogan, P.J. (2010). A modeling study of circulation and eddies in the Persian Gulf, Physical Oceanography, Vol. 40, No. 9, pp. 2122-2134, ISSN: 0022-3670.

Vethamony, P.; Reddy, G. S.; Babu, M. T.; Desa, E. & Sudheesh, K. (2005). Tidal eddies in a semi-enclosed basin: a model study, *Marine Environmental Research*, Vol. 59, No. 5, pp. 519–532, ISSN: 0141-1136.

11

Water Quality in Hydroelectric Sites

Florentina Bunea[1], Diana Maria Bucur[2],
Gabriela Elena Dumitran[2] and Gabriel Dan Ciocan[3]
[1]National Institute for R&D in Electrical Engineering ICPE-CA,
[2]Politehnica University of Bucharest,
[3]Université Laval, Laboratoire de Machines Hydrauliques,
[1,2]Romania
[3]Canada

1. Introduction

The most widely form of renewable energy is the hydropower, which produces electrical power using the gravitational force of falling or flowing water. Comparing to fossil fuel powered energy plants, hydropower plants are considered „green" energy source, because they do not produce direct waste and have almost no output level of greenhouse gas carbon dioxide. Hydropower is the most important source of renewable electricity generation – 86.3 %, and essential to operate the others sources of renewable energy that are random generation.

Hydro energy importance also comes from its own source – the water, an essential life resource. Therefore, to maintain the water quality is a main concern from ecological, economical and sustainable development point of view (Bunea et al., 2010).

Hydroelectric sites use the available head and flow rate of a water course. Sometimes, this is made using more natural configuration, but often it involves important construction works and arrangements. The most common hydroelectric sites are based on:

a. local rise of the water level by means of a dam, which creates a reservoir upstream the dam. The hydropower plant is usually placed next to the dam;
b. deviation of water course through a free surface channel or tunnel. At the downstream end of the channel or tunnel, the water is put under pressure and driven to the turbines;
c. other mixed arrangements, with surface or underground hydropower plant, specific to mountain areas. These sites have high dams and large reservoirs.

The water quality used in a hydropower plant depends on many elements such as: the size and depth of the lake, placement along water course, intake depth, hydropower releases, temperature variations, rain intensity and frequency, reservoir thermal stratification and hydropower plant operation regimes.

The ecological impact of hydropower plants will be presented. In the first part is analyzed the processes in the reservoir (fig. 1), then, the influence of water realized on the downstream river. Also, some methods and their efficiency for improving water quality will be presented. Finally a review of chemical and quality water evolution in time will be presented for a hydroelectric site of Romania.

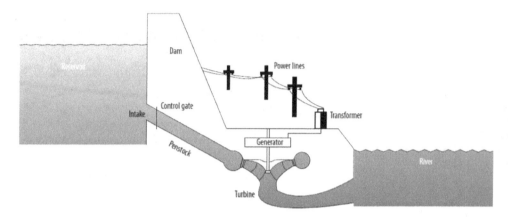

Fig. 1. Cross section of a hydroelectric power plant

During warm seasons the large reservoirs become subject of thermal stratification. Because the upper layers are close to the free water surface, they have a higher level of dissolved oxygen (DO). On the opposite side, the lower layers have a low level of DO, mainly because of the organic sediments at the bottom of the reservoir. When DO level goes under 5.0 mg/l, the aquatic life is endangered and large quantities of fish can die if the DO remains at 1÷2 mg/l for a few hours.

In hydropower plants, the water that goes to the turbines is taken from the lower layers of the reservoirs, sometimes with low DO content, which can affect the downstream water quality. The DO level from downstream water depends also on water head, periodic temperature variations, intensity and frequency of rain, hydropower plant design and its operation regimes.

Recently, the number of studies concerning water quality from hydropower releases increased. Many environment or ecological issues were reported, in different types of hydroelectric schemes. Scientist and engineers try to find solutions and mechanisms which will improve water quality, especially DO level. Generally, the low DO level is caused by organic sediments left on the reservoir bottom floor from the initial filling. When these organic sediments decompose, they absorb the oxygen from water, producing sulphuretted hydrogen, carbon dioxide and methane (like greenhouse gas). This pollution alters the local flora and fauna, even causing total extermination of some aquatic species.

Low DO level happens when the reservoir has a depth greater than 15 m and a volume bigger than $61 \cdot 10^6$ m³, the power output is more than 10 MW, and the retention time is longer than 10 days.

Romania has about 170 hydroelectric sites, a quarter of them having reservoirs larger than $61 \cdot 10^6$ m³ and deeper than 15 m, so they are susceptible for a low DO level (Bucur et al., 2010).

The usual methods used to increase the DO level in the hydropower plants downstream waters include selective intakes, air diffusers, hub and draft tube deflectors. These equipments are used in the hydropower plants with different success rates in the aeration process. Generally, in order to increase the DO with $1mg/l$, an air quantity of 1% from water volume is necessary (March, et al, 1992). A bibliographical revue is presented in this paper and recommendations are done for the implementation of aeration devices.

There is no legal support for DO level control downstream hydropower plants, but there are intense concerns regarding this issue. Usually, turbine aeration is made only in order to reduce turbine central vortex, so to increase the efficiency and reduce unsuitable pressure fluctuations and structure vibrations. The aeration made to increase the DO level downstream hydropower plant must be more consistent. Injection of a bigger air quantity can decrease the turbine efficiency; therefore air injection becomes an important factor for the balance between power output and ecology.

2. Hydroelectric reservoirs

Regarding the mean multiannual water flow, the surface water sources in Romania are much higher than ground water sources. Each type of water source has its own physic-chemical and biological characteristics, varying from region to region, depending on the mineralogical composition of the crossed areas, by the contact time, temperature, weather conditions, etc.

Water accumulated in reservoirs has the physical - chemical qualities significantly different from water flowing in the river, before the dam construction and the hydropower development. Thus, the processes occurring in lakes can have an important impact on the water quality. On one hand the stagnation of water leads to a natural settling of suspended materials which determinate a good transparency of the water and less sensitive to weather conditions. On the other hand, the stagnation of water leads to thermal and chemical stratification which excludes the water circulation on vertically direction.

2.1 Seasonal stratification of water in hydroelectrical reservoirs

Thermal structure of lakes varies by climate, by configuration of the lake basin, by the water intake surface and by the total mineralization of water. The most common structure is the direct stratification, which involves the higher temperatures at the water surface and lower temperatures to the bottom. For this kind of stratification the decrease of water temperature is not uniform with depth. Temperate regions are characterized by dimictic lake ecosystems. Most lakes in Romania are considered dimictic, meaning they mix twice a year - spring and fall. In the winter season, reverse stratification will be installed, while in the summer period a direct stratification will appear.

The lakes dynamic is characterized by energy and mass exchange processes. Dominant energy flow comes from the kinetic energy of wind and thermal energy produced by solar radiation. The vertical profile of temperature/density established in a lake results by superposing these two energy contributions (Dumitran and Vuta, 2011).

Thermal stratification consists in the existence of a vertical thermal gradient in the water mass. The low thermal conductivity of water contributes also, assuring that thermal energy is very slowly transferred to the bottom layers of the water. This transfer is accelerated by vertical turbulent mixing and convective cooling of the water body. In time, the cumulative effect of heat loss and convective cooling can be felt throughout the water column, reducing the lake water temperature and causing a full mixing between the water layers (Pourriot and Meybeck, 1995). The cumulative effect of heat loss and convective cooling can be felt in the entire water column, thus reducing the lake water temperature and producing a full mixing between the water near the surface and deeper layers. Turbulent mixing is a process that precludes stratification, tends to destabilize the water column and is caused by shear induced by wind action (Stevens and Imberger, 1996).

Convective cooling occurs only if the net heat flow from the lake surface is negative. Thus the lake is losing heat to the atmosphere and the water layers near the surface are cooling, becoming denser than deeper waters.

At this point thermal stratification becomes unstable, and the volumes of water near the surface descend to a water layer with same temperature. Because of friction, running water entails other volumes of water, producing a new vertical mixing. The movement is done without wind energy contribution and there is a destratification tendency of superior layer to the equilibrium depth (Fig. 2). In summer, a typical temperature/density profile for a temperate lake is composed from two layers of small temperature/density gradient (epilimnion, hipolimnion) divided by a layer of high temperature/density gradient (metalimnion).

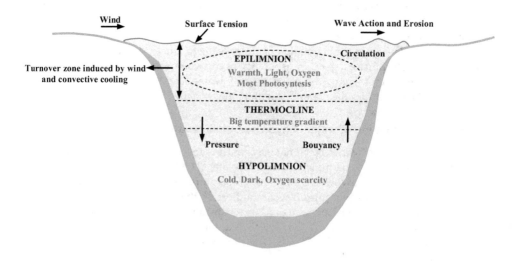

Fig. 2. Cross section of thermally stratified lake

Over the year, the lake water follows the cycle. In spring the ice melts into the lake, the wind picks up and the lake mixes. This is called spring turnover. Oxygen and nutrients get distributed throughout the water column as the water mixes. Then, as the weather becomes warmer, the surface water warms again and sets up summer stratification. During the summer the lake has a barocline structure, so at the surface a stable warmer layer of water overlies a colder water layer. The water movement due to wind and convection currents produces a mixing process which homogenize just the epilimnion, while the water temperature in the hypolimnion is kept at around 4 °C. In the fall the sunlight is not as strong as during summer and the nights become cooler. This change in season allows the epilimnion to cool off. As the water in the epilimnion cools, the density difference between the epilimnion and hypolimnion is not as great. Wind can then mix the layers. In addition, when the epilimnion cools it becomes denser and sinks to the hypolimnion, mixing the layers. This mixing allows oxygen and nutrients to be distributed across the whole water column. In winter, when surface water temperature drops below 4 °C, circulation ceases again and winter stagnation appear, characterized by an inverse thermal stratification. During this period the water mass is characterized by lower temperature at the surface and higher to bottom.

The lake stratification entails lower dissolved oxygen concentration in the bottom and the emergence of anaerobic oxidation processes. The stratification of lakes has a negative impact on trophic evolution of these ecosystems. Thus, the organic matter content and nutrients concentration will be increasing and sometimes even the hydrogen sulfide will appear at the bottom of lakes.

2.2 Day/night stratification of water in hydroelectrical reservoirs

In temperate regions the temperature differences between day and night are significant, so the water cooled during the night, goes down in a deeper layer. This depth is direct correlate with the reservoir size, so it can vary from 5 m up to 20 m (Read et al., 2011). In these conditions, the thermoclin layer appears which is characterized by a temperature drop of 10 to 15 °C (Fig.3).

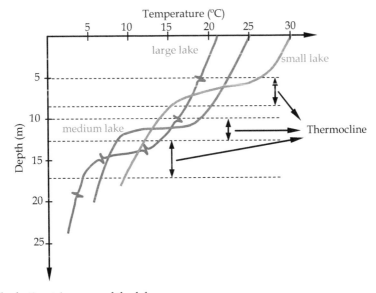

Fig. 3. The batimetric zones of the lake

2.3 Eutrophication of hydroelectrical reservoires

Biological quality of water is essentially affected by eutrophication, a phenomenon favored by building a hydropower plant. Eutrophication has proved to be one of the most widespread and serious anthropogenic disturbances to aquatic ecosystems. The major cause for eutrophication is the increased loading of nutrients, especially phosphorous. Increasing wastewaters, introduction of phosphorous containing detergents, use of fertilisers, and erosion in the watershed are the major reasons for increased loading of nutrients. The effects of the eutrophication phenomenon are negatively reflected on water quality, for reservoir ecosystem and also for river ecosystem (Fig. 4). Thus, eutrophication may lead in some cases even to the impossibility of using the water for certain uses.

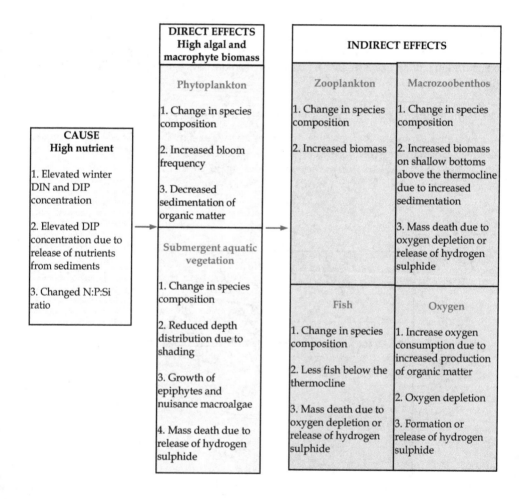

Fig. 4. Causes and effects of eutrophication

Effects of eutrophication emergence affect the lake ecosystem by:

- organoleptic changes of water (color, taste, odor and turbidity) by increasing the biomass of planktonic algae. Water may have a green color due to high content of green algae or diatoms, a red color in blue-green algae species presence, or even brown. This effect gives an unaesthetic aspect of water and leads to additional costs when water is used as a source of drinking water;
- premature clogging of filters and grids of treatment plants which are supply directly from the lake, due to increased phytoplankton biomass;
- biological clogging of the lake and therefore a reduction of its volume due to growth of organic matter content and organic detritus at the lake bottom;
- inability to release and to transform the organic matter due to its excessive quantity;
- pronounced decrease in the dissolved oxygen content, especially at the bottom of the lake, due to increased organic matter decomposition reactions;
- pronounced increase in the concentration of carbon dioxide, iron, manganese, ammonia and hydrogen sulfide due to occurrence of anaerobic decomposition conditions when dissolved oxygen is depleted;
- corrosion of water storage facilities due to the occurrence of precipitation reactions of iron and manganese. The same effect occurs in water in the presence of some *Cyanophyceae* species (*Oscillatoria*), which corrodes steel tanks in the presence of light;
- the appearance of toxic substances in water disposed by some *Cyanophyceae* species (*Microcystis* and *Anabaena flos-aquae aeroginosa*), causing human gastrointestinal disease;
- replacement of special fish species by common species due to changes in water quality.

From all the effects of water eutrophication, the most important consequence is the decrease of oxygen availability. During the day, plants produce oxygen, through photosynthesis, using sunlight. In the night, all organisms consume the oxygen dissolved in water by endogenous breathing.

When excessive amounts of biomass exist in the water body, decomposing organic matter will lead to higher oxygen consumption. Thus the oxygen in the water will be depleted, leading on the one hand to the impossibility of aquatic organisms breathing and on the other hand to the occurrence of anaerobic decomposition. Therefore all the biotic components of the aquatic ecosystem will suffer.

The heterotrophic organisms will be the first affected (fish and shellfish) because of their increased sensitivity to changes taking place in the chemical composition of water, like excessive alkalinity that occurs during intense photosynthesis processes and the lack of dissolved oxygen.

Eutrophication leads to changes in the populations of organisms that live in water. This is done through changes in ecological factors, which are becoming limiting factors for development of the aquatic organisms. Mostly are considered abiotic ecological factors, such as light, temperature, water movement or quantity of certain nutrients in water.

Since eutrophication involves a high input of allochthonous nutrients, often the light is the limiting factor for algae flourishing. Thus, due to changes in optical properties of water through cover of the surface water by vegetation, a high mortality of zoobenthos, nekton and zooplankton appear.

The nutrients demand vary widely from one species to another, both in terms of type and nutrient intake, so that a deterioration of the relationship between nutrients (nitrogen, phosphorus, silicon and iron) determine changes in qualitative and quantitative composition of the phytoplankton. From all the nutrients in the aquatic ecosystems, phosphorus is given the most attention, because is essential for all phytoplankton species. From the 5000 phytoplankton species with high abundance and wide geographical distribution, only 300 produce algae flourishing. Among the species that produce large biomass are many *Cyanobacteria*, which have the capacity to produce toxic substances in the water with effects over health. Changes in the phytoplankton composition in an aquatic ecosystem will cause major changes in the entire trophic chain. Thus, the composition of primary consumers (zooplankton and fish) will change (Cooke et al., 2005).

2.4 Accidental pollution of hydroelectrical reservoirs

The death of fish is mainly caused by the level of dissolved oxygen. There are also some situations of water pollution with toxic substances, but they will not be detailed in this paper. As an example, in river Târnava Mare (downstream Odorheiul Secuiesc, Romania) an historical pollution incident happened in 2002 (table 1) [Serban, 2005]. It caused fish morbidity, because of high temperature (over 20°C), water flow lower than annually average, low water velocity (0.3 ÷ 0.6 m/s) and overloading with some organic substances from an upstream wastewater treatment. All these made the level of DO lower than 4mg/l.

River	F [km²]	Q_{ma} [m³/s]	v [m/s]	Q [m³/s]	OD [mg/l]	T_{apa} [°C]	CCO-n [mg/l]	NH₄ [mg/l]
Crasna	1702	5.56	0.31	1.40	4.2	25	16.6	8.82
Someş	9753	82.60	0.60	33.9	4.5	28.4	15.8	4.2
Zalău	54	0.80	0.55	0.32	3.8	25	18.2	5.2
Someşul Mic	2954	17.50	0.38	6.88	1.03	28	32.4	6.2
Bistriţa	650	8.38	0.40	4.70	4.2	21	4.49	2.6
Târnava Mare (Od. Scuiesc)	646	5.70	0.35	1.31	4	23.9	32.0	4
Târnava Mare (Copşa Mică)	2960	12.10	0.33	3.12	2.6	22.4	13.9	0.8

Table 1. Fish morbidity on water courses in Romania in 2002 after water pollution incident (Q_{ma} - average multi annual water flow, Q - water flow at the incident time)

If the water pollution is limited, the reconstruction of original water quality is possible only by eliminating the accidental pollution sources.

2.5 Management of water quality in hydro electrical reservoirs

Because the lake stratification has a negative effect on water quality, the depth of reservoirs is a great disadvantage from water quality point of view. Therefore very deep lakes are not desired. For deeper reservoirs, one measure to combat the stratification is to locate water intakes at different depths.

The main effect of the stratification in the hypolimnion and sediments is the increased consumption of oxygen and appearance of anoxic conditions which impoverish the deep water fauna. This condition may also lead to a series of chemical and microbial processes like nitrate ammonification, denitrification, desulphurication and methane formation. The release of phosphorous from the sediments is extremely important as it accelerates eutrophication.

The following actions are required to maintain water quality in the reservoirs used by the hydroelectric development:

- the watershed management by river bed erosion control works, which will reduce the intake of silt in the lake;
- discharge of the effluent downstream the reservoir section;
- reducing water pollution;
- setting up sanitary protection perimeters around the lake and adjacent control of tourist areas;
- prevention of lake stratification and insurance of water vertical circulation through water intake at various depths and periodic use of bottom discharge system at a flow able to ensure hygiene riverbed downstream and hipolimnion renewal.
- discharges in reservoir mass is advantageous to be submerged, perpendicular to the surface of the lake for a maximum effect of aeration and movement of water layers.
- flows discharged from tailrace must have a minimum impact for downstream environment. Discontinuous discharge destroys the river bed, erodes the banks and can even break the roads and bridges. Such flows also have a stressful effect on fish.

In this way the water quality can be maintained in reservoirs without negative effects on water quality.

3. Aeration methods inside hydraulic turbines

As presented before, the water quality downstream hydropower plant depends mainly on the quality of water from the upstream reservoir. After water passes through the turbines, a supplementary degasification of water takes place, because of the low pressure in turbine draft tube, which lowers the DO level. This process happens mostly in Francis turbines at partial load operating regimes.

This is the main reason for developing and installing new aeration methods to increase DO level of turbined water. From hydraulic point of view, an air quantity injected downstream turbine runner, could affect turbine efficiency. For this reason, it is recommended that air inflow to be maximum 1÷3% of turbine water flow.

3.1 Existing solutions for aeration of hydraulic turbines

Measurement data are available for different technical solutions for water aeration:

- auto ventilation turbines, developed by Voith Hydro and Tennessee Valley Authority. The aeration can be made central, distributed or peripheral (through the outlet edge of blade) (Figure 5). Test were made for each aeration system individual and combined with the others. The air injection is made through new or existing passages (vacuum braking system and snorkel tubes), using air compressors or natural air admission (proffered for a lower cost).

- a system with air injection in turbine and another one with oxygen injection through porous hoses in penstock were installed at Tims Ford Dam (Harshbarger et al., 1999 and 1995).

The autoventing turbines (central, peripheral or distributed) were implemented for the first time at Norris Dam. These aeration systems can be used individually or combined. The justification for any solution depends on many parameters, characterizing each hydroelectric site. For autoventing turbines all combination were tested. When a group operates with all aeration systems the DO increased up to 5.5 mg/l. In this case the air absorbed in turbine is twice than the air absorbed by the original runners.

Depending on the operational conditions and the aeration system, the energetic efficiency decreased with 0 ÷ 4 %.

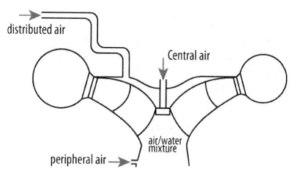

Fig. 5. Aeration methods for autoventing turbines

In other researches (March et al., 1992) the DO level in downstream water was up to 6 mg/l, trying to affect as little as possible the energetic efficiency. A few solutions were tested, like injecting air through runner, the design of a new deflector, low pressure edge blades, coaxial diffuser, injection in the conical part of the draft tube, or combination of them.

Aeration performance can be evaluated by measuring the DO level upstream and downstream turbine,

$$E = OD_u - OD_d \qquad (1)$$

where DO_u and DO_d are the OD concentration upstream and downstream turbine.

The effect of aeration over the hydraulic efficiency of the turbine is

$$\Delta\eta = \eta_a - \eta_0 \qquad (2)$$

where η_a is turbine efficiency with aeration system and η_0 is turbine efficiency without aeration system.

Two aeration systems were tested at Tims Ford Dam (Harshharger et al., 1995), in order to achieve a DO level of 6 mg/l, by injecting air in turbine and oxygen in penstock, through porous rubber pipes. For an upstream DO level of maxim 1 mg/l, if both aeration systems were operational, the DO level got to 5.2 mg/l and if only air system was on the DO level was 4.2 mg/l.

The air was injected with high pressure compressors under runner cover or in the draft tube. Also, porous line diffusers were installed in penstock, for oxygen injection, in case the desired DO level (6 mg/l) is not reached only with the air injection system. The cost of this oxygen injection system in 1995 was of 300'000 $.

Both systems were used during low DO level periods. In order to evaluate the DO level increase and turbine efficiency, the air, oxygen and water discharge were modified during tests. In any of the cases, the turbine efficiency decreased with maxim 1%, so the aeration didn't affect it too much. But this technology was rejected because of the rate between initial installing cost, long time operation and service cost.

Another research study, developed during two years, was made at la Bagnell Dam (Sullivan et al., 2006), over two turbines (with runner aeration orifices), an old one and a new one, with some changes. The tests were made for 51 combinations of water discharge, downstream water depth and aeration orifices diameters and were determined the water discharge through orifices, DO level and water temperature in sections upstream and downstream turbine. As a general conclusion, the oxygen transfer efficiency increases with the increase of air discharge and of downstream water level.

For the older turbine, it was noticed that for smaller openings the runner orifices are more efficient than draft tube orifices, and that when both aeration systems were operational and the water discharge was small, the DO level was over 5 mg/l.

Some researchers made studies concerning DO, temperature and fish growth downstream the hydro plants (Boring, 2005). For the Saluda River (U.S.) a model based on the historical data from 1990-2005 had been developed. In accordance with Environmental Protection Agency (EPA), the criteria established for 2006 are: for survival of trout – min. 3 mg/l, for growth protection – 6.5 mg/l for an average of 30 days, and for sensitive cold-water invertebrates – min. 4 mg/l.

The studies and researches continued with mathematical modelling of the flow (Rohland et al., 2010) for the three classical aeration methods in a Francis turbine. Each classical injection method has different characteristics and influence over the dimension and distribution of bubbles that flow through the draft tube and over the operating efficiencies.

The parameters that influence the efficiency are: the shape and length of draft tube, bubble retention time, quantity of air or the void fraction, air admission intake, bubble size and distribution. From quantity of air or the void fraction point of view, central aeration is the most efficient. The calculations for turbine efficiency and the aeration methods were used for optimization of aeration solution at Bridgewater plant, one of the first power plants designed in respect to aeration.

In Romania, environmental impact and water quality are main concerns, but the aspects of DO level are not taken in consideration. Even if the hydropower operators are preoccupied about environmental issues, there is no legal support for DO level. Turbine aeration (especially central zone) is made only in a few sites, but with hydraulic purposes (to reduce central vortex at partial load) in order to reduce pressure fluctuations.

3.2 Aeration efficiency and main parameters

As mentioned before, two main parameters must be considered in the aeration process: the DO transfer through air injection and the total energetic consumption to realize it.

The DO transfer necessary for water quality improvement, depends on many physical parameters: the quantity of injected air, the gas – liquid contact surface, time of contact, temperature, pressure gradient flow, DO level gradient, turbulence level of flow.

The energetic consumption necessary to introduce the air in liquid depends on the following operational parameters: the quantity of injected air, injection method (natural or induced), air influence over turbine efficiency (flow changes).

In order to obtain a good global balance the DO transfer must be done with a minim energetic consumption. For this purpose is necessary to generate an optimal dimension for the gas bubble – as small as possible (the positive effect is double, because of increasing contact surface and retention time), but with low hydraulic losses for decreasing energetic consumption.

The main parameters that must be considered to find the best compromise between the improvement of water quality and modification of turbine efficiency - see table no 2.

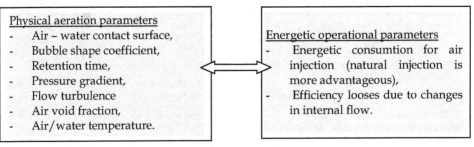

Table 2. Parameters considered for aeration efficiency balance

The air injected in turbine affects its efficiency in two ways: because of the flow perturbation caused by air introduced in water flow and because of energetic consumption necessary to introduce the air. Considering the above, the injected air flow is limited to 1÷3% of water inflow. Usually, this air quantity is not enough to have a good aeration and to obtain the minim reference DO level (5÷6 mg/l) in downstream waters. The natural absorption of atmospheric air is preferred, because it uses the existent turbine depression.

The hydrodynamics and the mass transfer of bubble columns are the subject of many research studies. The matter is analyzed from different aspects: the velocity induced by the bubbles to the liquid (Krishna et al., 1999), (Wiemann and Mewes, 2005), (Ekambara and Dhotre, 2010), bubbles generation regimes (uniform, transient and heterogeneous) (León-Becerril et al, 2002), (Pincovschi et al, 2007), size distribution of gas bubble gas (Katerina et al, 2004), (Polli et al, 2002), Laser Doppler Anemometry and Particle Image Velocimetry measurements (Mayur et al, 2010) (Laakkonen et al, 2005), (Becke et al., 1999), (Ciocan et al., 2011), void fraction and volumetric transfer coefficient (Krishna and van Baten, 2003), numerical simulation of mass transfer (Painmanakul et al. 2009), (Connie et al, 2003).

Mathematical models for turbulent two phase flow in complex configurations is one of the most difficult part in gas – liquid flow simulations, not solved until now. A more detailed representation of bubble movement and of their interaction with the liquid phase can get to development of turbulence models. In spite of consistent efforts for a correct description of closing rules for drag forces, lift forces and mass forces, the precisely model of interfacial forces remains an open matter in this kind of numerical simulations.

This paragraph is focused on correlation of aeration quality (oxygen transfer for a constant injected air flow), with the energetic efficiency of the transfer (energy consumption for introduction of the air flow).

Further on is presented the air bubble dimension influence on DO transfer, in relation to aerator required pressure, for different air discharges. Five metallic perforated plates were used (Bunea et al., 2010) with different orifice sizes (d) and identical geometries (Figure 6).

The total perforated surface area is equal in all configurations (12 mm²). The tests were made in same hydrodynamic conditions, in a tank with 79.2 l of water. For each plate were determined the DO level (C), water temperature (t) and pressure losses (Δp) on aeration system. The experimental results concerning DO level in time, are determined according ASCE regulations 2-91/1993.

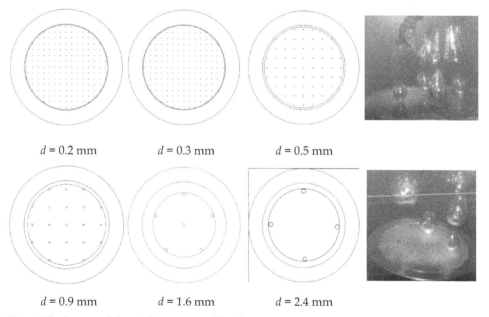

d = 0.2 mm d = 0.3 mm d = 0.5 mm

d = 0.9 mm d = 1.6 mm d = 2.4 mm

Fig. 6. The layout of the orifices on metallic plates

By keeping constant the following parameters: air flow rate Q_{air}, water volume and total area of perforations, the increasing of air-water interfacial area for the first layer of bubbles (A), standard oxygen transfer rate ($SOTR$) and standard aeration efficiency (SAE) with the decreasing of orifice diameter can be observed, while the pressure loss on perforated metallic plate (Δp) increases with maximum 27 mmH₂O (Table 3).

d (mm)	A (mm²)	Δp (mmH₂O)	SOTR (mg/min)	SAE (kg/kWh)
1.6	326	16	33.37	2.50
0.9	781	28	33.35	2.48
0.5	1524	30	49.74	3.67
0.3	3150	34	67.45	4.95
0.2	5233	43	77.08	5.59

Table 3. Evolution of mass transfer parameters with the orifice diameter for constant air discharge Q_{air}=360 l/h

Figure 7 shows the influence of orifice diameters on standard aeration efficiency provided by the metallic plates for different air flow rates (Bunea et al, 2010). Also, by increasing orifices diameter is necessary to increase the air flow rate, otherwise the plates generate bubbles from half of their surface. A bigger air flow rate leads to increased pressure losses on the aeration device and diminishes the standard aeration efficiency. Finally is obtained the efficiency of the oxygen transfer rate related to power needed to inject the air.

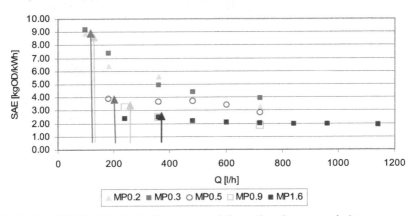

Fig. 7. Variation of SAE with the air flow rate and the orifice diameter of plates

A basic study of the air column in a water tank shows the importance of the quality of air injection, meaning: bubble size, pressure loss on the aeration device etc. for the oxygenation purpose. Different types of bubble aeration systems have been tested and compared. Starting with the experimental results, the influence of air bubbles dimension upon mass transfer is pointed out and the results are correlated with the pressure needed for the aeration device operation (pressure loss) at different air flow rates (Q = 180÷1160 l/h). Thus, efficiency SAE increases while air flow rate and the size of orifices in plates decreases.

4. Water chemical parameters in reservoirs of low head hydropower plants

It is important to study in time the water chemical parameters in large reservoirs. Big hydroelectric projects change the environment, creating retention lakes, the interaction between water and equipment materials causing different corrosion stages. In the same time, the equipments and their operation can affect the water quality (oil leaks, water degasification the through the turbines etc.).

The Lower Olt cascade from Romania (fig. 8) is an important and unique hydroelectric site, made of five identical plants (HPP). Each HPP has four bulb turbine-pump units, with a total installed discharge of 500 m³/s and 13.5 m net head. The corresponding reservoirs have a volume to normal retention level between 62 ·10⁶ m³ and 99 ·10⁶ m³.

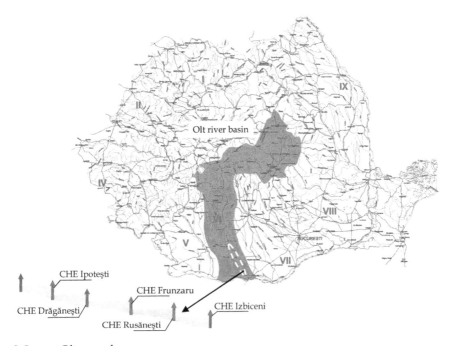

Fig. 8. Lower Olt cascade

A special particularity is that all five HPP operate both turbine and pump regime, which means important volumes of water are transported downstream and upstream in the same site. This could affect in a negative way the water quality by preserving for a longer time the accidental pollution effects.

Because water quality and environment protection are main concerns, the chemical parameters of water in Olt River, is periodically analyzed. For example, a few parameters recorded after ten years of operation and after twenty years, shows that the long time interaction between water and the equipment of HPPs, did not affect the quality of water.

Generally, all parameters remained at the same values, or even decreased, excepting sulphuretted hydrogen, but still remains in admitted values. During the analyzed period, there were no pollution accidents, neither in the hydropower plants in the Olt River and nor in its effluents.

It is also important to determine the effect of water on the equipment (corrosion). For this purpose, determinations in ten different measurement points along Lower Olt cascade were made for the level of pH indicator, chlorine content and manganese content. The results showed for pH indicator that five values are between 5.5÷6.5, and five of them are between 6.5÷7.0, all being between 6.5 ÷ 8.5, the reference limits for natural waters.

Parameter	Unit	Measured values	
		After 10 years	After 20 years
pH	u. pH	8.19-8.23	7.15
Dissolved oxygen	mg/l	4.8-5.9	5.1
CBO$_5$	mg/l	1.8-2.2	1.1
CCO-Mn	mg/l	5.7-6.2	2.203
Sulphuretted hydrogen	mg/l	<0.001	0.409
Phenols	µg/l	10	1.68
Phosphate	mg/l	<0.01	0.06
Chlorides	mg/l	85.3-92.1	73.4
Total hardness	°G	7.91-8.38	10.53

Table 4. Water chemical parameters in Lower Olt cascade reservoirs

The chlorine content is between 20 mg Cl/l and 106.5 mg Cl/l and the total manganese contents are up to 0.004 mg Mn/l, but mostly is zero. The total manganese content being so low the possibility of strong oxidizer appearance on the steel surface can be excluded (Bucur et al., 2010).

These results show that after twenty years of continuous, the water quality in this complex hydroelectric site was preserved from physical – chemical point of view. The dissolved oxygen quantity has a level that allows the preservation of aquatic life.

5. Conclusion

The hydraulic energy is an essential green energy source and important for the integration of the others energy sources in the energy system. However, as an essential live source – water - is used for energy generation, to conserve the green character of power generation, accompaniment measures are to be implemented on the hydropower plants sites.

The quality of water in a hydro electrical site depends on both natural factors – like temperature variations, precipitation intensity and frequency, thermal stratification reservoirs, and operational parameters and components of the hydro electrical site. The behaviour of the water in the lake is to be studied and considered in operation. To not take in consideration this behaviour, the complete eutrophication of the lake can happen, with huge consequences of the ecological system.

Another essential topic is related to the released water. The required parameters to preserve the aquatic life are to be insured; the main parameter is the DO, and 5 mg/l are needed. Aeration devices can be implemented to improve the DO content. The implementation has to consider the compromise between the positive effect of air injection and the inconvenient of energy losses due to air injection (energy needed for injection and energy losses due to flow perturbation by the air injection). Aeration devices can be implemented on new facilities or in the refurbishment process. Actual studies are done to improve the efficiency of the aeration.

The environmental impact of a hydroelectric power plant should be minim, and the water parameters should be as close as possible to natural water course values. The necessity of supervising of the water quality is a reality that should be a main concern for all hydropower users and will permit to prevent the degradation of the ecological system or to implement the needed system to improve the water quality.

6. Acknowledgment

This paper has been elaborated with the support from the Romanian National Council for Scientific Research in Education, IDEAS program, contract no. 705/2009, ID 1701, and by the Sectoral Operational Programme Human Resources Development 2007-2013 of the Romanian Ministry of Labour, Family and Social Protection through the Financial Agreement POSDRU/89/1.5/S/62557.

7. References

ASCE (American Society of Civil Engineers) standard ANSI/ASCE 2-91/1993, *Measurement of Oxygen Transfer in Clean Water - 2nd Edition*, ISBN 087262885X, 45 p.

Becke S., De Bie H. and Sweeney J. (1999). *Dynamic flow behaviour in bubble columns*, Chemical Engineering Science, V. 54, Issue 21, pp. 4929-4935

Boring, S., 2005, *Lower Saluda Site Specific DO Standard*, Joint Meeting with Fish & Wildlife RCG, Presentation, Kleinschmidt. (http://www.saludahydrorelicense.com)

Bucur D.M., Bunea F., Ciocan G.D., Băran G., Isbășoiu E.C. (2010). *Water parameters evolution in a hydroelectric site*, Environmental Engineering and Management Journal, Vol.9, No. 11, 469 – 472

Bunea F., Houde S., Ciocan G.D., Oprina G., Băran G., Pincovschi I. (2010). *Aspects concerning the quality of aeration for environmental friendly turbines*, 25th IAHR Symposium on Hydraulic Machinery and Systems, IOP Conf. Series: Earth and Environmental Science, vol. 12, 012035

Ciocan G.D., Bunea F., Houde S., Deschênes C., (2011). *Measurement technique for air-water disperse system study*, 5TH International Conference on Energy and Environment CIEM2011, Bucharest, Romania

Connie D. DeMoyer, Erica L. Schierholz, John S. Gulliver, Steven C. Wilhelms, (2003). *Impact of bubble and free surface oxygen transfer on diffused aeration systems*, Pergamon, Water Research 37 1890–1904

Cooke D, Welch E., Peterson S., Nichols S. (2005). *Restoration and management of lakes and reservoirs*, Boca Raton Florida, Taylor&Francis Group

Dumitran G.E., Vuță L. (2011). *Study on Lake Izvorul Muntelui rehabilitation*, Simulation Modelling Practice and Theory, vol. 19, pp 1235-1242

Ekambara K., Dhotre M.T. (2010). *CFD simulation of bubble column*, Nuclear Engineering and Design 240 963–969

Harshbarger, E.D., Herrold, B., Robbins, G., Carter, J. (1999). *Turbine Venting for Dissolved Oxygen Improvements at Bull Shoals, Norfork and Table Rock Dams*, Waterpower '99 - Hydro's Future: Technology, Markets, and Policy, CD-ROM

Harshharger, E.D., Mobley, M.H., Brock, W.G. (1995). *Aeration of hydroturbine discharges at Tims Ford Dam, San Francisco"*; ASCE, 9 pp., Waterpower '95 - Proc.of the Conf.on Hydropower, San Francisco, 1, 11-19

Katerina A. Mouza, Nikolaos A. Kazakis & Spiros V. Paras, (2004). *Bubble column Reactor design using a CFD code*, 1st International Conference "From Scientific Computing to Computational Engineering" 1st IC-SCCE, Athens, 8-10 September

Krishna R., Urseanu M. I., van Baten J. M. & Ellenberger J. (1995). *Influence of scale on the hydrodynamics of bubble columns operating in the churn-turbulent regime: experiments vs. Eulerian simulations,* Chemical Engineering Science, V. 54, Issue 21, pp. 4903-4911

Krishna R., van Baten J.M. (2003). *Mass transfer in bubble columns,* Elsevier, Catalysis Today 79–80 67–75, pp. 67-75

Laakkonen M., Honkanen M., Saarenrinne P., Aittamaa J. (2005). *Local bubble size distributions, gas–liquid interfacial areas and gas holdups in a stirred vessel with particle image velocimetry,* Chemical Engineering Journal 109, pp. 37–47.

León-Becerril E., Cockx A. and Liné A. (2002). *Effect of bubble deformation on stability and mixing in bubble columns,* Chemical Engineering Science, V. 57, Issue 16, pp. 3283-3297

March P.A., Brice T.A., Mobley M.H., Cybularz J.M. (1992). *Turbines for solving the DO dilemma,* Hydro Review, 11, 30-36.

Mayur J.S., Iqbal H.T., Tyson E.S., Jyeshtharaj J. (2010). *Advanced PIV/LIF and shadowgraphy system to visualize flow structure in two-phase bubbly flows,* Chemical Engineering Science 65 2431–2442

Painmanakul P., Wachirasak J., Jamnongwong M. & Hebrard G. (2009). *Theoretical Prediction of Volumetric Mass Transfer Coefficient (k_La) for Designing an Aeration Tank,* Engineering Journal V. 13, No 3, ISSN 0125-8281

Pincovschi I., Oprina G., Bunea F. & Băran G. (2007). *Methods for determining flow regimes in bubble columns with fine pore diffusers,* Proceeding of the 5th International Conference Management of Technological Changes, August 2007, v. I, ISBN 978-960-8932-1-2, Alexandropoulis, Greece,

Polli M., Di Stanislao M., Bagatin R., Bakr E. A. & Masi M. (2002). *Bubble size distribution in the sparger region of bubble columns,* Chemical Engineering Science, V. 57, Issue 1, pp. 197-205

Pourriot R., Meybeck M. (1995). *Limnologie générale,* Paris, Collection d'écologie N° 25, Masson

Read, J.S., et al. (2011),, *Derivation of lake mixing and stratification indices from high-resolution lake buoy data,* Environmental Modelling & Software, doi:10.1016/j.envsoft.2011.05.006

Reasearch contract (2006). *Evaluarea micropotenţialului hidroenergetic românesc, sursă regenerabilă de energie, în vederea identificării de amplasamente pentru dezvoltarea investiţiilor în acest sector,* Romanian Ministry of Economy, Commerce and Business.

Rohland K., Foust J., Lewis G. & Sigmon J. (2010). *Aeration Turbines for Duke Energy's New Bridgewater Powerhouse,* Hydro-Review, pp. 58-63, ISSN 0884-0385

Serban A, Analysis of the fish mortatlity phenomena caused by reduced oxygen dilution in walter, Rom Aqua, nr. 11/2005, V. 39p. 17-22

Stevens C., Imberger J. (1996). *The initial response of a stratified lake to a surface shear stress,* Journal of Fluid Mechanics, 312, pp 39-66 doi: 10.1017/S0022112096001917

Sullivan, A., Bennet, K. (2006). *Retrofit Aeration System (RAS) for Francis Turbine,* Final Report, Ameren UE and MEC Water Resources Inc., contract FC36-02ID14408, US

Wiemann D., Mewes D. (2005). *Calculation of flow fields in two and three-phase bubble columns considering mass transfer,* Chemical Engineering Science, V. 60, issue 22, p. 6085-6093

12

Removal Capability of Carbon-Soil-Aquifer Filtering System in Water Microbiological Pollutants

W.B. Wan Nik[1]*, M.M. Rahman[2], M.F. Ahmad[1], J. Ahmad[3] and A. M Yusof[4]

[1]*Dept. of Maritime Technology, Universiti Malaysia Terengganu,*
[2]*Dept. of Pharm. Chemistry, Faculty of Pharmacy, Int. Islamic University Malaysia*
[3]*Dept. of Engineering Science, Universiti Malaysia Terengganu,*
[4]*Department of Chemistry, Faculty of Science, Universiti Teknologi Malaysia*
Malaysia

1. Introduction

1.1 Definition and factors

Water can be defined as a clear colourless, nearly odourless and tasteless liquid, H_2O and is very essential for most plant, animal life and the most widely used of all solvents. It can be obtained in many forms such as rain, lake water, column, river, etc. Household used water and other types that are related to our body is very important from the viewpoint of bacterial and microbiology. Microbiology is the study of organisms that are usually too small to be seen by the unaided eye; it employs techniques such as sterilization and the use of culture media that are required to isolate and grow these microorganisms (Prescott et al., 2005).

In rural area of developed, developing and underdeveloped countries, untreated ground and surface waters are used as the sole source of drinking and cooking water. This is due to the general ignorance about water quality and its treatment and also due to relatively poor economy. Ground and surface water are protected from pollution so that raw water can be directly used for drinking and household purposes. The main sources of pollution are municipal and domestic wastewater, industrial as well as irrigation flows, animals wastes; pesticides, fertilizer and human excreta which contaminate ground as well as surface water in rural areas. It affords an opportunity for certain species of flies to lay their eggs, breed, feed on the exposed materials and carry infectious diseases. It is responsible for the incidence of certain diseases, which include *paratyphoid*, *cholera*, *typhoid*, *dysentery*, infant *dicholera* as well as other similar intestinal infection, parasitic infections and chronic disease.

1.2 Principles of microbiology

Biology is the science of life. It has three major divisions: zoology - the study of animals, botany - the study of plants, and microbiology - the study of microbes. These primary

---

*Corresponding Author

partitions may be divided further into specialities. For instance, *algology* and *mycology*, the study of algae and fungi respectively, are subdivisions of botany. *Protozoology*, the study of unicellular animals, is a division of zoology whilst bacteriology and virology, the study of bacteria and viruses, are subdivisions of microbiology. Knowledge of the behaviour of micro-organisms is heavily dependent upon biochemistry although knowledge about macroscopic animals and plants may be acquired through studies of anatomy and morphology as well as structure and form.

Life does not have simple definition. It may be characterized by a list of properties which are shared by all living organisms, with the exception of the viruses, and which discriminate them from non-living matter:

Movement: It is characteristic of organisms that they, or some part of them, are capable of moving themselves. Even plants, which at first sight appear to be an exception, display movements within their cells.

Responsiveness: All organisms, including plants, react to stimulation. Such responses range from the growth of a plant towards light to the rapid withdrawal of one's hand from a hot object.

Growth: Organisms grow from within by a process which involves the intake of new materials from outside and their subsequent incorporation into the internal structure of the organisms. This is called assimilation and it necessities some kind of feeding process.

Feeding: Organisms constantly take in and assimilate materials for growth and maintenance. Animals generally feed on ready-made organic matter (*heterotrophic* nutrition) whereas plants feed on simple inorganic materials which they build-up into complex organic molecules (*autotrophic* nutrition).

Reproduction: All organisms are able to reproduce themselves. Reproduction involves the replication of the organism's genetic "blueprint" which is encoded in a nucleic acid. Generally this is *deoxyribonucleic acid* (DNA) but in some viruses it may be *ribonucleic acid* (RNA).

Release of energy: To sustain life an organism must be able to release energy in a controlled and usable form. This is achieved by breaking down *adenosine triphosphate* (ATP). The energy to generate ATP is obtained by the breakdown of food by respiration. The occurrence of ATP in living cells appears to be universal.

Excretion: The chemical reactions that take place in organisms result in the formation of toxic waste products which must be either eliminated or stored in a harmless form.

1.3 Global impact of waterborne disease

Throughout the world, many people do not have access to safe drinking water. As a consequence, there is significant morbidity and mortality due to disease-causing organisms in water. It is estimated that nearly one-fourth of all hospital beds in the world are occupied by patients with complications arising from infection by waterborne organisms (Gerba, 1996). Citing the *WHO/UNICEF Global Water Supply and Sanitation Assessment 2000 Report*, it is estimated that nearly 6000 people, mostly children, die every day because of water related diseases. Even in the United States, an estimated US$20 billion per year in lost productivity has been attributed to diseases caused by waterborne pathogens (Gerba, 1996).

1.4 Microbiological aspects

Drinking water should not contain any microorganisms known as pathogens. It should be free from bacteria indicative of excremental pollution. The primary bacterial indicator recommended for this purpose is the coliform group of organisms as a whole. Although as a group they are not exclusively of faecal origin, they are universally present in large numbers in the faeces of man and other warm-blooded animals, and thus can be detected even after considerable dilution. The detection of faecal (*thermotolerant*) coliform organisms, in particular *Escherichia coli* or *E.coli* provides definite evidence of faecal pollution.

The tiny and invisible microorganisms pose a serious threat to the safety of the world's drinking water. Water when contaminated by microorganisms, particularly by the pathogenic ones, can become a growing peril with the potential to cause significant outbreaks of various types of infectious disease. The list of potentially pathogenic microorganisms transmitted by water is increasing significantly each year. Indeed the distribution of safe drinking water to the home can no longer be taken for granted, not even in the United States and Western Europe. The average person consumes about 2-4 liters of water per day through food and drinks. All these deliberations suggest an urgent need for supply of safe drinking purposes. Pilot and mini scale carbon-soil-sand filtering system has successfully produced potable drinking water in the developed countries for a century. Cost of fabrications, usefulness of local raw materials, ease of operation and maintenance, and low energy requirements make slow carbon-soil-sand filtration system a water treatment technology that is particularly well suited for developing countries.

In the midpoint of international drinking water supply and sanitation decade (1985) only 42% of the world's rural populations and 77% of the world's urban population had access to safe drinking water excluding the Peoples of Republic of China (Rotival, 1987). Extensive research has been performed on the efficiency of carbon-soil-sand filters for treatment of water in normal climates. When source waters are relatively low in temperature and turbidity, this process has been found to be effective in the removal of toxic elements, organics and bacteria (Bellamy et al., 2006, Slezak and Sims , 1984). Tropical rivers and other surface waters, which serve as receiving waters for waste discharges and runoff, also serve as water supplies for downstream communities. Low flow rates of surface waters during dry seasons will result in high levels of toxic and microbial contamination from upstream human waste, agricultural runoff and industrial processing facilities. As concentrations of faceal coliforms in untreated water supplies vary widely in developing countries (Vsscher et al., 1986), polluted source waters were simulated by maintaining a filter influent concentration of approximately 10^6 $E.$ $coli$ cells per 100 mL.

Adsorption with carbon-soil-sand filter has been one of the most useful techniques in water treatment. In the past, activated carbon was predominantly used to remove odor and colour producing molecules in water (Suffer and McGuire, 1980). We have previously reported the removal of toxic elements (heavy metals) using the same filtering system (Yusof et al., 2002 and Rahman et al., 2011).

In these papers, the competitive separation and adsorption microbiological pollutants (Coliform, total count etc.) are reported. One of the problems related to groundwater is the reddish colour caused by the presence of iron and manganese. This colour can be seen after it has been exposed to the air, the oxidation of groundwater will promote the precipitation of iron(III) and manganese ions. A significant removal of iron and manganese was reported by Ahmad et al. (2005). While a fixed bed column or continuously flow study to remove heavy metal specifically cadmium and lead by using granular activated carbon has been successfully removed up to 99 percent (Ahmad et al., 2007).

A culture medium is a solid or liquid preparation used to grow, transport and store micro-organisms also based on Prescott, et. al (2005). The medium must contain all the nutrients the microorganism requires for growth. The isolations and identifications of microorganisms need special media which is essential for testing the sensitivities of water. Sources of energy, *carbon, nitrogen, phosphorus* and various minerals are required for the growth of microorganisms and the precise composition of a satisfactory medium will depend on the species one is trying to cultivate because nutritional requirements vary so greatly.

Indicator organisms are bacteria that are used as a sign of quality or hygienic status in water. The definition of indicator is the concept of the indicator organisms which is so strictly associated with particular conditions that its presence is indicative of the existence of these conditions. The minimum requirement for an indicator is that it must be a biotype that is prevalent in sewage and excreted by humans or warm-blooded animals. Historically, these conditions have been related to insanitation and public health concerns. Over the years, however, the use of indicator organisms has been extended to provide evaluation of the quality, in addition to the safety, of particular commodities. In addition, the indicator should be present in greater abundance than pathogenic bacteria, incapable of proliferation or at least not more capable than enteric bacteria, more resistant to various disinfectants than the pathogenic bacteria, and qualified by simple and rapid laboratory procedures. To ensure of value in evaluating the risk of disease as well as water quality, the indicator should satisfy the following criteria:

i.   The indicator should always be present when the source of the pathogenic microorganisms of concern is present and absent in clean uncontaminated water.
ii. The indicator must present in numbers much greater than the pathogen or pathogens it is intended to indicate.
iii. The indicator should respond to natural environmental conditions and water as well as wastewater treatment processes in a manner similar to the pathogens of interest.
iv. The indicator should be easy to isolate, identify and enumerate.

1.5 Bacteria

The bacteria (blue green bacteria) formerly known as blue-green algae, constitute in the kingdom *Procaryotae*. They are single-celled organisms which use soluble food. From all organisms, this group is the most significant to the public health engineer, since biological wastewater treatment processes rely almost exclusively on the activity of bacteria. Bacteria are relatively endurant in many habitats on earth, growing in soil, acidic hot springs, water, deep in the earth's crust, in organic matter and the live bodies of plants and animals. They constitute the highest population of microorganism in wastewater.

Bacteria can be classified into two major groups: heterotrophic and autotropic depending on the source of nutrients. Heterotrophs utilize organic matter as energy as well as a carbon source for their synthesis. Whereas autotropic bacteria use oxidizing inorganic compounds for energy and carbon dioxide as a carbon source (Hammer, 2008).

Bacteria that cause bacterial infection are called pathogenic bacteria. Pathogenic bacteria are a major cause of human death and disease and cause infections such as cholera, tetanus, typhoid fever, diphtheria, syphilis, foodborne illness, leprosy and tuberculosis. A pathogenic cause for a known medical disease may only be discovered many years after, as was the case with *Helicobacter pylori* and peptic ulcer disease. Bacterial diseases are also important in farm animals such as mastitis, salmonella and anthrax as well as in agriculture, with bacteria causing leaf spot, fire blight and wilts in plants. Some of commonly found bacteria are shown in Figure 1. They are *Bacillus, Bordetella, Clostridium, Escherichia, Spirilina, Staphylococcus, Streptococcus* and *Salmonella.*

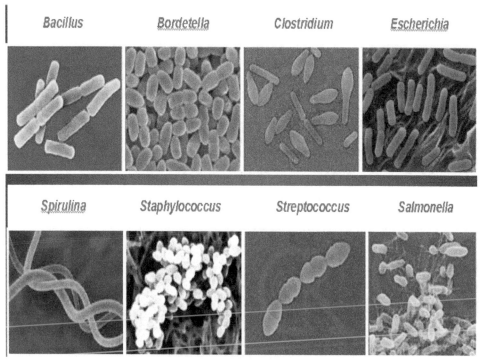

Fig. 1. Commonly found bacteria (Source http://agrobacter wikispaces com/file/list)

It is well known that soil washes into natural bodies of water, particularly after heavy rains. Many different kinds of bacteria also will be carried into water, including soil bacteria and bacteria from the faeces of animals, such as tiny invertebrates, insects, birds, and other "lower animals." In addition, water will often have its own bacterial flora, contributed by its resident species of animals. Bacterial flora from literally hundreds of different species of animals enters natural waters each week. The normal flora of man has been thoroughly studied. The stool flora of man alone may comprise over a hundred different species of bacteria. Although less well studied, other animals have their own characteristic bacterial flora. Many of the bacteria species which form the normal flora of man probably evolved from bacterial species which form the normal flora of lower animals. In the cases that have been carefully studied, it is easy to tell the difference between closely related species with a sensitive technique, such as DNA-*hybridization*. Often, however, it is not easy to tell the difference between closely related species with the simple tests samples taken from insects, birds, and mammals.

1.6 Coliform

The coliform group was the mainstay of the sanitarian's tools for detecting the presence of faecal contamination in aquatic environments. The broad general characteristics which define this group have allowed it to be one of the most useful of bacterial indicators and at the same time have been responsible for its displacement as an indicator of faecal contamination.

The coliform group is made up of bacteria with defined biochemical and growth characteristics that are used to identify bacteria that are more or less related to faecal contaminants. The total coliforms represent the whole group, and are bacteria that multiply at 37°C. The thermotolerant coliforms are bacteria that can grow at a higher temperature (44.2°C) and *E.coli* is a thermotolerant species that is specifically of faecal origin. A finding of any coliform bacteria, whether thermotolerant or not, in water leaving the treatment works requires immediate investigation and corrective action. There is no difference in the significance of total coliforms, thermotolerant coliforms and *E. coli* in water leaving a treatment works, as they all indicate inadequate treatment, and action should not be delayed pending the determination of which type of coliform has been detected. Upon detection in a distribution system, investigations must be initiated immediately to discover the source of the contamination.

Tests for detection and enumeration of indicator organisms, rather than of pathogens, are used. The cultural reactions and characteristic of this group of bacteria have been studied extensively. Coliform group density is a criterion of a degree of pollution. Membrane filter technique, which involves direct plating for detection and estimation of coliform densities, is as effective as the multiple-tube fermentation test for detecting bacteria of the coliform group. Modification of procedural details, particularly of the culture medium, has made the results comparable with dose given by the multiple-tube fermentation procedure. Although there are limitations in the application of the membrane filter technique, it is equivalent when used with strict adherence to these limitations and to the specified technical details.

Thus, two standard methods are presented for the detection of bacteria of the coliform group. It is customary to report results of the coliform test by the multiple-tube fermentation procedure as a Most Probable Number (MPN) index. This is an index of the number of coliform bacteria that, more probably than any other number, would give the results shown by the laboratory examination; it is not an actual enumeration. By contrast, direct plating methods such as membrane filter procedure permit a direct count of coliform colonies. In both procedures coliform density is reported conventionally as the (most probable number) MPN or membrane filter count per 100 mL. Use of either procedure permits appraising the sanitary quality of water and the effectiveness of treatment process. *E. coli* infection often causes severe bloody diarrhoea and abdominal cramps; sometimes the infection causes none bloody diarrhoea or no symptoms. Usually little or no fever is present, and the illness resolves in 5 to 10 days. To some people, particularly children under 5 years of age and the elderly, the infection can also cause a complication called hemolytic uremic syndrome, in which the red blood cells are destroyed and the kidneys fail. About 2%-7% of infections lead to this complication. In the United States, hemolytic uremic syndrome is the principal cause of acute kidney failure in children, and most cases of hemolytic uremic syndrome are caused by *E. coli*.

1.7 Fungi

Fungi are multi-cellular, non-photosynthetic, heterotrophic organisms. Fungi are obligate aerobes that are reproduced by a variety of methods including fission, budding, and spore formation. Their cells require only half as much nitrogen as bacteria so that in a nitrogen-deficient wastewater, they predominate over the bacteria. Fungi are plants that are unable to do photosynthesis such as yeast and moulds. Yeast is normally used for fermentation in making bread, cake and alcohol. Moulds are filament in shape that lives in acidic condition. They reduce the efficiency of secondary sedimentation tank and cause unpleasant smell and taste.

Few members of the group are readily visible when they are present. It is their fruiting body which is seen rather than the vegetative body of the organisms. Perhaps the most familiar is the mushrooms, whose Greek name *mykes* gives rise to the term applied to the scientific study of fungi and mycology.

The fungi have highly distinctive biological organization. Although some aquatic fungi and the yeasts are unicellular they are readily distinguishable from bacteria by their large cells and membrane-bound nuclei. Some aquatic fungi do show resemblances to flagellate protozoa. Fungi occupy a wide variety of habitats including the sea and fresh waters. However, the majority occupy moist habitats on land and are abundant in soil. At least 100,000 species are known. Although some produce macroscopic fruiting bodies (e.g. mushrooms) the overwhelming majority are microscopic. The fungi are heterotrophic organisms acquiring organic materials for their nutrition. Those which feed on dead organic materials are described as *saprophytic*. Saprophytes bring about the decomposition of plant and animal remains and in doing so release simpler chemical substances into the environment. In soil this is vital significance in maintaining fertility by recycling essential plants nutrients.

Fungi are present in, and have been recovered from, diverse, remote, and extreme aquatic habitats including lakes, ponds, streams, estuaries, marine environments, wastewaters, sludge, rural and urban storm-water runoff, well waters, acid mine drainage, asphalt refineries, jet fuel systems, and aquatic sediments. Fungi are widely distributed and are found wherever moisture is present. Glycogen is the primary storage polysaccharide in fungi. Most fungi use carbohydrates (preferably glucose or maltose) and nitrogenous compounds to synthesise their own amino acids and proteins. Identification of fungi which are larger than bacteria is dependent on colonial morphology on a solid medium, growth as well as reproduction morphology and for yeasts, physiological activity in laboratory cultures.

Increasing numbers of fungi usually indicate the increasing organics loadings in water or soil. Large numbers of similar fungi suggest excessive organic load while a highly diversified mycobiota indicates populations adjusted to the environmental organics. Despite their wide occurrence, little attention has been given to the presence and ecological significance of fungi in aquatic habitats. The relevance of fungi and their activities in water is emphasized by increasing knowledge of their pathogenicity for humans, animals, and plants; their role as food or energy sources; their activity in natural purification processes; and their function in sediment formation. Quantitative enumeration of fungi is not equivalent like the unicellular bacteria because a fungal colony may develop from a single cell (spore), an aggregate of cells (a cluster of spores or a single multi-celled spore), or from a mycelial or pseudo-mycelial fragment (containing more than one viable cell). It is assumed that each fungal colony developing in laboratory culture originates from a single colony-forming unit (CFU) which may or may not be a single cell.

The advantages of fungi are for food and food preparation such as edible fungi and fermentation of bread, wines and beers; for medicine e.g. Penicillin is the best known antibiotic and it is actually made from a mould; and decomposers of organic material. The disadvantages of fungi are poisonous (i.e. wild fungi can be both delicious and deadly poisonous. The high concentration of metals such as arsenic, cadmium, copper and lead in wild fungi) causing fungal diseases to plants (i.e. mildew, smuts, ruts, etc).

1.8 Algae

Algae are plant-like organisms that usually photosynthetic and aquatic but do not have true root, stem, leaf, vascular tissues and have simple reproductive structures. They range from tiny single cells to branched forms of visible length that appear as attached green slime. They may be either unicellular or multi-cellular. There are a wide variety of algal species in various shades of commonly green, brown and red. They are distributed worldwide in sea, in freshwater, waste water, marine water, and non-marine water such as mud and sand. Typical green algae are *Oocystis* and *Pediastrum*. Whereas the blue green algae that associated with polluted water are *Anacystis*, *Anabaena*, and *Aphanizomenon*.

Algae are photoautotrophic that use carbon dioxide or bicarbonate as carbon source and inorganic nutrients of nitrogen as ammonia or nitrate and phosphate. Algae have chlorophyll that can manufacture their own food through the process of photosynthesis. They conduct photosynthesis within membrane bound structure called chloroplast. In the presence of sunlight, the photosynthetic production of oxygen is greater that the amount used in respiration. At night they use up oxygen in respiration. If the daylight hours exceed the night hours by a reasonable amount, there is a net production of oxygen. They produce more oxygen than all the plants in the world.

The good of algae are: form important food source for many animals. e.g: little shrimps and huge whales; most important at the bottom of food chain with many living things depend upon them; and have an economic importance because they are a source of carotene, glycerol, and alginates and can be converted into a food source for aquaculture. While the bad of algae are: too much algae does suffocate the lake, so it will kill many fish; blue green-algae are very toxic and algal toxin can seriously affect human and animal; and toxic blue-green algal blooms cause a rash known as "swimmer's itch", while powerful neuromuscular toxins released by other cyanobacteria (blue green algae) can kill fish or animals that drink the water.

1.9 Filtration

Filtration is most often a polishing step to remove non-settleable flocs remaining after chemical coagulation and sedimentation or precipitant particles of softened water. Under certain conditions, filtration may serve as the primary turbidity removal process especially in direct filtration of raw water.

The normally used filtration process involves passing the water through a stationary bed of granular medium. Solids particles in the water are retained in the interstices of the filter media. Several modes of operation are possible in granular medium filtration. These include upflow, downflow, pressure and vacuum filtration. The most common practice is downward gravity flow filtration, with the weight of the water column above the filter providing the driving force.

The solids removal operation with granular medium filter involves several complicated processes. The most obvious processes include straining, flocculation and sedimentation. The straining process occurs at the interface between filter media and water. Initially, materials larger than the pore openings at the interface are strained. In the filtration process, conditions within pores of a filter bed promote flocculation. Flocs grow in size and become trapped in the interstices. Other processes are also important since most of the solids existence in settled water is too small to be completely removed by straining. Removal of particles and flocs in the filter media depends on transport mechanisms that carry the solids through the water to the surface of the filter grains, and on retention of the solids by the medium once contact has occurred. Transport mechanisms include gravity settling, inertial impaction, diffusion of colloid, Brownian movement and van der Waals forces (Kim and Whittle, 2006). Retention of solids once contact has occurred can be attributed primarily to electrochemical forces, van der Waals force, and physical adsorption (Yaroshevskaya, 2007).

With chemical preconditioning of the water, a well-designed and operated filter should remove virtually all solids down to the submicron size. Removal begin at the top portion of the filter. As pore opening are filled by the filtered material, increased hydraulic shear sweeps particles farther into the bed. The ideal filter media should be coarse enough for large pore openings to retain huge quantities of particles or floc however sufficiently fine to avoid passage of small floc. It should has an optimum filter depth to produce relatively long operation filter run and graded to permit effective cleaning during backwash. Dual-media filter of coarse burnt oil palm shell granules or anthracite overlaying the sand media provide higher porosity at the upper layer as well as higher filter run of more than three times than the conventional sand filter as recorded by Ahmad et al. (2009).

2. Methodology

2.1 Sampling protocols

Sampling site was whole area of Terengganu Darul Iman, one of the east coast states, which was divided by 7 districts and these are Kemaman, Dungun, Marang, Kuala Terengganu, Hulu Terengganu, Setiu and Besut. Two samples were taken randomly from every single district, except Kuala Terengganu, and from there 4 samples were taken. All together 16 samples were taken in the whole Terengganu area.

Chosen places are in Kemaman, Kampong (Kpg.) Baharu and Esso petrol pump. Two samples were taken from there. Kpg. Kemenyer and Kpg. Pasir were chosen in Dungun districts and two samples were taken from there to be analyzed. Kuala Terengganu districts which consists of Tanjung, Pantai Tok Jembal, MengabangTelipot 1 and MengabangTelipot 2 were places of sampling where 4 samples were taken. Hulu Terengganu, there were two samples taken that was in Kpg. Bt. Gemuroh and Kpg. Nibong. Marang district, two samples was taken in Kpg. Lubok Perah and Wakaf Tapai. In Setiu, Kpg. Guntung Luar and Kpg. Tembila were chosen and two samples were taken there. In Besut, Kpg. Bt. Bunga and Kpg. Pasir Aka were chosen and two samples were taken there.

For the field method, refer to Figure 2 and the explanation in sampling procedure for the well.

Fig. 2. Sampling Map

2.2 Filtration system

The setup of the filtration system consisting of the carbon-soil-aquifer filtering system is shown in Figure 3. The experimental set-up consists of two identical columns, though only final setup was diagramed. Filter column height is 18 cm, external diameter is 8.5 cm, internal diameter is 7.3 cm and its constructed material is polyethylenetetrathelate (PET). Granulated activated carbon, modified activated carbon and red soil particles were prepared from locally available raw materials. Sand and silica particles were collected from locally available sources.

2.3 Sampling procedure and preserved for microbiological test

In Figure 3, firstly, the metal bucket was cleaned until it is free from earth and rubbish to avoid contamination. Next, a little methylated spirit was poured into the bucket, light it and allowed the burning alcohol to run over the walls of the bucket so as to sterilise the inner surface. After that, the bucket was lowered into the well and natural water sources, and making sure that the rope does not enter the water or the inside of the bucket, as well as the bucket should not touch the side of the well. When the bucket is full of water, it was carefully raised. Next, the string fixing the protective cover was untied and the stopper removed. Then, he sample bottle was filled with the water from the bucket. Finally, the bottle was capped and the protective sheet fixed in place with string.

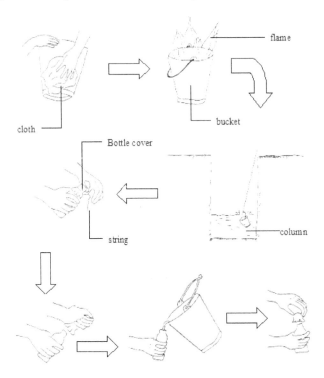

Fig. 3. Sampling method for dug well

The sequence of the components, its thickness and quantity inside the filtration system is shown in Figure 4. The Teflon plastic bottles were cleaned until free from earth and rubbish to avoid contamination. Next, a little methylated spirit was poured into the bottle, light it and allowed the burning alcohol to run over the inside wall of the bottle so as to sterilise the inner surface. When the bottle is full of water, it is carefully transferred, and then bottles are capped and protective sheet fixed in place with a piece string.

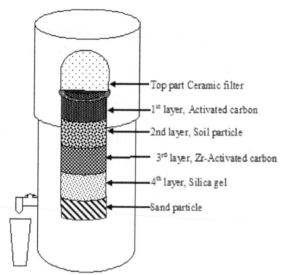

Fig. 4. Constructed filter unit (Yusof et al., 2002 and M. M. Rahman et al., 2011)

2.4 Preparation of media for coliform and total count

Endo broth: About 2.4 g of (7.5 g agar 1.25 g Yeast Extract + 2.5g Peptone) MF Endo Broth mixture was dissolved in 50 mL of distilled water containing 1.0 mL ethanol. Then this was sterilized in an autoclave at 121⁰C, 15 psi for 20 minutes. The broth was cooled to room temperature before incubation. This medium was freshly prepared to ensure accurate results used for coliform count.

Exactly 1.25 g of tryptone was added to 50 mL of deionized water and to it was added 3.75 g agar. The mixture was then gently heated. The pH was adjusted to 8.6 to 9.0 and then 1.25 g of glucose was added and then autoclaved. Yeast and mould count was then done.

Appropriate volume of mixture was filtered through a sterile 47-mm, 0.45µm, grid membrane filter (cellulose nitrate membrane filter), under partial vacuum. The funnel was rinsed with three 20 to 30ml portions of sterile dilution water. An exact amount of 100 ml of water sample was poured through the funnel. If the water was heavily polluted, the water sample will be diluted using a dilution bottle 3 to 5 times dilution with 90 ml sterile dilution water. The filter was placed on the agar in the petri dish. Dishes were placed in close fitting box containing moistened paper towels. Incubation at 44.5± 0.5°C for 24 hours was done if using the EMB Agar medium. Duplicate plates may be incubated for other time and temperature conditions as desired. Membrane filter technique is shown in Figure 5.

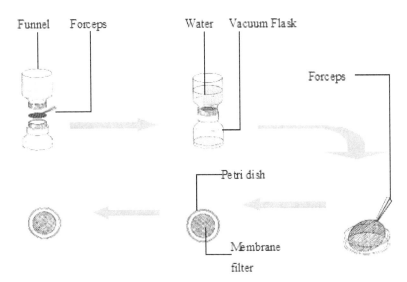

Fig. 5. Membrane filter technique

2.5 *E. coli* (faecal)

Colonies on membrane filters were counted by viable count or using a stereoscopic microscope at 10 to 15 x magnification. Preferably, the petri dish was placed on the microscope stage slanted at 45^0 and light source was adjusted vertically to the colonies. Optimal colony density per filter is 20 to 200. If colonies are small and there is no crowding, a higher limit is acceptable.

All colonies were counted on the membrane when there were 1 to 2 or fewer, colonies per square. For 3 to 10 colonies per square count 10 squares and obtain average count per square. For 10 to 20 colonies per square count 5 square and obtain average count per square. Multiply average count per square by 100 times reciprocal of the dilution to give colonies per millilitre. If there are more than 20 colonies per square, record count as >2000 times the reciprocal of the dilution. Report averaged counts as estimated colony-forming units (CFU). Make estimated counts only when there are discrete, separated colonies without spreaders

2.6 Fungi – Membrane filter method

Sample preparation: Each plate was marked with a sample number, dilution, date, and any other necessary information before examination. Duplicate plates were prepared for each volume of sample or dilution examined. All samples were thoroughly mixed or dilutions done by rapidly making about 25 complete up-and-down movements. Optionally, the use of a mechanical shaker to shake samples or dilution blanks for 15 second would be useful.

Filtration: Appropriate volumes of well shaken samples or dilution were filtered in triplicates, through membrane filter with a pore diameter of 0.45 or 0.80μm. The filters were then transferred and incubated at 15°C for 5 days in a humid atmosphere. Alternatively, incubate-ion can also be performed at 20°C for 3 days, or longer depending on the fungi present. Using a binocular dissecting microscope at a magnification of 10×, all colonies on each selected plate were then counted. If counting must be delayed temporarily, plates are held at 4°C for not longer than 24 hours. Ideal plates should have 20 to 80 colonies per filter.

2.7 Total count of bacteria – Membrane filter method

Appropriate volume of water samples were filtered through a sterile 47-mm, 0.45μm, grid membrane filter, under partial vacuum. The funnel was rinsed with three 20 to 30ml portions of sterile dilution water. The 100 ml of water sample was poured into the funnel. Dilution is required if the water was heavily polluted, and done with a dilution bottle 3 to 5 times dilution with 90 ml sterile dilution water. The filter was placed on the agar in the petri dish. The dishes were later placed in close fitting box containing moistened paper towels and incubated at 35± 0.5°C for 24 hours if the nutrient agar medium was used. Duplicate plates may be incubated for other time and temperature conditions as desired.

Counting the colonies was done on membrane filters by viable count or using a stereoscopic microscope at 10 to 15x magnification. Preferably, the petri dish was placed on a microscope stage slanted at 45o and the light source adjusted vertical to the colonies. Optimal colony density per filter is 20 to 200. If the colonies are small and there is no crowding, a higher limit is acceptable.

Counting was done on all colonies on the membrane when there are 1 to 2 or fewer, colonies per square. For 3 to 10 colonies per square count 10 squares an average count per square was obtained. For 10 to 20 colonies per square, counting was done on 5 squares and an average count per square obtained. Multiplying the average count per square by 100 times reciprocal of the dilution will give colonies per millilitre. If there are more than 20 colonies per square, the count was recorded as >2000 times the reciprocal of the dilution. The average count was reported as estimated colony-forming units. Estimated counts are made only when there are discrete, separated colonies without spreaders.

3. Discussion

Environmentally polluted water samples were taken from various sources such as rivers, columns, tube well, sea and water falls. This naturally polluted water was filtered continuously through the filtering column for 10 days. The results of coliform and TC in raw and treated shown in Table 1 and Table 2. Coliform count is the most important microbiological count for drinking water. Coliforms are quantified to assess water treatment effectiveness and the integrity of the distribution system. They are also used as a screening test for recent feacal contamination. Treatment that provides coliform-free water should also reduce the pathogens to a minimal level. A major shortcoming is that coliforms under certain circumstances may proliferate in the bio-films of water distribution systems, clouding their use as an indicator of external contamination.

From the results in Table 1 it was observed that all raw water samples contain coliform except raw water from locations 10, 13 and 16. The sample taken from location 16 was observed to contain uncountable coliforms but treated water using filter column contained no coliforms. Therefore, it could be confirmed that the filtering column has produced coliform free water and safe for drinking. From Table 2, it was found that the total count includes three types of microorganisms; bacteria, yeasts and moulds which presence in the raw water samples. Filtrate water supplies were selected at random from the stock. After 24 and 72 hours of observation the results showed that the yeasts and moulds could be distinguished from each other. The 24 hour observation could not distinguish between the yeast and mould but with the 72 hours observation yeast and mould were detected.

Sample ID	WHO Std.	Raw Water	Treated Water
1	0/100 ml	02	0
3	0/100 ml	01	0
6	0/100 ml	01	0
7	0/100 ml	20	0
8	0/100 ml	01	0
9	0/100 ml	23	0
10	0/100 ml	0	0
11	0/100 ml	11	0
12	0/100 ml	01	0
13	0/100 ml	0	0
14	0/100 ml	04	0
15	0/100 ml	25	0
16	0/100 ml	Uncountable	0

Table 1. Coliform count in the raw and treated water samples

Sample ID	WHO Std	Raw water/10 ml		Treated water/10 ml	
		24 hours	72 hours	24 hours	72 hours
1	<25/10mL	28	36/14	09	11/1
3	<25/10mL	38	40/38		
6	<25/10mL	02	4/1		
7	<25/10mL	06	20/2		
8	<25/10mL	25	30/18	13	18/12
9	<25/10mL	21	32/6		
10	<25/10mL	15	15/3		
11	<25/10mL	70	78/12	11	12
12	<25/10mL	Uncountable	/6	09	10/9
13	<25/10mL	0	13/3		
14	<25/10mL	18	22/88		
15	<25/10mL	Uncountable	/25	02	12/3
16	<25/10mL	08	8/8		
utm tap	<25/10mL			00	00

Table 2. Total count in raw and treated water samples

4. Conclusion

A newly fabricated filter unit as shown in Figure 4 has been found to be significantly improved for its removal capability of microbiological hazardous materials from the filtered water. It provides a safe drinking water in the point of view of microbiological, toxicity, softening hardness and pH value. All kinds of parameters for drinking water have fulfilled the requirements as provided by this water filter unit.

5. Acknowledgment

The authors wish to express their gratitude to the "**Ministry of Science, Technology and the Environment (MOSTE)**" (Fund no. ERGS55064) and Ministry of Higher Education (Fund no. FRGS59210) for funding this research.

6. References

Ahmad, J.; Cheng, W.H.; Wan Nik, W.B.; Noraaini, A.& Megat Mohd Noor, M.J. (2005). Study on Removal of Iron and Manganese in Groundwater by Granular Activated Carbon. *Desalination*. 182, pp. 347 – 353.

Ahmad, J.; Shiung LS.; Ali N.& Noor M.J.M.M. (2007). Assimulation Study of the Removal Efficiency of Granular Activated Carbon on Cadmium and Lead. *Desalination*. 206, pp. 9 – 16.

Ahmad, J.; Nora'aini A.;Halim A.G. & Wan Nik, W.B. (2009). Effective Depth, Initial Headloss and Backwashing Criteria as the Key Factors of Burnt Oil Palm Shell (BOPS) Granular Filtration. *Journal Desalination and Water Treatment*. 10. pp. 80 – 86.

Bellamy, W.D.; Hendricks, D.W. & Lonsdon, G.S. (2006). Slow Sand Filtration: Influences of Selected Process Variables. *J.W.W.W.A.* (77) pp. 62-66.

Gerba, C. P. (1996). Pathogens in the Environment. In: Pepper, I. L., Gerba C. P., Brusseau, M. L. (Eds.), Pollution Science Academic Press, New York, pp. 279 – 299.

Hammer M.J. (2008). *Water and Wastewater Technology*. Pearson Prentice Hall.London U. K. Sixth Edn., pp. 49 – 61.

Kim, Y.S. & Whittle, A.J. (2006). Filtration in a Porous Granular Medium: Simulation of Pore Scale Particle Deposition and Clogging. *Journal of Transport in Porous Media*. 65, pp. 53 – 87.

Rahman M.M.; YusofA.M. & Wan Nik, W.B. (2011). Increase of Purification Capacity and the Performance of Slow Filtering in the Removal of Bacteria in an Activated Carbon-Soil-Sand Filter Unit. *International Journal of Applied Science and Technology (IJAST)*. 1 (5), pp.127-130.

Prescott, Harley & Klein, "*Microbiology*" 6th Edition, International Edition (2005). McGraw Hill.

Rotival, A.H. (1987). "State of the International Drinking Water Supply and Sanitation Decade: in *Resource Mobilization for Drinking Water and Sanitation in Developing Nations*." American Society of Civil Engineers, NY.

Slezak, L.A. & Sims, R.C. (1984)."The Application and Effectiveness of Slow Sand Filtration in the United States."*J.W.W.W.A.* (76) pp. 38-43.

Suffer, I.H. & McGuire, M.J. (1980)."Activated carbon adsorption of organics from aqueous phase."Ann. Arbor., MI.; *Ann. Arbor. Science*,**1**.

Vsscher, J.T.; Paramasivam, R. & Santacruz, M. (1986). "IRC's Slow Sand Filtration Projects." *Waterlines*, (4) pp. 4-27.

Yaroshevskaya, N.V. (2007). The Method of Computing a Water-treatment Filter with the Account of Media Granulometric Composition. *Journal of Water Chemistry and Technology*. 29 (3), pp. 152 – 162.

Yusof A.M.; Rahman M.M.; Wood A.K.H.; Hamzah S. & Shamsiah A. (2002). Development and water quality parameters assessment of a filter unit using local raw materials for use in safe drinking water, *Proceedings of SKAM-15/EXPERTS 2002*, Bay view Beach Resort, Pulau Pinang, Malaysia, September 10-12, 2002.

13

An Overview of the Persistent Organic Pollutants in the Freshwater System

M. Mosharraf Hossain[1], K.M. Nazmul Islam[1] and Ismail M.M. Rahman[2]*
[1]*Institute of Forestry and Environmental Sciences, University of Chittagong, Chittagong,*
[2]*Department of Applied and Environmental Chemistry, University of Chittagong,*
Chittagong,
Bangladesh

1. Introduction

Organic contamination can be viewed as the secondary and tertiary dispersions of organic compounds from various sources into the global circulation across different spheres of the environment *viz.* hydrosphere, lithosphere, atmosphere and biosphere - mediated primarily by human activities [1, 2]. Freshwater contamination occurs in diverse ways, of which, contamination by organic compounds of various kinds and traits from a wide range of sources is a major concern due to their persistent nature and harmful biological impacts, especially on human beings [3]. All over the world, supply of freshwater is shrinking in quantity and dwindling in quality [4] - forcing the modern city planners to recycle potable water from raw water coming out of waste treatment facilities [5]. Contaminants are coming into the fresh water streams from different natural and anthropogenic sources of which the anthropogenic contribution is getting larger in volume each day as do their diversity [6]. These contaminants are either attenuated naturally into less toxic or non-toxic forms [7], or they persist in the fresh water ecosystem for long time, enter the food chain through bioaccumulation and bio magnified to cause cascading effect on terrestrial and aquatic bio-diversity altogether [8, 9]. Presence of organic compounds and their harmful derivatives makes water unfit to consumption. Maintaining a supply of pure water for ever increasing population is already a daunting challenge all over the world while the organic contaminants are aggravating the challenges further [10, 11]. Hence, we need a clear understanding of the classes of organic compounds, which find their way into the fresh water system, their sources, and their transformation through physicochemical and biological processes in the fresh water system in order to control their entry and undesired transformation thereof. Moreover, we need to know the chemistry of these contaminants and their effects on the environment in general and on human health in particular while detecting their presence in fresh water in a quick and easy manner. Better understanding of these pollutants is inevitable for developing better techniques of purification of water from persistent organic contaminants. Substantial improvement has so far been achieved in all these aspects. This chapter is an effort to get a contemporary picture of our understanding about the organic contaminants in the freshwater system. After introducing the issue of fresh water pollution in the second section, we shall intensely focus on different aspects of contamination of fresh water by persistent organic compounds in section three. In this section, we shall describe the categories of organic compounds which contribute to fresh water contamination with their sources, fate and effects. This section will also include a review of detection techniques available for the detection of organic contaminants in the freshwater system. The section will be concluded with two sub sections – one highlighting our understanding of the natural assimilation of persistent organic contaminants through normal aquatic ecological processes and the other focusing on the techniques that we have at our discretion for the removal of such contaminants from water in order to purify water. Different conventions and protocols are in place to reduce and control the issue of organic contamination of water. Before concluding the chapter, we shall try to shed some light on the current status of such protocols and conventions regarding organic contamination of fresh water.

2. Pollution and contamination of fresh water

Contamination of fresh water is a natural process; in the passage of the normal cycling of water through the water cycle, during precipitation, different organic substances of varied origins, both soluble and insoluble, get transported into fresh water streams. Among the natural contributors of organic contaminants, volcanic eruption, forest fires, natural decomposition of organisms, are substantial sources [4]. Organic substances of biological origin also result from either excreta or wastes from of living organisms, from municipal waste decomposition, etc. Among the dissolved organic substances in water, humic and fulvic substances, polycyclic aromatic hydrocarbons (PAHs), halogenated aromatic and aliphatic hydrocarbons, phthalate esters are the major classes [12]. However, the input of organic contaminants from natural sources does not increase the way they are increasing from anthropogenic sources. Human activities are increasing the consumption of the already known organic compounds all over the world for meeting diverse needs while organic pharmaceuticals, pesticides, paints and coloring materials, cosmetics, etc with novel properties are synthesized every day for many purposes. Consequently, the quantity and diversity of anthropogenic organic contaminants that enter the fresh water system and polluting it are accumulating. According to a recent study [6] on the presence organic contaminants in untreated drinking water sources in USA, for surface sources, natural sterol – cholesterol, herbicide – metolachlor, nicotine metabolite – cotinine, natural plant sterol – β-sitosterol and caffeine metabolite – 1,7-dimethylxanthine were the five most frequently detected organic contaminants. In ground water sources, solvent – tetrachloroethylene, pharmaceutical – carbamazepine, plasticizer – bisphenol-A, caffeine metabolite – 1,7-dimethylxanthine, and fire retardant – tri (2-chloroethyl) phosphate were the five respective major contaminants. However, they could not establish any specific seasonal trends in the distribution patterns of contaminants.

3. Organic contaminants in freshwater system

In recent decades there has been escalating concern about the potential effects of the occurrence of a diverse array of organic contaminants in the fresh water system on the human and environmental health [12, 13]. Organic contaminants are mostly human-induced chemicals entering into natural fresh water through pesticide use, industrial chemicals, and as by-products of degradation of other chemicals and persist long enough in the environment to cause harmful effects. They tend to accumulate in reservoirs such as water, soil, sediments etc. From these reservoirs, they are remobilized through various processes, switch form or speciation and become available to the biological food chain. In this way, these contaminants tend to bio accumulate and bio magnify exhibiting toxicity and other related outcomes – mutagenicity, carcinogenicity and teratogenicity - resulting into chronic and acute disorders [14].

3.1 Common organic contaminants in freshwater

Organic contaminants are composed basically of hydrocarbons both from anthropogenic and natural sources. Many contaminating hydrocarbons exist in the environment but carcinogens, mutagens and teratogens among them are the most closely monitored [15]. These contaminants enter the fresh water system both from point and non-point sources. Point sources are defined in a spatially explicit manner both in terms of chemical residues from organic contaminants they contribute and by epidemiological factors thereof, like morbidity, mortality or community disruption [16]. Effluents from municipal sewage-treatment plants, industrial sources, storm sewer systems, mining and construction sites, etc. are examples of point sources. In contrast, organic contaminants from non-point-sources are diffused over broad geographical scale in a relatively uniform environmental concentration explicitly delineated into spatial or temporal patterns. Consequently, the management of non-point organic contaminants is difficult. Among non-point-sources are agricultural runoff, urban runoff and atmospheric wet and dry depositions [17]. Based on how they are formed, organic contaminants of fresh water systems can be of two categories - natural and synthetic. Organic contaminants of biological origin are natural organic contaminants of which sugars, alkaloids and terpenoids are prominent. Synthetic organic contaminants are generated through reaction among different chemical species, which subsequently get discharged into the fresh water - examples include DDT, Polychlorinated biphenyls (PCBs), Chlordane, and Dieldrin, etc. [18]. Some organic contaminants are easily broken down upon entry into the environment, but others are very persistent, popularly termed as, Persistent Organic Pollutants (POPs) which are of particular concern because of the long term risks they pose. POPs, due to their persistent nature, become widely distributed geographically to pose adverse effects to human health and the environment [2]. The *Stockholm Convention* on POPs entered into force in 2004, and it identified some POPs potential for damaging environment and human health (Table 1).

Since the mid-twentieth century, pesticides became ominous in agriculture for saving crop yields from pastes both in the field and in storage. Pesticides are vital for crop protection and human health in many parts of the world [19, 20], yet due to their detrimental effects on natural ecosystems [21-23], we are becoming increasingly worried about their short and long term harms on the environment. A number of pesticides are in use throughout the world, but the four most important groups are *insecticides*, *rodenticides*, *herbicides* and *fungicides* – each group contains a rapidly accumulating list of persistent organic compounds capable of contaminating fresh water. Upon application, a substantial amount of these chemicals found their ways into fresh water systems. Most of the pesticides tend to bioaccumulate and exert ruinous effects on environment and human health [24, 25]. Among the twelve toxic substances listed for being phased out as per the *Stockholm Convention*, nine are pesticides. The most common pesticide among the fresh water organic contaminants are DDT, Diedrin, Aldrin, Chlordane, Endrin, Mirex, Toxaphene, and Heptachlor [25, 26] - all of which has a long half-life, can break down to intermediate species and have detrimental health effects.

POPs	Class
Heptachlor	Pesticide
Endrin	Pesticide
Dieldrin	Pesticide
Aldrin	Pesticide
Chlordane	Pesticide
Hexachlorobenzene	Pesticide / industrial chemical / by-product
Dioxins	By-product
Furans	By-product
Polychlorinated biphenyls (PCBs)	Industrial chemical / by-product
DDT	Pesticide
Mirex	Pesticide
Toxaphene	Pesticide

Table 1. POPs scheduled to be phased out and eliminated under the Stockholm Convention, 2004.

3.2 Sources of persistent organic contaminants in the freshwater system

3.2.1 Industrial sources

Poly Aromatic Hydrocarbons (PAHs) are produced through burning of common fossil fuels, such as coal, petroleum and natural gas as well as from biomass fuels such as fuel wood, animal excreta etc. Naphthalene, an ingredient in dyeing industry, aluminium smelting industry and lubricant oils as well as wood procession industry, enters into water, mainly through discharges and spills during the storage, transportation and disposal of fuel oil and coal tar and incomplete combustion of organic compounds [27]. Anthracene releases from dye and pesticide manufacture, from exhaust of engines, from the incomplete combustion of organic compounds [28]. Burning of gasoline, garbage or biomass causes release of benzopyrene into the environment. It also releases from burning of tar during road construction, from wood preservative creosote and from glues used in electrical components [29]. The environmental release of benzene occurs from industries such as rubber processing, dyeing and washing, pharmaceuticals, and agrochemicals. Underground gasoline storage tanks also leak benzene into ground water [30]. Xylene is a common solvent, thinning and cleaning agent used widely in printing, rubber, wood processing, plastic and leather industries. Being liquid in form, xylene easily leaks into surface or ground water [31]. Dioxins are another class of organic contaminants released from pulp and paper mills, wood preservatives, etc. Trichloroethene is used as solvent during cleaning engine parts from grease and spillage of it to environment occurs during cleaning process or from the wastes of such cleaning facilities.

3.2.2 Agricultural and farming sources

Agrochemicals are predominantly organic in nature and diverse in classes, nature and applications. They constitute the major anthropogenic source of organic contaminants into the fresh water system [32]. Usually, agrochemicals when applied in agricultural field or agro-product storage or processing, come in direct contact with flow of fresh water and can contaminate gradually both the surface and ground sources of fresh water. As they enter through flow water, they can reach unimaginable distance from the place of their applications.

3.2.3 Natural sources

The background concentration of organic contaminants in the environment is negligible, however, a number of them exist in the environment. Some of the organic contaminants enter the environment through various natural processes [1] such as volcanic eruptions, forest fires, biological decomposition and microbial activities. PAHs may enter the fresh water system naturally from thermal geological reactions, *i.e.* volcanic eruptions, forest fires, etc. Anthracene releases from natural sources such as coal and tar, can seep into the ground and surface water from coal piles. Among seventy five dioxins, originating naturally from volcanic eruption or forest fires, 2,3,7,8- tetrachlorodibenzo-*p*-dioxin (TCDD), is the most contaminating to the environment. A deeper treatment of contaminant contribution from natural sources has been discussed elsewhere [32].

3.2.4 Domestic and municipal sources

Burning of biomass fuels, such as wood, leaves, cow dung, etc., lead to the release of organic contaminants such as PAHs, anthracene and benzopyrene, which ultimately reaches water or soil [33]. Burning of coal for cooking and home heating is another source of anthracene and benzopyrene, which find their way into the fresh water system. One of the major concerns of modern city or municipality management is the vast quantity of solid wastes produced every day in big municipal cities. The solid waste stream contains a huge amount of organic substances - human and animal excreta and wastes, vegetable and food remaining, healthcare wastes, etc. Upon their disposal into land filling or open dumping ground, the leachates thereof easily finds their way into surface or ground water [34]. Rich in organic contaminants for fresh water system, the leachates either contaminates surface water in the short run and ground water gradually, in the long run. Incineration of municipal waste also releases different organic contaminants into the air which ultimately pollute fresh water. In 37 rivers of Japan, presence of many pharmaceutical organic compounds has been found of which crotamiton, carbamazepine, ibuprofen and mefenamic acid were positively correlated to population in the respective catchment areas indicating the contribution of these compounds from sewage sources [35].

3.2.5 Atmospheric deposition

Organic contaminants from different sources easily enter to the atmosphere and subsequently adsorbed with atmospheric particles or moisture. Such deposition of contaminants from the atmosphere ultimately reaches the aquatic ecosystem. An example of atmospheric movement of organic contaminants to fresh water ecosystem has been clearly demonstrated in case of two high arctic lake systems - Lake Ellasjøen and Lake Øyangen - which are 500 km away from known point source [36]. In sediments and biota from both the lakes, high levels of POPs especially PCB and *p,p* 9-DDE, have been detected. However, the higher levels of these contaminants in Lake Ellasjøen has been linked to the higher precipitation it receives, and that it receives POPs from birds that use the lake as resting ground.

3.3 Fate of organic contaminants

Organic contaminants usually break down easily in the environment through natural assimilative processes, but there are compounds highly resistant and these needs special attention from the standpoint of pollution of fresh water systems. Popularly known as *Persistent Organic Pollutants* (POPs), they remain in the environment from many months to even years [1].

3.3.1 The media involved in the aquatic transport of organic chemicals

Some of the persistent organic contaminants, including DDT, move from the point of application through the atmosphere and translocate from relatively warm regions to get condensed at colder, higher latitudes through a process known as *global distillation effect*. This explains the deposition of organic compounds at high concentrations in the Arctic region which is free from usage of such compounds [9, 37]. Benzene is water soluble, to some extent, and can seep into groundwater as well. Among the POPs, polychlorinated biphenyles (PCBs) were wonder materials with a wide range of applications. However, they entered the fresh water system during their manufacture and use. Even after the ban on them, they are still entering into the environment from the waste products where PCBs were used in the past. In fresh surface water or ground water, most of the PCBs adhere, weakly or strongly, to suspended sediment particles and remain as such for years [38]. Sediments containing PCB settle down to the bottom of fresh water reservoirs. Such sediments become the source of PCBs and gradually but continuously release PCBs in scanty amounts for years.

3.3.2 Bio-transformation and bio-accumulation

POPs are less soluble in water and more in lipid or fat resulting in their higher accumulation into the fatty tissues of living organisms. These organic contaminants enter living organisms from water through drinking of contaminated water or through ingestion of foods either processed using such water or foods from sources, which already have accumulated these contaminants. Bioaccumulating POPs include pesticides, PCBs, dioxins, and furans [1]. Repeated intake of organic contaminants or food contaminated thereof may cause bio-magnification of these contaminants leading to the enhancement of the harm caused by these contaminants. The example of this is DDT, which shows higher concentration in fish fat tissue compared to its concentration in water where fishes grow. Different derivatives of DDT, for example, DDE or DDD forms inside the organisms and they are excreted off through excreta, but they may excrete through milk of mammals leading to transfer of them to offspring from parents. These compounds are stored most readily in fatty tissue. The tendency of these substances to persist in the environment and to be build-up in plant and animal tissues poses the greatest risk to human health and the environment. Usually, the organic contaminants function as mutagenic, carcinogenic or teratogenic agents. Benzene, in fresh water, breaks down quickly and does not build-up in plants or animals. Fat soluble dioxins bioaccumulate easily in fatty tissue as well as in skin, muscle and other organs of most of the exposed animals, including fishes grown in dioxin rich surface water sources. The risk is higher for fishes from streams where affluent from paper and pulp industry is released.

3.4 Effect of fresh water organic pollutants

DDT causes sperm decline, eggshell thinning of birds and birth defects in many animals, which have been linked to near elimination of some species of animal. Bald eagle, a carnivore, is an example of what damage a POP like DDT can do [1]. On the other hand, PCBs from fresh water bioaccumulate into fish to a concentration hundreds or thousands of times higher than their levels in water. Benzene, if ingested, is broken down into secondary metabolites in the liver and bone marrow which are suspected to have to link with liver and bone marrow tissues. Exposure to moderate or high toluene levels has potential adverse effects on their liver, kidneys, and lungs. Aquatic animals absorb only a minimal amount of the xylene available in water contaminated with it. However, methylbenzaldehyde, a breakdown product from xylene cause damage to lungs of some animals. Exposure of animals to moderate to high levels of trichloroethane (TCE) causes liver and kidney damage.

Among the organic contaminants, PCBs, DDT, DDE, DDD, naphthalene, benzopyrene, dioxins, etc., are reported or suspected carcinogens for humans. Benzopyrenes have been related to the production of metabolites doubted as a carcinogen precursor. PCBs enter the human body through eating of PCB containing meat or fish and get converted to other metabolites some of which excreted naturally, but others stay in fat and in the liver for months or years. Inclusion of PCBs into breast milk fat and their subsequent entries into babies have been reported [8, 39, 40]. In Western Japan, the consumption of rice bran oil contaminated with some thermal degradation products of PCBs, *i.e.* furans and quaterphenyls, which are more toxic than PCBs, led to a severe form of acne called chloracne followed by fatigue, nausea, and liver disorders [41, 42]. Short-term low human exposure to naphthalene in mild concentration has been linked to eye and skin irritation while at elevated exposure levels, it causes headaches, fatigue and nausea. However, ingestion of naphthalene may cause hemolytic anemia, damages to kidneys and liver. Decreasing fertility, fatal damage, lung and skin tumors are also evident in the case of chronic exposure to naphthalene. Trace amount of dioxins at parts per trillion levels may cause hormone disruption; it also causes numbness, fluctuations in liver enzymes levels, nausea, etc. Exposure to large quantity of TCE creates dizziness to senselessness and even death. Range of concentrations of different pharmaceutical organic contaminants of fresh water in water bodies around the world, their lowest predicted no-effect concentrations (PNEC), their health effects at higher doses along with natural attenuation have been summarized by Pal *et al.* [2].

In a high-latitude freshwater food web of a remote lake in the Canadian Arctic, bioaccumulation of atmospherically deposited organochlorides such as PCBs, DDT, chlordane (CHL)-related compounds and hexachlorocyclohexane (HCH)- isomers were found at ng L^{-1} concentrations in water and sediment samples, while in fish samples, 6- to 10-fold higher concentrations of these POPs compared to water samples were observed [9].

Every day, the flow of organic contaminants into the fresh water system is increasing with the increase in the industrial activities and intensification of agricultural practices alongside the expansion of big municipalities all over the world. The rapid growth of population and the changes in lifestyle is enhancing the demand for global freshwater consumption. Hence, the necessity of treating larger volume of water is increasing and the task of meeting this demand is becoming complicated in terms of technology and costlier in terms of capital investment due to increased flow of contaminants in a larger variety into the water. The challenges that the water treatment will face in near future have been discussed in a nice review by Shannon et al. [43].

3.5 Detection of organic contaminates in fresh water system

3.5.1 Instrumental and laboratory techniques

The ubiquitous presence of persistent organic pollutants in surface and ground water sources necessitates their detection at trace level in order to ensure safe drinking water standards for public health in one hand and to develop appropriate technology for the removal or avoidance of these contaminants, on the other hand. Significant efforts are evident to the development of novel techniques and systems for the detection of the presence of organic fresh water contaminants in sub parts per billion levels to sub parts per trillion levels along with the refinement of existing detection capabilities for these organic compounds in the environment [6]. Pressurized solvent extraction, solid phase extraction, and capillary-column gas chromatography/mass spectrometry techniques has been developed for the determination of organic contaminants [44, 45]. Gas chromatography–high resolution mass spectroscopy (GC–HRMS) has been used in Canada to detect estrogenic organic fresh water contaminants in municipal wastewater as well as in effluents from bleached kraft pulp mill at ng L^{-1} level [46]. Natural organic matter and the anthropogenic organic contaminants have distinct fluorescence signatures, which have been used to develop a technique for early detection of contamination in a groundwater based drinking water supply plant [47]. Very recently, low cost calibration free methods for determination of organic contaminants have been reported [48].

3.5.2 Bio indicators

Bio monitoring of water contamination using bio indicators is becoming a cost-effective mode of contamination detection in aquatic systems [49, 50]. Some species of aquatic biota helps the depuration of aquatic resources contaminated by organic contaminants through bioaccumulation. The presence and concentration of contaminants can therefore be bio-monitored by using these species as biological indicators. Bivalves are usually preferred as a bio indicator and Goldberg in 1975 introduced the concept of "Mussel Watch" followed by the wide use of bivalves as an ideal bio-monitor in routine monitoring for National and Regional programmes [51]. The current results from the implementation of this approach succeeded to identify hot spots of contaminants and in following up their spatial distribution in the marine environments. "Mussel Watch" approach is based simply on using the characteristic feeding habits of bivalves as filter feeders which can accumulate tremendous quantity of contaminants in their tissues; reflecting the present quality and quantity of bio-available chemicals in their surrounding waters. The approach has been used to monitor PCBs in water [52], organotin contamination of fresh water lakes [53], and for other organic contamination issues [54, 55]. Diatoms are also used in bio monitoring of water contamination [56].

3.5.3 Assimilative capacity of water and organic contaminants

Once into the fresh water system, the natural assimilative capacity of water can attenuate majority of the organic compounds in a relatively shorter period of time using natural processes such as biotransformation, photolysis, sorption, volatilization, and dispersion, or a combination thereof [57]. The completion of natural breakdown of PCBs in the water may take several years, or even decades. Benzopyrenes, once get into the fresh water system, is either photo-degraded at the surface of waters or biodegraded gradually, however, their adsorption to particulates slows down the rate of natural assimilation. After entry, xylene may remain in fresh water for months or more but it gradually breaks down into relatively less harmful or harmless organics [58]. Trichloroethane and other halogenated hydrocarbons degrade naturally by photo-oxidation and biodegradation. In surface water, natural breakdown occurs in weeks while in groundwater, the breakdown is much slower. However, the slow rate of desorption and resorption of trichloroethane in subsurface soil make their long persistence in soil from where they may find their way into ground water for a quite long time [59]. Reports on the natural attenuation of pharmaceutical organic compounds in the fresh water system have been reviewed in an excellent review by Pal *et al.* [2]. New approaches are emerging to assess the natural attenuation of organic contaminants in the freshwater system. Plume-scale electron and carbon balances have been used in UK in a Permo-Triassic Sandstone aquifer contaminated with phenolic compounds [7].

3.6 Treatment options for organic contaminants in freshwater system

The entry of organic contaminants into ground and surface fresh water source is increasing with industrialization and rapid urbanization. Accordingly, efforts towards the enhancement of treatment capabilities are increasing [60]. Bolong *et al.* [61] has reviewed the use and efficiencies of activated carbon, oxidation, activated sludge, nanofiltration and reverse osmosis membranes in removal of persistent organic contaminants. The current options for water treatment for the removal of organic contaminants are based on continued development of activated sludge method, which is basically a bio transformation method [62, 63], the membrane filtration methods [60, 64] and reverse osmosis [65-67]. Activated sludge with zeolite improves organic contaminants biodegradability during anaerobic oxidation of municipal waste water by increasing the number of heterotropic bacteria as high as 55 times compared to control activated sludge [68]. With the use of zeolite, the COD removal rate reached to about 90% while the TOC removal rates with 0.22µm and 0.45µm filter membrane reached respectively at 97% and 92%. The sludge with zeolite can depredate further organic contaminants, which was difficult to degrade using organisms. Removal of organic contaminants are also enhanced by using novel biodegradable coagulants like chitosan, poly-lactic acid derivatives, etc. [69], and adsorption into porous materials such as the activated charcoal [70], or granular carbon [71], etc. The focuses of these techniques are basically on the removal of POPs from freshwater entering into the environment through industrial or municipal effluents, since it became a binding requirement for business entities in almost all the countries of the world to release industrial effluents after making it free from such contaminants, which made the development of modern and effective removal technology an active field of research with adequate funding. However, there is a grim picture even after all these developments, as a recent report from South-East Queensland, Australia reported the presence of a total of 15 organic contaminants, including NDMA and bisphenol in potable water recycled from raw water coming off waste treatment plants with reverse osmosis equipped advanced water treatment facility [5]. On the other hand, the need for removal of organic contaminants puts tremendous pressure on the water-treatment facilities in terms of increased cost. However, some alternatives are emerging in this case as well to bring the cost of treatment down. Low-cost alternative adsorbents (LCAs), which comprise of both natural and synthetic materials, are in use for removal of POPs. LCAs have shown fast kinetics and appreciable adsorption capacities in removing organic dyes from the contaminated fresh water [72].

4. Convention/protocol on organic contaminant of fresh water

There are a number of conventions and protocols dealing with the issue of organic contaminants of the fresh water system, however, none of these protocols specifically deals only with this issue. The progress is in place, and we can expect that in the years to come, specific global convention covering the range of organic pollutants of the fresh water system will be adopted and ratified by all. Among the current conventions and/or protocols, Stockholm Convention on Persistent Organic Pollutants comprises of 173 parties, and it entered into force in 2004. The convention deals with POPs and it has identified 12 highly hazardous organic compounds, mostly agrochemicals and outlined the phasing out of these twelve gradually by ratifying parties [61]. Convention on the Protection and Use of Trans-boundary Watercourses and International Lakes, adopted at Helsinki, in 1992 by United Nations Economic Commission for Europe (UNECE), outlines the protocols for controlling trans-boundary movement of contaminants [73]. There are many country or region-specific regulations and directives related to water quality, which also covers the issue of organic contaminants of the freshwater system.

5. Conclusion

Our understanding of the extent to which organic contaminants are entering into the freshwater system is still sketchy, as do our knowledge of the subsequent chemistry of these contaminants – their fate, effect and mediation. The reasons of non-comprehensive understanding are many. The first is the impossibility, in practice, to keep a clear accounting of the entry of persistent organic contaminants into the aquatic system since the entry is taking place through plethora of means from innumerable sources. Secondly, the rapid increment of the pool of persistent organic fresh water contaminants in number and diversity; the interaction among the organic contaminants themselves, between contaminants and aquatic environment mediated by biological, physical and chemical forces leading to the formation of many intermediates and derivatives thereof. However, new information is added to our knowledge base each day, and we are becoming more capable of addressing the issue – locally and globally.

6. References

[1] T. Navratil and L. Minarik, Trace elements and contaminants, In: V. Cilek and R. H. Smith (Eds.), Encyclopedia of Life Support System, Vol. IV, pp. 1184–1213, 2005.

[2] A. Pal, K.Y.-H. Gin, A.Y.-C. Lin and M. Reinhard, "Impacts of emerging organic contaminants on freshwater resources: Review of recent occurrences, sources, fate and effects", *Science of the Total Environment*, Vol. 408, pp. 6062–6069, 2010.

[3] I. Rostami and A.L. Juhasz, "Assessment of persistent organic pollutant (POP) bioavailability and bioaccessibility for human health exposure assessment: A critical review", *Critical Reviews in Environmental Science and Technology*, Vol. 41, pp. 623–656, 2011.

[4] R.P. Schwarzenbach, T. Egli, T.B. Hofstetter, U. von Gunten and B. Wehrli, "Global water pollution and human health", *Annual Review of Environment and Resources*, Vol. 35, pp. 109–136, 2010.

[5] D.W. Hawker, J.L. Cumming, P.A. Neale, M.E. Bartkow and B.I. Escher, "A screening level fate model of organic contaminants from advanced water treatment in a potable water supply reservoir", *Water Research*, Vol. 45, pp. 768–780, 2011.

[6] M.J. Focazio, D.W. Kolpin, K.K. Barnes, E.T. Furlong, M.T. Meyer, S.D. Zaugg, L.B. Barber and M.E. Thurman, "A national reconnaissance for pharmaceuticals and other organic wastewater contaminants in the United States–II) Untreated drinking water sources", *Science of the Total Environment*, Vol. 402, pp. 201–216, 2008.

[7] S.F. Thornton, D.N. Lerner and S.A. Banwart, "Assessing the natural attenuation of organic contaminants in aquifers using plume-scale electron and carbon balances: model development with analysis of uncertainty and parameter sensitivity", *Journal of Contaminant Hydrology*, Vol. 53, pp. 199–232, 2001.

[8] N.M. Tue, A. Sudaryanto, T.B. Minh, T. Isobe, S. Takahashi, P.H. Viet and S. Tanabe, "Accumulation of polychlorinated biphenyls and brominated flame retardants in breast milk from women living in Vietnamese e-waste recycling sites", *Science of the Total Environment*, Vol. 408, pp. 2155–2162, 2010.

[9] K.A. Kidd, R.H. Hesslein, B.J. Ross, K. Koczanski, G.R. Stephens and D.C.G. Muir, "Bioaccumulation of organochlorines through a remote freshwater food web in the Canadian Arctic", *Environmental Pollution*, Vol. 102, pp. 91–103, 1998.

[10] C. Chen, X. Zhang, W. He, W. Lu and H. Han, "Comparison of seven kinds of drinking water treatment processes to enhance organic material removal: A pilot test", *Science of the Total Environment*, Vol. 382, pp. 93–102, 2007.

[11] J.D. Wilcox, J.M. Bahr, C.J. Hedman, J.D.C. Hemming, M.A.E. Barman and K.R. Bradbury, "Removal of organic wastewater contaminants in septic systems using advanced treatment technologies", *Journal of Environmental Quality*, Vol. 38, pp. 149–156, 2009.

[12] H.F. Al-Mudhaf, F.A. Alsharifi and A.S. Abu-Shady, "A survey of organic contaminants in household and bottled drinking waters in Kuwait", *Science of the Total Environment*, Vol. 407, pp. 1658–1668, 2009.

[13] R. Loos, B.M. Gawlik, G. Locoro, E. Rimaviciute, S. Contini and G. Bidoglio, "EU-wide survey of polar organic persistent pollutants in European river waters", *Environmental Pollution*, Vol. 157, pp. 561–568, 2009.

[14] K.K. Barnes, D.W. Kolpin, E.T. Furlong, S.D. Zaugg, M.T. Meyer and L.B. Barber, "A national reconnaissance of pharmaceuticals and other organic wastewater contaminants in the United States–I) Groundwater", *Science of the Total Environment*, Vol. 402, pp. 192–200, 2008.

[15] M.S. Evans, D. Muir, W.L. Lockhart, G. Stern, M. Ryan and P. Roach, "Persistent organic pollutants and metals in the freshwater biota of the Canadian Subarctic and Arctic: An overview", *Science of the Total Environment*, Vol. 351-352, pp. 94–147, 2005.

[16] A. Kronimus, J. Schwarzbauer, L. Dsikowitzky, S. Heim and R. Littke, "Anthropogenic organic contaminants in sediments of the Lippe river, Germany", *Water Research*, Vol. 38, pp. 3473–3484, 2004.

[17] I. Baranowska, H. Barchańska and A. Pyrsz, "Distribution of pesticides and heavy metals in trophic chain", *Chemosphere*, Vol. 60, pp. 1590–1599, 2005.

[18] Z.L. Zhang, H.S. Hong, J.L. Zhou, J. Huang and G. Yu, "Fate and assessment of persistent organic pollutants in water and sediment from Minjiang River Estuary, Southeast China", *Chemosphere*, Vol. 52, pp. 1423–1430, 2003.

[19] D.G. Hela, D.A. Lambropoulou, I.K. Konstantinou and T.A. Albanis, "Environmental monitoring and ecological risk assessment for pesticide contamination and effects in Lake Pamvotis, northwestern Greece", *Environmental Toxicology and Chemistry*, Vol. 24, pp. 1548–1556, 2005.

[20] P. Mineau, "A review and analysis of study endpoints relevant to the assessment of "long term" pesticide toxicity in avian and mammalian wldlife", *Ecotoxicology*, Vol. 14, pp. 775–799, 2005.

[21] C.A. Laetz, D.H. Baldwin, T.K. Collier, V. Hebert, J.D. Stark and N.L. Scholz, "The synergistic toxicity of pesticide mixtures: Implications for risk assessment and the conservation of endangered Pacific salmon", *Environmental Health Perspectives*, Vol. 117, pp. 348–353, 2008.

[22] M. Olsson, A. Bignert, J. Eckhéll and P. Jonsson, "Comparison of temporal trends (1940s-1990s) of DDT and PCB in Baltic sediment and biota in relation to eutrophication", *AMBIO: A Journal of the Human Environment*, Vol. 29, pp. 195–201, 2000.

[23] A. Baun, N. Bussarawit and N. Nyholm, "Screening of pesticide toxicity in surface water from an agricultural area at Phuket Island (Thailand)", *Environmental Pollution*, Vol. 102, pp. 185–190, 1998.

[24] W. Lopes Soares and M. Firpo de Souza Porto, "Estimating the social cost of pesticide use: An assessment from acute poisoning in Brazil", *Ecological Economics*, Vol. 68, pp. 2721–2728, 2009.

[25] L.C.M. Cuyno, G.W. Norton and A. Rola, "Economic analysis of environmental benefits of integrated pest management: A Philippine case study", *Agricultural Economics*, Vol. 25, pp. 227–233, 2001.

[26] J. Fernandez-Cornejo, S. Jans and M. Smith, "Issues in the economics of pesticide use in agriculture: A review of the empirical evidence", *Review of Agricultural Economics*, Vol. 20, pp. 462–488, 1998.

[27] J.K. Rosenfeld and R.H. Plumb, "Ground water contamination at wood treatment facilities", *Ground Water Monitoring & Remediation*, Vol. 11, pp. 133–140, 1991.

[28] D.H. Thomas and J.J. Delfino, "A gas chromatographic/chemical indicator approach to assessing ground water contamination by petroleum products", *Ground Water Monitoring & Remediation*, Vol. 11, pp. 90–100, 1991.

[29] G. Chao and H. Ting-lin, *Comparative research on petroleum composition between pro-release and post-release at the low petroleum contaminant sediments*, In: International Symposium on Water Resource and Environmental Protection (ISWREP), Xi'an, 2011.

[30] O. Adeyemi, J.O. Ajayi, A.M. Olajuyin, O.B. Oloyede, A.T. Oladiji, O.M. Oluba, I.A. Ololade and E.A. Adebayo, "Toxicological evaluation of the effect of water contaminated with lead, phenol and benzene on liver, kidney and colon of Albino rats", *Food and Chemical Toxicology*, Vol. 47, pp. 885–887, 2009.

[31] R. Kandyala, S.P.C. Raghavendra and S.T. Rajasekharan, "Xylene: An overview of its health hazards and preventive measures", *Journal of Oral and Maxillofacial Pathology : JOMFP*, Vol. 14, pp. 1–5, 2010.

[32] S. Vedal, Natural sources – wildland fires and volcanoes, In: S. Tarlo, P. Cullinan and B. Nemery (Eds.). *Occupational and Environmental Lung Diseases*, John Wiley & Sons, Ltd: Chichester, pp. 389–404, 2010.

[33] A. Asa-Awuku, A.P. Sullivan, C.J. Hennigan, R.J. Weber and A. Nenes, "Investigation of molar volume and surfactant characteristics of water-soluble organic compounds in biomass burning aerosol", *Atmospheric Chemistry and Physics*, Vol. 8, pp. 799–812, 2008.

[34] D. Barceló and M. Petrovic, *Emerging contaminants from industrial and municipal waste: Occurrence, analysis and effects*. Springer Verlag: Berlin, 2008.

[35] N. Nakada, K. Kiri, H. Shinohara, A. Harada, K. Kuroda, S. Takizawa and H. Takada, "Evaluation of pharmaceuticals and personal care products as water-soluble molecular markers of sewage", *Environmental Science & Technology*, Vol. 42, pp. 6347–6353, 2008.

[36] A. Evenset, G.N. Christensen, T. Skotvold, E. Fjeld, M. Schlabach, E. Wartena and D. Gregor, "A comparison of organic contaminants in two high Arctic lake ecosystems, Bjørnøya (Bear Island), Norway", *Science of the Total Environment*, Vol. 318, pp. 125–141, 2004.

[37] G.A. Stern, E. Braekevelt, P.A. Helm, T.F. Bidleman, P.M. Outridge, W.L. Lockhart, R. McNeeley, B. Rosenberg, M.G. Ikonomou, P. Hamilton, G.T. Tomy and P. Wilkinson, "Modern and historical fluxes of halogenated organic contaminants to a lake in the Canadian arctic, as determined from annually laminated sediment cores", *Science of the Total Environment*, Vol. 342, pp. 223–243, 2005.

[38] G. Welfinger-Smith and D.O. Carpenter, Addressing sources of PCBs and other chemical pollutants in water, In: J. M. H. Selendy (Ed.). *Water and Sanitation-Related Diseases and the Environment*, Wiley-Blackwell: Hoboken, N.J., pp. 359–384, 2011.

[39] L.-M.L. Toms, L. Hearn, K. Kennedy, F. Harden, M. Bartkow, C. Temme and J.F. Mueller, "Concentrations of polybrominated diphenyl ethers (PBDEs) in matched samples of human milk, dust and indoor air", *Environment International*, Vol. 35, pp. 864–869, 2009.

[40] P.O. Darnerud, "Toxic effects of brominated flame retardants in man and in wildlife", *Environment International*, Vol. 29, pp. 841–853, 2003.

[41] C.J. George, G.F. Bennett, D. Simoneaux and W.J. George, "Polychlorinated biphenyls a toxicological review", *Journal of Hazardous Materials*, Vol. 18, pp. 113-144, 1988.

[42] G. Ross, "The public health implications of polychlorinated biphenyls (PCBs) in the environment", *Ecotoxicology and Environmental Safety*, Vol. 59, pp. 275-291, 2004.

[43] M.A. Shannon, P.W. Bohn, M. Elimelech, J.G. Georgiadis, B.J. Marinas and A.M. Mayes, "Science and technology for water purification in the coming decades", *Nature*, Vol. 452, pp. 301-310, 2008.

[44] M.R. Burkhardt, *Determination of wastewater compounds in sediment and soil by pressurized solvent extraction, solid-phase extraction, and capillary-column gas chromatography/mass spectrometry [electronic resource]*. U.S. Dept. of the Interior, U.S. Geological Survey: Reston, Va., 2006.

[45] J.D. Cahill, E.T. Furlong, M.R. Burkhardt, D. Kolpin and L.G. Anderson, "Determination of pharmaceutical compounds in surface- and ground-water samples by solid-phase extraction and high-performance liquid chromatography-electrospray ionization mass spectrometry", *Journal of Chromatography. A*, Vol. 1041, pp. 171-180, 2004.

[46] M.P. Fernandez, M.G. Ikonomou and I. Buchanan, "An assessment of estrogenic organic contaminants in Canadian wastewaters", *Science of the Total Environment*, Vol. 373, pp. 250-269, 2007.

[47] C.A. Stedmon, B. Seredyńska-Sobecka, R. Boe-Hansen, N. Le Tallec, C.K. Waul and E. Arvin, "A potential approach for monitoring drinking water quality from groundwater systems using organic matter fluorescence as an early warning for contamination events", *Water Research*, Vol. 45, pp. 6030-6038, 2011.

[48] A. Radu, S. Anastasova, C. Fay, D. Diamond, J. Bobacka and A. Lewenstam, *Low cost, calibration-free sensors for in situ determination of natural water pollution*, In: IEEE Sensors, Kona, HI 2010.

[49] W. Liu, L.Z. Shan, Q.L. Xie and others, "Application of biomonitoring in water pollution", *Journal of Environment and Health*, Vol. 25, pp. 456-458, 2008.

[50] Q. Zhou, J. Zhang, J. Fu, J. Shi and G. Jiang, "Biomonitoring: An appealing tool for assessment of metal pollution in the aquatic ecosystem", *Analytica Chimica Acta*, Vol. 606, pp. 135-150, 2008.

[51] E.D. Goldberg, V.T. Bowen, J.W. Farrington, G. Harvey, J.H. Martin, P.L. Parker, R.W. Risebrough, W. Robertson, E. Schneider and E. Gamble, "The mussel watch", *Environmental Conservation*, Vol. 5, pp. 101-125, 1978.

[52] A.A. Raeside, S.M. O'Rourke and K.G. Drouillard, "Determination of in situ polychlorinated biphenyl elimination rate coefficients in the freshwater mussel biomonitor Elliptio complanata deployed in the huron-erie corridor, southeast Michigan, USA, and southwest Ontario, Canada", *Environmental Toxicology and Chemistry*, Vol. 28, pp. 434-445, 2009.

[53] J. Yang, H. Harino, H. Liu and N. Miyazaki, "Monitoring the organotin contamination in the Taihu Lake of China by bivalve mussel *Anodonta woodiana*", *Bulletin of Environmental Contamination and Toxicology*, Vol. 81, pp. 164-168, 2008.

[54] S. Mishra, R.K. Mishra, B.K. Sahu, L. Nayak and Y. Senga, "Differential growth of the freshwater mussel, *Lamellidens marginalis* in relation to certain drugs", *Environmental Toxicology*, Vol. 23, pp. 379-386, 2008.

[55] H. Liu, J. Yang and J. Gan, "Trace element accumulation in Bivalve Mussels *Anodonta woodiana* from Taihu Lake, China", *Archives of Environmental Contamination and Toxicology*, Vol. 59, pp. 593–601, 2010.

[56] A. Beyene, T. Addis, D. Kifle, W. Legesse, H. Kloos and L. Triest, "Comparative study of diatoms and macroinvertebrates as indicators of severe water pollution: Case study of the Kebena and Akaki rivers in Addis Ababa, Ethiopia", *Ecological Indicators*, Vol. 9, pp. 381–392, 2009.

[57] C.J. Gurr and M. Reinhard, "Harnessing natural attenuation of pharmaceuticals and hormones in rivers", *Environmental Science & Technology*, Vol. 40, pp. 2872–2876, 2006.

[58] J.P. Barker, G.C. Patrick and D. Major, "Natural attenuation of aromatic hydrocarbons in a shallow sand aquifer", *Ground Water Monitoring & Remediation*, Vol. 7, pp. 64–71, 1987.

[59] N.S. Thom, A.R. Agg and A.L. Downing, "The breakdown of synthetic organic compounds in biological processes", *Proceedings of the Royal Society of London. Series B. Biological Sciences*, Vol. 189, pp. 347–357, 1975.

[60] N. Tadkaew, M. Sivakumar, S.J. Khan, J.A. McDonald and L.D. Nghiem, "Effect of mixed liquor pH on the removal of trace organic contaminants in a membrane bioreactor", *Bioresource Technology*, Vol. 101, pp. 1494–1500, 2010.

[61] N. Bolong, A.F. Ismail, M.R. Salim and T. Matsuura, "A review of the effects of emerging contaminants in wastewater and options for their removal", *Desalination*, Vol. 239, pp. 229–246, 2009.

[62] G.M. Shaul, T.J. Holdsworth, C.R. Dempsey and K.A. Dostal, "Fate of water soluble azo dyes in the activated sludge process", *Chemosphere*, Vol. 22, pp. 107–119, 1991.

[63] M. Petrovi, S. Gonzalez and D. Barceló, "Analysis and removal of emerging contaminants in wastewater and drinking water", *TrAC Trends in Analytical Chemistry*, Vol. 22, pp. 685–696, 2003.

[64] S. Gonzalez, M. Petrovic and D. Barcelo, "Removal of a broad range of surfactants from municipal wastewater–Comparison between membrane bioreactor and conventional activated sludge treatment", *Chemosphere*, Vol. 67, pp. 335–343, 2007.

[65] T.A. Peters, "Purification of landfill leachate with reverse osmosis and nanofiltration", *Desalination*, Vol. 119, pp. 289–293, 1998.

[66] C.Y. Tang, Q.S. Fu, A.P. Robertson, C.S. Criddle and J.O. Leckie, "Use of reverse osmosis membranes to remove perfluorooctane sulfonate (PFOS) from semiconductor wastewater", *Environmental Science & Technology*, Vol. 40, pp. 7343–7349, 2006.

[67] J. Radjenovic, A. Bagastyo, R.A. Rozendal, Y. Mu, J. Keller and K. Rabaey, "Electrochemical oxidation of trace organic contaminants in reverse osmosis concentrate using RuO_2/IrO_2-coated titanium anodes", *Water Research*, Vol. 45, pp. 1579–1586, 2011.

[68] G.-W. Cheng, W.-Y. Wei, Z.-C. Wu, F.-J. Zhang and X.-F. Huang, *A study on degradation of organic contaminants of the A/O process of denitrogenation enhanced by zeolite and chemical phosphorus removal*, In: International Conference on Energy and Environment Technology, Guilin, Guangxi 2009.

[69] S. Zodi, J.-N. Louvet, C. Michon, O. Potier, M.-N. Pons, F. Lapicque and J.-P. Leclerc, "Electrocoagulation as a tertiary treatment for paper mill wastewater: Removal of non-biodegradable organic pollution and arsenic", *Separation and Purification Technology*, Vol. 81, pp. 62–68, 2011.

[70] D. Stalter, A. Magdeburg, M. Wagner and J. Oehlmann, "Ozonation and activated carbon treatment of sewage effluents: Removal of endocrine activity and cytotoxicity", *Water Research*, Vol. 45, pp. 1015–1024, 2011.

[71] D. Mohan, K.P. Singh and V.K. Singh, "Wastewater treatment using low cost activated carbons derived from agricultural byproducts–A case study", *Journal of Hazardous Materials*, Vol. 152, pp. 1045–1053, 2008.

[72] V.K. Gupta and Suhas, "Application of low-cost adsorbents for dye removal – A review", *Journal of Environmental Management*, Vol. 90, pp. 2313–2342, 2009.

[73] S.M.A. Salman, "The Helsinki Rules, the UN watercourses convention and the Berlin rules: Perspectives on international water law", *Water Resources*, Vol. 23, pp. 625–640, 2007.

14

Fluxes in Suspended Sediment Concentration and Total Dissolved Solids Upstream of the Galma Dam, Zaria, Nigeria

Y.O. Yusuf, E.O. Iguisi and A.M. Falade
Department of Geography, Ahmadu Bello University, Zaria
Nigeria

1. Introduction

Soils are formed from the weathering of the earth's surface and it is loose thereby subject to washing away by various erosive agents including water. Earth materials can therefore be transported in water bodies as either sheet or channel erosion. Water is the transporting medium through which sediments entering the stream from the catchment area are carried down stream. As water moves, its potential energy is transformed into kinetic energy and part of the latter is consumed for transporting the sediments. Mineral materials of many different shapes and particles sizes erode and contribute to the overall stream load. Stream load is broken into three types: dissolved load, suspended load and bed load (Ritter, 2006).

Dissolved loads are materials carried by rivers in true chemical solution. Consist mainly of materials, organic or inorganic, carried in solution by moving water. This type of load can result from mineral alteration chemical erosion or may even be the result of ground water seepage into the stream. Materials comprising the dissolved load have the smallest particle size of the three load types. The quantity of the dissolved solids in the stream is referred to as the Total Dissolved Solids (TDS) and it is important in the assessment of water quality and pollution (Leopold *et al*, 1964; Smith and Stopp, 1978 and Degens *et.al.*, 1991).

Suspended load comprises of fine sediment particles suspended and are transported through the stream. They are transported with no direct contact with the channel floor. In other words, they are too large to be dissolved, but too small to lie on the stream bed; they are rather buoyed by the flow of the stream. They account for the largest majority of the stream load in many rivers and consist of materials such as clay and silt. Suspended Sediment Concentration (SSC) is the volume of suspended sediments at successive depths along the vertical profile of a river, from the water surface to close the river bed (Colby and Hubbell, 1961). Walling and Kleo (1979) added that information on the magnitude of the suspended loads of rivers has many practical applications ranging from geomorphologic studies of denudation rates and patterns of landform development to problem of upstream soil loss and downstream channel and reservoir sedimentation.

Bed load are materials that remain in contact with the bed of the stream and moves by a combination of rolling, sliding and skipping (Knighton, 1998).

The product of the SSC and TDS with the stream discharge gives the suspended sediment discharge (load) and the dissolved sediment discharge (load) respectively. The SSC and the load computed from it is said to increase downstream, while the TDS with the load computed from it is said to decrease downstream, all things being equal (the reverse should be true for an upstream consideration) (Leopold *et al*, 1964; Smith and Stopp, 1978 and Degens *et al.*, 1991). However, since all things are not equal in nature, the study of SSC and TDS variations upstream of the Galma Dam is therefore necessary. Furthermore, an idea of the upstream fluxes of these parameters will help in the development and maintenance of the water infrastructure, which will be of great relevance to the inhabitants of Zaria and Kaduna State at large.

2. Study area

The Galma River is mainly situated in Zaria, with some portion of it extending to Kano and Katsina states. The Galma River originates from the Jos plateau in the South Western area of the Shetu hills, which is some 350km away from Zaria. The river then flows from there, Northwest towards longitude 8⁰E and Southwest towards Zaria. It flows from Zaria to join the Kaduna River (Abdulrafiu, 1977). Galma dam is located on the Galma River, at a distance of about 10km Northeast of Zaria town (see fig.1a, 1b and 1c). The dam consists of a 550m long earthfill embankment with a maximum height of about 14 metres, and a spillway that is 91.5 metres. (NUWSRP, 2004)

The study area lies within the tropical wet and dry climatic zone, exhibiting a strong seasonality of rainfall. The area is characterized with 6 months of rain and 6 month of dryness, denoted as the Aw climatic type. The rainy season starts around May and terminates around October; on the other hand, the dry season lasts between November and April (Iguisi and Abubakar, 1998). The mean monthly temperature is about 27⁰C, temperature varies and it is highest between the months of March and May, which represents the hot and dry period. It is lowest in December/January reaching about 22⁰C.

The basement complex rocks primarily underlie the Zaria region. The region is an area within the Zaria plain, a dissected part of the Zaria-Kano portions, an extensive peneplain that had developed in crystalline metamorphic rock. The geology is constituted of three basic rock types: gneissic complex, metasediments and older granites. Younger granites can occur mostly as the ring complexes of Jurassic age. Alluvial deposits are quite extensive in Galma area, traversing wide open, shallow and gently sloping channels that had been cut mainly into the thick mantle of the overburden with some bedrock exposures. Shallowness of the stream channel enables the formation of extensive flood plain along the river and hence is often flooded during the rainy season (Nassef and Olugboye, 1979).

The vegetation type of the northern Guinea Savanna, which is mainly characterized by herbs and grasses, with few deciduous trees widely scattered, characterizes the area. The herbs and grasses grow very tall, in some places, along the perennial rivers and streams, riparian vegetation consisting of evergreen trees are found. The common grass and tree communities in the study area are mostly *Andropogon spp*, Mango trees, *Parkia clappertonia*, *Azfelia africana* and *Daniella oliveri* used for making mortar and *Acacia balanites* are found in the area.

The soil type in the area is highly leached ferruginous tropical soils that develop on weathered regolith overlain by a thin deposit of wind blown silt from the tropical continental air mass into the area (McCurry and Wright, 1970 and Torkarski, 1972). The physical conditions of the soils are generally poor. The soil aggregates are very small and unstable with tendency to compact under wet condition (Kowal and Kassam, 1978).

The rural inhabitants living around the river are mainly subsistent and peasant farmers, hence the land is subjected to intensive seasonal farming. In addition, along the Galma River and its tributaries, irrigation farming is being practised on a small scale, where farmers divert flowing water into their farmlands, or they can use machines or pumping device to supply water to their farmland. Rising population however, has led to massive deforestation in the area to create space enough for cultivation, thus posing a serious environmental threat, in terms of upstream soil loss and downstream sedimentation. The available grassland encourages grazing of cattle and other pasturing activities in the area, likewise, the river supports fishing activities, which also improves the economy of the area.

Fig. 1a. Map of Nigeria showing Kaduna State

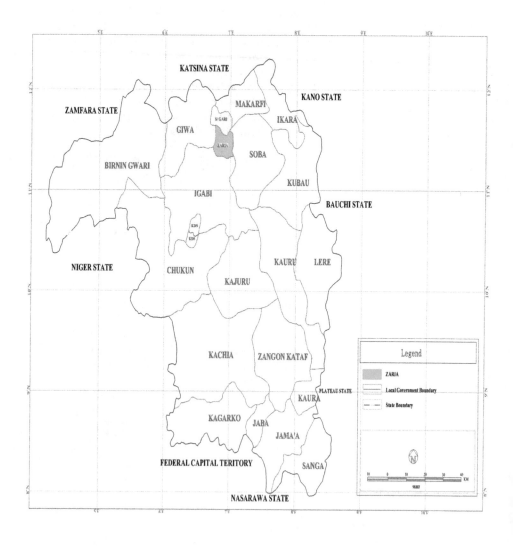

Fig. 1b. Kaduna State map showing Zaria local government

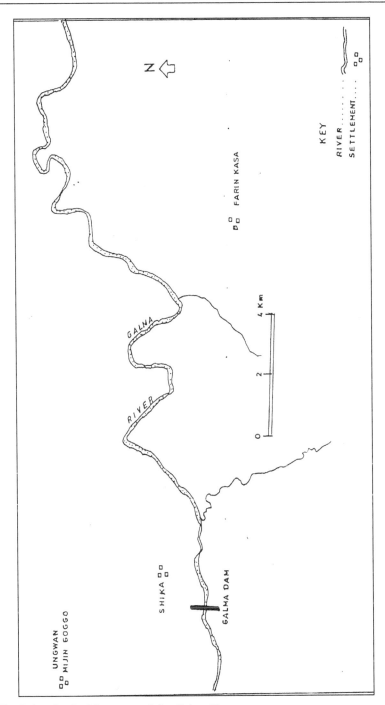

Fig. 1c. The Galma basin: Upstream of the Galma Dam

2. Methodology

2.1 Techniques of data collection

Data were collected using a systematic sampling approach, from the upstream section of the Galma dam. Starting point for the sampling was determined by moving 2km upstream away from the dam's embankment. From this starting point, 2km away from the dam's embankment, data on SSC, TDS and the stream discharge were obtained at 20 sampling points, over a distance of 100 metres each.

The velocity-area technique of discharge measurement was employed for this study, because it's relatively easier, accurate and less costly; here discharge is by definition the product of velocity and Cross Sectional Area (CSA) of flow, and this procedures evaluate these two terms for a stream section at a particular point in time. The width of the river and its depth was measured at each sampling point, the product of which gives the Cross Sectional Area. The velocity of flow of each sampling point was also read on a current metre, which is a digital equipment consisting of a graduated rod and a valeport, a propeller and the digital machine attached to it. The discharge therefore is calculated as the product of its flow velocity and Cross Sectional Area.

Sediment samples were collected at each sampling point using the USDH 48 sediment sampler, which is a depth-integrated sediment sampler. It was used because concentrations vary in a stream's cross-section and therefore, a sampling device is required to provide a representative sample of the concentration at a particular point within the channel. The sediment sample collected was then transferred into a plastic container and tightly corked. The samples were taken to the laboratory for onward analysis to determine the SSC and TDS values of the various samples.

The analysis of the sediment samples to determine their SSC and TDS values was carried out in the laboratory. The SSC was obtained by fetching 250ml of the sediment samples, which had been vigorously shaked to resuspend some of the settled particles. A membrane filter, which had been weighed, was attached to the filtration apparatus, which is equipped with a vacuum pump to make the filtration faster. The filter paper and the residue on it are reweighed. The difference between the initial weight and the new weight gives the SSC.

In getting the TDS, the aliquot from the filtration process for each sample is poured into a weighed crucible and oven-dried. On complete drying, the crucible is reweighed with the dissolved sediments that did not evaporate. The difference in the initial weight of the crucible and the weight after oven-drying gives us the TDS. However, for both the SSC and the TDS, the values gotten was multiplied by four (4), this is done to make the values up to mg/litre, since the sample taken is just 250ml.

The values of the SSC and that of the TDS multiplied by the stream discharge gives the Qs (suspended sediment discharge) and Qd (dissolved sediment discharge) respectively.

2.2 Techniques of data analysis

All generated data were logarithmically transformed. The log-transformed data were plotted against distance upstream on a line graph to reveal their upstream trends. Data on Suspended Sediment Concentration and Total Dissolved Solids were separately correlated with distance upstream to show their correlations. Also data on suspended and dissolved solids discharge were separately correlated with distance upstream to observe their correlations.

Student's t was also used to test the significance of the correlation co-efficient at the 95% significance level (α= 0.05). Mean, standard deviation, co-efficient of variation and ranges for each variable were also calculated. All analyses were carried out by the use of Microsoft excel and SPSS statistical Package.

3. Results and discussions

Table1 shows summary of the descriptive statistics data on sediment concentration (suspended sediments and dissolved solids) and discharge of the upstream section of the Galma Dam.

	N	Range	Minimum	Maximum	Sum	Mean	Standard error	Standard dev	Coefficient of variation
SSC	20	72.00	180.00	252.00	4416.00	220.80	5.09	22.76	8%
TDS	20	1280.00	120.00	1400.00	10160.00	508.00	86.48	386.75	76%
Qs	20	3.03	0.65	3.68	39.83	1.99	0.16	7.16	36%
Qd	20	15.15	0.42	15.57	88.15	4.41	0.81	3.62	82%

Table 1. Table showing Summary of Descriptive Statistics for SSC, TDS, Qs and Qd

The results from the summary statistics show that Suspended Sediment Concentration (SSC) and Suspended Discharge are moderately variable while the Total Dissolved Solids (TDS) and dissolved discharge have higher variability. SSC varied from a concentration of 180mg/l to a maximum concentration of 252 mg/l, giving a range of 72 mg/l with a standard deviation of 19.04 mg/l and a very low co-efficient of variation of 8.65%. Suspended discharge varies between 0.65g/s to 3.8g/s, with a range of 3.03g/s, mean of 1.99g/s standard deviation of 0.72g/s and co-efficient of variation of about 36%.

Total Dissolved Solids varies between 120 mg/l and 1400 mg/l, with a wide range of 1280 mg/l. TDS has a mean of 508 mg/l, standard deviation of 386.75 mg/l and co-efficient of variation of 76%. Solute discharge on the other hand ranges between 0.42g/s and 15.57g/s, giving a range of 15.15g/s mean solute discharge is 4.41g/s with a standard deviation of 3.62g/s and co-efficient of variation of 82%.

The higher mean values of TDS (and solute discharge) above SSC (and suspended discharge) indicates the dominance of chemical denudation within the basin. This confirms findings elsewhere (Dole and Stabler, 1909; Morisawa, 1968; Smith and Stopp, 1978 and Bowale, 2007)

Table 2 below shows that the correlation co-efficient of SSC with distance upstream is -0.16, while those of TDS, Qs and Qd are 0.330, -0.32 and 0.346 respectively. All the correlation coefficients are not significant at 95% significance level (α = 0.05), because the r – values are all greater than 0.05. This means that the relationship of SSC and TDS (with their respective sediment loads) with distance upstream is not well defined as to say that an increase in distance upstream will connote an increase in TDS and decrease in SSC. This is largely attributed to the fluxes experienced in the SSC and TDS as one move upstream, giving an irregular, haphazard and indiscernible pattern relating to the findings of Bowale, 2007.

Variables	Correlation with Distance (r)	t – values	Significance at 0.05 level
SSC (mg/l)	- 0.016	0.948	Insignificant
TDS (mg/l)	0.330	0.155	Insignificant
Qs (g/s)	- 0.320	0.893	Insignificant
Qd	0.346	0.135	Insignificant

Table 2. Correlation of SSC, TDS, Qs and Qd with distance upstream

4. Upstream fluxes of sediments

A more detailed and careful examination of the upstream patterns of SSC, TDS, Qs and Qd as shown in fig. 2-5 reveals that they show some pulses upstream; the pulsations are similar to waves having an amplitude and a pulse-length. SSC has four (4) pulses (fig. 2.); this is same for TDS, Qs and Qd (fig. 3-5). Nonetheless, despite showing some pulsations, the logarithmic line in figures 2-5, show that the variables increases generally upstream, except for the suspended discharge, Qs (fig. 4). The patterns exhibited by the four variables as revealed by the graphs are summarized in tables 3-6.

4.1 Fluxes of suspended sediment concentration

From fig. 2 and the summary in table 4, the first pulse of SSC started from 100m, which is the first point of measurement as well as the minimum SSC of 180mg/l, it peaked at a concentration of 232mg/l at a distance of 400m from starting point of measurement. The first pulse of SSC therefore has amplitude (range) of 52mg/l, representing 22.4% of maximum concentration, over a pulse length (distance) of 400m. The second pulse started immediately at the end of the first at 501m, got to a peak at a distance of 1000m and then declined to another minimum at a distance of 1200m from starting point. Pulse 2 had SSC amplitude of 56 mg/l, representing about 22.2% of maximum concentration and a pulse length of 700m. Pulse 3 starting from 1201m got to a maximum at 1400m and then declined to another low at 1700m, giving amplitude of 40mg/l, representing just 16.1% of maximum concentration, over a pulse length of 500m. Pulse 4 of SSC starting from 1701m reached a maximum at 2000m, the last point of measurement to give amplitude of 68mg/l, representing 27% of maximum concentration and a pulse length of 300m.

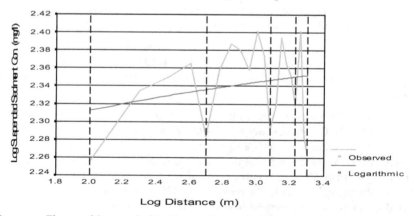

Fig. 2. Upstream Fluxes of Suspended Sediment Concentration

Pulse Parameter	Pulses			
	1	2	3	4
Distance of peak pulse concentration (m)	400	1000	1400	1800
Beginning of pulse (distance (m)	100	501	1201	1701
End of pulse (distance) (m)	500	1200	1700	2000
Pulse length (m)	400	700	500	300
Maximum concentration, (mg/l)	232	252	248	252
Minimum concentration (mg/l)	180	196	208	184
Amplitude (range) of concentration (mg/l)	52	56	40	68
Amplitude as % of maximum concentration (%)	22.4	22.2	16.1	27.0

Table 3. Upstream Fluxes (Pulses) of SSC, produced from fig. 2

4.2 Fluxes of total dissolved solids

Total Dissolved Solids, similar to the SSC also has four pulses (fig.3 and Table 4). Beginning from 100m, the first pulse of solute concentration peaked at a concentration of 200mg/l at the 100m point of measurement and thereafter declined to a minimum at 300m to give solute concentration amplitude of 80mg/l, representing about 40% of maximum concentration, over a pulse length of 300m. The 2nd pulse starting at 401m, got to a maximum concentration of 1000mg/l at 1100, thereafter, declining to a minimum about 100m from the point of maximum concentration, over a pulse length of 800m. Pulse 3 starting at 1201m, got to a maximum solute concentration at a distance of 1400m, and then declining to minimum solute concentration at 1600m. It has amplitude of 600mg/l, representing 75% of maximum solute concentration, over a pulse length of 400mg/l. The fourth pulse starting at 1601m got to a maximum solute concentration at 1800m and then declined to a minimum at a distance of 200m, the last point of measurement. Pulse 4 has an amplitude solute concentration of 1280mg/l, representing 91.4% of maximum solute concentration, over a pulse length of 400m.

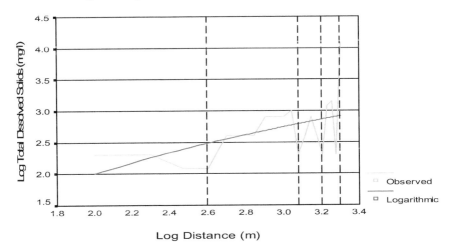

Fig. 3. Upstream Fluxes of Total Dissolved Solids

Pulse Parameter	Pulses			
	1	2	3	4
Distance of peak pulse concentration (m)	100	1100	1400	1800
Beginning of pulse (distance (m)	100	401	1201	1601
End of pulse (distance) (m)	400	1200	1600	2000
Pulse length (m)	300	800	400	400
Maximum concentration, (mg/l)	200	1000	800	1400
Minimum concentration (mg/l)	120	200	200	120
Amplitude (range) of concentration (mg/l)	80	800	600	1280
Amplitude as % of maximum concentration (%)	40	80	75	91.4

Table 4. Upstream Fluxes (Pulses) of TDS, produced from fig. 3

4.3 Fluxes of suspended sediment discharge

The suspended sediment discharge also has four pulses (fig.4 and table 5). The first pulse beginning from 100m reached the maximum suspended load of 2.39 g/s at a distance of 400m, after recording the minimum sediment discharge for the pulse at the first point of measurement of 100m, giving amplitude of 1.1g/s, representing 46% of the maximum sediment discharge over a pulse length of 400m. The second pulse beginning from 501m reached peak sediment discharge at 700m and declined to a minimum at 900m. It has an amplitude sediment discharge of 1.46, representing 50.3% of maximum sediment discharge over a pulse length of 400m. pulse 3, beginning at a distance of 901m peaked at 1000m and declining thereafter to a minimum at 1300m, with an amplitude of 1.33g/s, representing 51.8% of maximum sediment discharge, over a distance of 400m. Pulse 4, beginning from 1301m, reached peak sediment discharge at 1900m and declining to a minimum at 100m further upstream at 2000m, with amplitude of 3.03g/s, representing 82.3% of maximum sediment discharge over a pulse length of 700m.

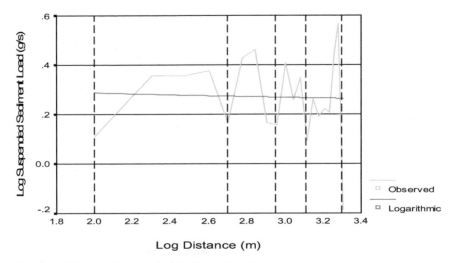

Fig. 4. Upstream Fluxes of Suspended Sediment Load

Pulse Parameter	Pulses			
	1	2	3	4
Distance of peak pulse concentration (m)	400	700	1400	1900
Beginning of pulse (distance (m)	100	501	901	1301
End of pulse (distance) (m)	500	900	1300	2000
Pulse length (m)	400	400	400	700
Maximum concentration, (mg/l)	2.39	2.90	2.57	3.68
Minimum concentration (mg/l)	1.29	1.44	1.24	0.65
Amplitude (range) of concentration (mg/l)	1.1	1.46	1.33	3.03
Amplitude as % of maximum concentration (%)	46.0	50.3	51.8	82.3

Table 5. Upstream fluxes (Pulses) of Qs produced from fig. 4

4.4 Fluxes of solute discharge (load)

The upstream fluxes of solute discharge (fig 5. and Table 6) closely replicate those of Total Dissolved Solids (fig. 3 and Table 5). Consequently, solute discharge has four pulses, same for TDS with the points of maximum discharge, discharge amplitude, pulse length and beginning and end of each pulse resembling those of TDS, except for the lower numerical values of minimum, maximum and amplitude of solute discharge.

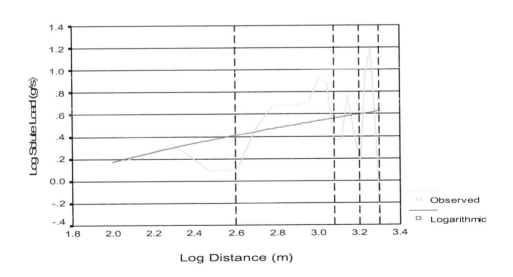

Fig. 5. Upstream Fluxes of Solute Load

Pulse Parameter	Pulses			
	1	2	3	4
Distance of peak pulse concentration (m)	200	1000	1400	1800
Beginning of pulse (distance (m)	100	401	1201	1601
End of pulse (distance) (m)	400	1200	1600	2000
Pulse length (m)	300	800	400	400
Maximum concentration, (mg/l)	2.11	8.15	5.90	15.57
Minimum concentration (mg/l)	1.22	2.28	1.48	0.42
Amplitude (range) of concentration (mg/l)	0.89	5.87	4.42	15.15
Amplitude as % of maximum concentration (%)	42.2	72	75	97.3

Table 6. Upstream Fluxes (Pulses) of Qd, produced from fig. 5

5. Comparison of sediment variables

5.1 Relationship between SSC and TDS

Suspended Sediment Concentration (SSC) in the Galma basin is lower than Total Dissolved Solids (TDS); the mean SSC is 220.8 mg/l, while mean TDS is 508 mg/l (table 2), a ratio of about 1:2.3. Looking at fig. 2 and fig. 3 in relation to each other, especially their logarithm lines, it can be observed that there is a general upstream increase in both variables, implying that a positive association exists between both variables. The test of association between both variables, shows a significant correlation between them, r=0.535, a moderately positive correlation (table 7). Therefore, generally on the average, despite the fluxes, as we move upstream, both the SSC and TDS tends to increase.

Compared Variables	Correlation values (r)	t- values	Significance at 0.05 level
SSC/ TDS	0.535	0.015	Significant
SSC/ Qs	0.321	0.168	Insignificant
TDS/ Qd	0.927	0.000	Significant
Qs/ Qd	0.317	0.174	Insignificant

Table 7. Correlations of SSC, TDS, Qs and Qd with each other

5.2 Relationship between SSC and Qs

In Table 7, the correlation between the SSC and sediment discharge (Qs) had an r-value of 0.321, which is poorly and insignificantly correlated. The correlation though positive is not strong, this can also be observed by viewing closely together the graphs of SSC and Qs (fig. 2 and 4), and they show a similar though not perfect trend upstream. Therefore, high levels of SSC correspond to high levels of suspended sediment discharge and vice versa.

However, when the logarithm line for both variables graph are viewed, there is a difference, which is reflected in the general upstream increase in concentration for the SSC, whereas there is a general upstream decrease for suspended sediment discharge, this could be a reason for the insignificant correlation between the two variables.

5.3 Relationship between TDS and Qd

TDS and Qd show a very strong, positive and significant relationship, with an r-value of 0.927 (table 6). This correlation is strongly positive indicating a perfectly direct relationship between both variables i.e. as the TDS increases, so does the solute discharge and vice-versa. This relationship is also reflected in the similar pattern and pulsations, the graphs of both variables with distance upstream gives, and in the general upstream increase as depicted by the logarithm line of both variables (fig. 3 and 5).

A downstream study of both variables by Walling and Webb (1986), however, revealed that while solute discharge increased with increase in river discharge downstream, solute concentration decreased, it can therefore be said that for an upstream scenario, the relationship will be positive, all things being equal. The decrease in the downstream case of the solute discharge can be attributed to the dilution effects of solutes, which increases as river discharge increases downstream.

5.4 Relationship between Qs and Qd

Suspended Sediment Discharge and Solute Discharge showed a poor and insignificant correlation, with an r-value of 0.317 (refer to fig. 4 and 5). At the start of measurement, especially the first and second pulse for both variables, there was an increase in suspended discharge that was associated with a decrease in the solute discharge, however, pulses 3 and 4 for both variables show a great correspondence, in that as the suspended discharge increases, the solute discharge increases and vice-versa. This kind of association is found between suspended and solute discharges due to the moderating effect of river discharge.

6. Discussion

6.1 Fluxes of SSC and Qs

The fluxes in SSC and Qs can be attributed to sand mining at irregular points along the river channel, to animals grazing around the channel banks, and the effect of cultivation near the riverbanks. Sand mining within the channel, most especially for construction works, promotes turbulence, which entrains sediments already settled on the riverbed. These sediments go into suspension to become part of suspended sediment load thereby increasing SSC. In addition, the release of sediments into the channel during mining operations enhances the concentration of suspended sediments. The entrained or released suspended sediments travel from point of release for a short distance before they settle back on the riverbed; the distance travelled by these sediments before resettling depends on the nature (weight and shape) of sediments, degree of turbulence eddies generated at the point of entrainment or release, flow characteristics and channel configuration among others. As sand mining occurs in a few irregular spots along the channel, sediment concentration and discharge show fluxes in the form of irregular pulses along the river channel. Dearing (1992) and Bowale (2007) have observed similar effect of sand mining on SSC and load in Wales and Samaru respectively.

Grazing of animals around the riverbanks has caused bank collapse in many parts of the channel by their trampling effects. Some of the bank materials slumped into the channel go into suspension and are transported in suspension before settling down on the riverbed further downstream. Likewise, human activities such as cultivation practices close to the riverbanks can lead to the intermittent addition of soil particles to the river and forthwith carried in suspension. Cultivation of lands along the river channel generally increases the level of SSC in a water body, as the soil becomes more susceptible to erosion, because of tillage.

6.2 Fluxes of TDS and Qd

The cause of the fluxes of Total Dissolved Solids and solute discharge (Qd) in the river channel is uncertain. However, fertilizer application by farmers cultivating the land along the river channel may be thought of as a possible cause. Excess chemical fertilizer not used up by plants fined their way into the river through various routes to increase the solute load of the river.

7. Conclusion

This study shows that chemical denudation seems more dominant within the basin than mechanical denudation; this is reflected in the higher mean value of dissolved solids over the suspended loads. Furthermore, sediment concentrations and discharges show wave like fluxes upstream. Using the logarithm line, the suspended sediment concentration, total dissolved solids and discharges showed an increasing trend upstream except for the suspended sediment discharge. The fluxes were attributed to certain physical and anthropogenic factors, implying that certain elements capable of creating disturbances within the river channel can lead to a diversion from the ideal patterns.

8. Recommendation

The study of sediment concentration and discharge fluxes is very important and useful in municipal water supply. This study therefore is of great importance to the Federal Ministry of Health as well as the Kaduna State Water Authority (K.S.W.A). The study was carried out in the upstream section of the Galma dam, which serves as the major reservoir for municipal water supply in Zaria and environs. The study was necessitated considering the effects of suspended and dissolved sediments on water quality and quantity in the reservoir, it also provides information on the rate at which sediment are being deposited into the reservoir.

Institution of an effective watershed management policy is recommended to help curtail the general increase for sediment released into the reservoir. The watershed management authority should be mandated to moderate or stop practices of sand mining at irregular spots on the channel; cultivation practices in the river basin; and cultivation requiring use of inorganic fertilizer should be moderated in the river basin. This will reduce the dissolved sediment sources in the basin thereby increasing the quality of water going into the reservoir.

Grazing of cattle on the riverbank should also be discouraged, to reduce sediment generated because of loosening of the soil by trampling effect of the cattle. The authority should also embark on massive afforestation programmes to reduce the rate of erosion, consequently, reducing sediment generation in the basin. Planting trees and cover crops that will protect the soil from raindrop impact and erosive work of wind and water is important. The authority should also carry out regular studies on sediment concentration and discharges to obtain data that gives an idea of the reservoir's lifespan; it can also be used to determine the cost of water treatment. This information will help the Kaduna State Water Authority, Nigeria to effectively carry out their function of municipal water supply.

9. References

Abdulrafiu, B.G. (1977). Land use changes Association with the New Galma Dam in Zaria. *Unpublished B.Sc. Dissertation*, Geography Department, Ahmadu Bello University, Zaria, Nigeria.

Bowale, A.R. (2007) "Downstream Fluxes in Suspended Sediment Concentration and Total Dissolved Solids, Upper Kubanni River Channel, Zaria, Nigeria". Paper Presented at the 49th Annual Conference of the Association of Nigerian Geographers, University of Abuja, Abuja, Nigeria.

Colby, B.R. and Hubbell, D.W. (1961). Simplified methods for computing total sediment discharge with the modified Einstein Procedure. U.S. *Geological Survey Water Supply Paper*. 1593.

Dearing, J.A. (1992). Sediment yields and sources in Welsh upland lake catchment during the past 800 years. *Earth surface Processes and Landforms*. 17, 1 – 22

Degens, E.T., Kempe, S and Richey, J.E. (1991). Summary: Biogeochemistry of major world rivers. In Degens, *et al* (eds), *Biogeochemistry of Major World Rivers*.

Dole, R.B. and Stabler, H. (1909). Denudation. *U.S. Geological Survey Water Supply Paper*, 234.

Iguisi, E.O. and Abubakar, S.M. (1998). "The Effect of Land use on Dam Siltation" Paper presented at 41st Annual Conference of Nigeria Geographical Association, University of Uyo, Akwa Ibom.

Knighton, D. (1998). *Fluvial Forms and Processes*. Arnold Hodder, Headline Group, London.

Kowal, J.M. and Kassam, A.H. (1978). Drought in the Guinea and Sudan Savanna Area of Nigeria. *Savanna*, Vol. 3. No. 1.

Leopold, L.B., Wolman, M.G. and Miller, J.P. (1964). *Fluvial Processes in Geomophology*. W.H. Freeman and Company, San Francisco.

McCurry, P.M. and Wright, J.B. (1970) Geology of Zaria. In Mortimore, M.J. (edition). *Zaria and its Region*. Occasional Paper No. 4. Department of Geography.

Morisawa, M. (1968). Streams: *Their dynamics and Morphology*. McGraw – Hill Company, New York.

Nassef, M.O. and Olugboye, M.D. (1979). The Hydrology and Hydrogeology of Galma Basin. *Federal Department of Water Resources*. File Report.

National Urban Water Sector Reform Project (NUWSRP) (March, 2004). *Report on Dam Safety Measures*. Federal Ministry of Water Resources.

Ritter, M.E. (2006). *The Physical Environment: An Introduction to Physical Geography*. Methuen & Co. Ltd. London.

Smith, D.I. and Stopp, P. (1978). *The River Basin. An Introduction to the Study of Hydrology.* Cambridge University Press, Cambridge.

Torkarski, A. (1972). Heterogeneous Terrace Arrangement of Ahmadu Bello University Zaria. *Occasional Paper*, No. 2. Dept. of Geography. A.B.U.

Walling, D.E. and Kleo, A.H.A. (1979). Sediment yield of Rivers in of areas of low precipitation: A global view in: The hydrology of areas of low precipitation, proceedings of the Canberra Symposium, 1979, *IAHS-AISH Publication.* 12, 479 – 93.

Walling, D.E. and Webb, B.W. (1986). Solute in River System in Trudgill, S.T. (ed), *Solute Processes.* Chichester. Wiley, 251 – 327.

Rainwater Harvesting Systems in Australia

M. van der Sterren[1,2], A. Rahman[1] and G.R. Dennis[1]
[1]University of Western Sydney,
[2]NSW Office of Environment and Heritage,
Australia

1. Introduction

The Australian continent has an extremely variable climate, as a result of the different oceanic currents and atmospheric variation. Australia has regular cycles of droughts and floods resulting in highly variable storage volumes in its major dams. The population in Australia is nearly 23 million (Australian Bureau of Statistics (ABS), 2011) of which the majority lives in the South-East coast of Australia. The largest cities (see Figure 1) are Sydney (4.58 million people.), Melbourne (4.08 million people), Brisbane (1.07 million people) and Canberra (358,600 people) (ABS, 2011). The water supply storage for these cities is located in the nearby mountain ranges and brought to the metropolitan areas through large distribution water pipes. The urban fringe areas, rural locations and the outback have limited reticulated water supply and often rely on capturing roof water, farm dam water and bore water for their water supply. The roof water in these regions provides the principal potable water supply, whilst farm dam and bore water are often used to meet non-potable requirements and for livestock (ABS, 2010). Historically, this has been different in the urban areas where potable and non-potable supply demands are met with a reticulated water supply.

A shift has occurred in the Australian Water industry as a result of population growth, the worst drought in living memory (Horstman, 2007) and a desire to become more sustainable. Total Water Cycle Management has gained momentum in Australia and new property developments must consider all aspects of the water cycle, including water supply, waste water treatment, stormwater control and water quality control of all discharges and supplies (Argue, 2004; Argue & Pezzaniti, 2009; Barton & Argue, 2009; Hardy, 2009; Hardy et al., 2003; Wong, 2006b, c; Wong et al., 2008; Wong & Brown, 2009). Rainwater tanks are being installed in urban areas, resulting in an increase resilience of the cities to droughts and a reduction of mains water demand. These rainwater tank installations are encouraged in various Development Control Plans (DCPs), through state legislation, such as the NSW Building and Sustainability Index (BASIX Sustainability Unit, 2009), and by providing rebates (Blacktown City Council, 2006; Blue Mountains City Council, 2005; Gardiner & Hardy, 2005; Ku-ring-gai Council, 2005; Penrith City Council, 2010). The reasons for installing a rainwater tank in Australia include reducing mains water costs, helping the environment, irrigating the garden and because it was mandatory when the house was built (ABS, 2010; Blackburn et al., 2010; White, 2010).

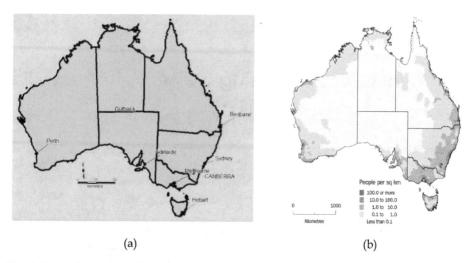

(a)   (b)

Fig. 1. Australia with the location of major urban centres (a) and Population density (b) (ABS, 2011)

2. Legislation

2.1 Australian guidelines

The water quality requirements in Australia have been developed using a fit-for-purpose approach. The guidelines for water quality in Australia are supported by the Department of Sustainability, Environment, Water, Population and Communities of the Australian Commonwealth Government. The Natural Resource Management Ministerial Council (NRMMC) released the guidelines and, in some cases, with input from the National Health and Medical Research Council (NHMRC) and the Australian Health Ministers Conference. These National Councils update and publish a number of guidelines applicable to rainwater tank installation in Australia. These guidelines are discussed below and are dependent on the point of discharge, users, use and general application of the water in the rainwater tank. The guidelines are summarised below:

- Australian and New Zealand guidelines for fresh and marine water quality (AFWG) (Australian and New Zealand Environment and Conservation Council & Agriculture and Resource Management Council of Australian and New Zealand (ANZECC & ARMCANZ), 2000a, b);
- Australian drinking water guidelines (ADWG) (NHMRC & NRMMC, 2004);
- Guidelines for urban stormwater management (ANZECC & ARMCANZ, 2000c); and
- Australian guidelines for water recycling: managing health and environmental risks - Phase 2 (NRMMC et al., 2009c), including
    - Australian Guidelines for Water Recycling: Stormwater Harvesting and Reuse (NRMMC et al., 2009a) or Recycled Water Guidelines (RWG); and
    - Australian Guidelines for Water Recycling: Augmentation of Drinking Water Supplies (NRMMC et al., 2009b).

2.2 Stormwater harvesting and reuse

Stormwater harvesting and reuse guideline have been prepared as part of the National Water Quality Management Strategy (NWQMS) by NRMMC, Environment Protection and Heritage Council (EPHC) and the NHMRC (2009a). The guideline uses a risk-based approach to stormwater harvesting and can be adopted for any stormwater harvesting technique. The management and treatment techniques for the harvested stormwater depend on the type of catchment area (roof, drainage system, driveway) and its intended end-use (potable, non-potable or irrigation only). The guideline has a specific section for the management of harvested roof water and identifies these systems as low risk. In particular, the guideline indicated that roof water harvesting for residential dwellings should meet the criteria and management techniques specified in the Australian Standards (AS) (Standards Australia, 2008) and *Guidance on use of rainwater tanks* (enHealth Council, 2004). For non-residential buildings or for a larger user group (nursing homes), the stormwater harvesting guideline (NRMMC et al., 2009a) is recommended to be applied and a risk assessment is to be conducted for the proposed end-uses of the rainwater harvesting system

The end-uses for harvested roof water are also dependent on other guidelines and policies. Most council Development Control Plans (DCP) specify that water from a rainwater tank is not to be used for drinking (Blue Mountains City Council, 2005; Hawkesbury City Council, 2000; Upper Parramatta River Catchment Trust (UPRCT) et al., 2005) even though the New South Wales (NSW) Department of Health does not prohibit its use. The NSW Department of Health does recommend using the water only for non-potable purposes when town water is available (enHealth Council, 2004; Environmental Health Branch, 2008). In contrast to this, it was concluded by Heyworth, Maynard & Cuncliffe (1998, p. 9) that 'little is known about the associated health risks' in regards to using water from a rainwater tank. It should be noted that a large number of Australians still rely on their rainwater tank as their sole supply of drinking water. They are advised by various government bodies to refer to Cuncliffe (1998) and enHealth (2004) to determine how to maintain a rainwater tank and how to improve the quality of drinking water from their rainwater tank.

In addition to these guidelines, the Australian Standards (AS) specify how these systems should be constructed. The AS 3500 (Standards Australia, 2006) specifies connections, backflows prevention systems and roof gutter design. The handbook (Standards Australia, 2008) shows how the systems can be designed and constructed to ensure a low risk, high quality fit-for-purpose water supply. In Australia, all designs and constructions of roof water harvesting and drainage systems need to adhere to minimum standard specified in the Australian Standards. Stormwater harvesting can also be used augment drinking water supplies and the guideline divides the augmentation of drinking water supplies in direct augmentation and indirect augmentation (NRMMC et al., 2009b). Indirect augmentation includes storages, such as dams or rivers prior to treatment and distribution. Direct augmentation includes storages such as rainwater tanks and recycled water treatment plants and requires the treatment and adequate management of the system to protect users.

2.3 Australian Drinking Water Guidelines

The Australian Drinking Water Guidelines (NHMRC & NRMMC, 2004) give the minimum standards to which water supply authorities need to treat drinking water to protect the users, especially those who are very young, elderly or immuno-compromised. Rainwater tanks on private property are not required to meet these guidelines for low risk use. A risk assessment approach is recommended for larger harvesting systems to ensure the risk associated with the use of the rainwater is managed (NRMMC et al., 2009a). Adopted management techniques can include fit-for-purpose use of the harvested water or treating the harvested water to the ADWG (NHMRC & NRMMC, 2004).

2.4 Fresh and marine water quality guidelines

The AFWG (ANZECC & ARMCANZ, 2000a, b) are used to determine objectives for in stream water quality of fresh and marine waters from high conservation areas to heavily modified systems. The overflow from harvesting systems is commonly directly diverted to the receiving waters through a drainage system. So although the AFWG (ANZECC & ARMCANZ, 2000a, b) are ambient in-stream water quality objectives, it is important to understand the impact that rainwater tank overflows might have on the receiving waters. Mixing of flows will occur within the drainage system thereby affecting the final concentrations of the water quality parameters.

2.5 Stormwater management

The current guideline for stormwater quality, *Australian Guidelines for Urban Stormwater Management* (ANZECC & ARMCANZ, 2000c) is based on a risk management approach, which is appropriate for catchment wide management, but difficult to implement on a lot scale. Council-wide assessments are made by local councils to develop a stormwater management plan. These plans have been developed by some councils and Catchment Management Authorities to identify management scenarios and pollutant hot spots, and are developed to reduce the pollutant levels in the run-off discharges from the catchment. Some councils have implemented water quality targets or water quality objectives for new developments, based on published water quality targets (see Table 1).

Pollutant	Percentage Detained
Suspended Solids	80% retention of average load
Total Phosphorus	45% retention of average annual load
Total Nitrogen	45% retention of average annual load
Litter	Retention of litter (> 50 mm) up to the 3-month ARI peak flow
Coarse Sediment	Retention of sediment (> 0.125 mm) up to the 3-month ARI peak flow
Oil and Grease	No visible oils up to the 3 month ARI peak flows

Table 1. Guidelines for stormwater pollutant discharges (data from Wong, 2006, p. 1-6).

Water quality and quantity control for redevelopment in many council areas is required to include an On Site Detention system (OSD) and a trash rack in front of an orifice screen. This is often combined with an in-line commercial pollutant filter. The trash screen is a mesh screen bolted to the OSD pit wall to remove litter, vegetation and sediment (Nicholas, 1995).This screen removes the gross pollutants and may also trap other pollutants (Goyen et al., 2002; UPRCT2005). In most cases, a commercially available pollutant filter is added to the design to reduce the contaminants below the threshold values shown in Table 1. OSD systems by themselves may not be able meet the standards shown in Table 1. Brisbane City Council (2008) developed a framework for water quality requirements that can be specifically applied to a site, based on the urban stormwater point of discharge to the catchment (for example a specific stormwater outlet or receiving stream). Other councils such as Blacktown City Council (BCC) (2006) and Blue Mountains City Council (BMCC) (2005) require the designs to protect specific receiving waters and these councils refer designs to the respective Catchment Management Authorities (CMA).

Often council requirements do not take into consideration any improvement in discharge water quality as a result of rainwater tank on-site. A number of contrasting results have been reported in regard to the water quality of rainwater tank. Evans et al. (2006, p. 37) reviewed the current literature on water quality of rainwater tank and found that 'a clear consensus on the quality and health risks associated with rainwater has not been reached' for water use. Furthermore, research has also indicated that the internal processes in the rainwater tank, such as sedimentation and micro-layer flocculation, can improve the water quality in the tank (Spinks et al., 2003).

3. Construction and typical Australian rainwater tanks

The application and size of rainwater tanks in Australia is highly varied. The tanks come in all shapes and sizes, but are also made from different materials. Australian rainwater tanks are often constructed above ground (see Figure 2(a)), partially in-ground (see Figure 2(b)), in-ground (see Figure 2(c)) or under house and the tanks can be made of polypropylene (see Figure 2(a)), concrete (see Figure 2(b)), and coated corrugated iron (Zincalume®, Colorbond®) (see Figure 2(e)). The size of the tank is dependent on roof area, water demand, local rainfall characteristics, and availability of mains water including required security of supply. Each of these characteristics has a different effect on the size of the tank to meet demand (Barry & Coombes, 2006, 2007; Coombes, 2002; Standards Australia, 2008). The size of the tank also has an impact on the materials used as the forces associated with water and ground pressure have a direct impact on the total stress on the tank wall (Standards Australia, 2008). The structural integrity of a large tank can be difficult to maintain for the polypropylene tanks (PVC) and therefore often concrete is used for larger tanks. The smaller tanks are often chosen for urban areas and need to be able to be installed in confined spaces, such as small side setbacks or little alcoves. The shape, size and general visual appeal of small tanks are therefore very different than the larger concrete tanks (Standards Australia, 2008). Finally, the cost of the tank material can often also become a deciding factor in the selection of size and material.

All rainwater tanks are required to have as an absolute minimum an inlet, an outlet to meet demand and an overflow point (see Figure 3). Rainwater tanks can also be fitted with a pump for ease of use (Standards Australia, 2008), a top-up system for areas where mains water is available (Barry & Coombes, 2006, 2007; Standards Australia, 2008), a first flush (Cuncliffe, 1998; enHealth Council, 2004; Standards Australia, 2008) and a water treatment system (Föster, 1996; Standards Australia, 2008). Furthermore in some urban areas stormwater is also detained in the tank, adding a third outlet in the tank to separate detention[1] and retention[2] of run-off from the roof. These additional items are dependent on

---

[1]Detention of run-off is to keep the water on the site for a period of time and slowly release the flow to the downstream system.
[2]Retention of run-off is to keep the water on the site for an unknown period of time and reuse the captured volume on the site.

the local legislation, but a first flush is recommended to be installed everywhere to improve the water quality on the tank.

Fig. 2. Rainwater tank site locations and materials

In a rainwater tank study in Western Sydney, it was identified that topping-up of a rainwater tank with mains water can cause a change in the harvested water quality (van der Sterren et al., 2010a; 2009; 2010b). A larger value of hardness and conductivity was observed when mains water was added to the tank. The tank had reached a low volume as a result of high levels of water use and top-up of the water in the tank was needed. Mains water in the area has a typical value between 35 and 62 mg L^{-1} $CaCO_3$, which increased the hardness of water within the rainwater tank after, which created similar concentration in the tank as in the mains water supply (Sydney Water, 2007). The results were significantly higher than the average conductivity in the rainwater tank and the equilibrium was not re-established until 3 months after the mains water was added (see Figure 4).

A first-flush device diverts the first part of the run-off (e.g. first 5 minutes)and it is allowed to drain overland. Griffin et al. (1980) showed that the first 30% runoff generated by a storm has a higher contaminant concentration and contains nearly 70% of the total pollutant load. Bucheli et al. (1998) and Föster (1999) found that the run-off from the first 2 mm of rain contains most of the total pollutants on a number of different roof types. The remaining flows for the roofs have a lower contaminant concentration and are treated by the natural processes within the rainwater tank itself. Goonetilleke et al. (2005, p. 33) disagrees with

[3]Reproduced with permission from Rocla Pty Ltd (www.rocla.com.au)

these findings as the first-flush is 'never precisely defined' and argued that the pollutant load, rather than the concentration, should be the governing factor for the design.

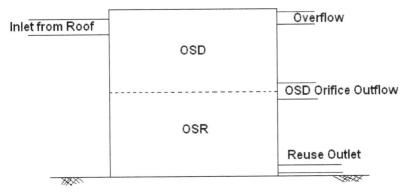

Fig. 3. Rainwater tank with inlet, overflow, outlet and optional third outlet for detention control.

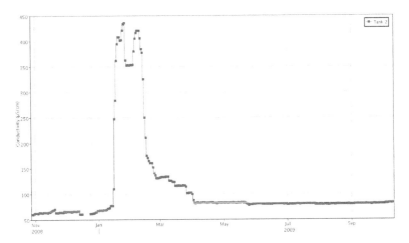

Fig. 4. Conductivity measurements for a rainwater tank in Western Sydney showing the effect of the mains water added to the tank.

Brodie (2007) indicated that the volume required for an effective first-flush is hard to quantify when duration of the storm event is similar to the time of concentration (t_c) of the roof catchment and this can cause difficulties in designing a first-flush. Recently a new technique was suggested by Bach et al. (2010) to determine the first-flush based on statistical analysis of the concentration of pollutants in the first-flush versus the background concentrations rather than a set amount of rainfall as the criteria. This could provide a volume-based design approach suitable for industry application. The removal of pollutants by collecting the first-flush and separating it from the water in the rainwater tank can result in better water quality in the tank. Mendez et al. (2011), however, did not find a significant difference in the water quality in rainwater tank with or without a first-flush. A recent study in Western Sydney showed elevated levels of microbes, nutrients and heavy metals in the first-flush, which were significantly higher than the samples of the harvested water (van der Sterren et al., 2009; 2010b). In addition, the *E.coli* and *Enterococcus* enumeration exceeded secondary contact guidelines, which indicates that pathogens can be present in the first-flush. The data collected in the Western Sydney study supports the emptying of the first-flush after a storm event, which often is not conducted on a regular basis. This defeats the purpose of the system, highlighting the importance of having a first-flush device that can operate without human intervention, especially to prevent the inhalation of pathogens when cleaning is conducted. Recently some councils (e.g. Penrith City Council, 2008, 2010) have added the requirement to connect the overflow from the rainwater tank directly to the street drainage; recent research has found that the overflow quality is not significantly cleaner than the harvested tank water due to settlement of pollutants in the rainwater tank (van der Sterren et al., 2010a; 2009; 2010b). The impact of the overflow quality and comparison of overflow water quality to tank water quality is included in Section 4.2.

4. Water quality of harvested water in Australia

4.1 Processes within the rainwater tank

During storage of water in a rainwater tank, various chemical and biological processes take place. Well known processes are those of settlement and re-suspension of particles (Davis & Cornwell, 2006), conservation of energy (Moeller et al., 1984; Young et al., 2004) and conservation of mass (Moeller et al., 1984). Each of these processes plays a vital role in improving the water quality in a rainwater tank and in turn ensures that the water in the tank ages and is purified through natural processes.

Settlement and the monolayer/multilayers at the air/liquid interface are two critical processes contributing to water quality in rainwater tanks (Spinks et al., 2003). The settlement could significantly reduce Total Solid (TS) concentrations in the overflow and tank water quality, whilst a biofilm and/or chemical gradients could potentially be critical contributors to overflow contamination. A number of pollutants have been shown to adhere to Suspended Solids (SS) and TS. The principal law that governs the settling of particles is known as Stoke's Law and is a function of particle size and gravitation forces through the medium (Davis & Cornwell, 2006). The particle density and radius vary within the solution and are dependent on the pollution run-off characteristics. In Australia, the average particle size of SS is larger than 125 µm (Melbourne Water, 2005). Miguntanna et al. (2010a; 2010b) found that the 75 to 150 µm particle range had the strongest influence on the Total Nitrogen (TN) build-up, whilst the 1 to 150 µm particle range had the strongest influence on the Total Phosphorous (TP) build-up. Brodie (2007), on the other hand, investigated sediment size and found that over 50% was smaller than 63 µm in roof run-off. The particle density varies from 1600 kg m^{-3} for sand, 1720 kg m^{-3} for coarse materials and 1210 kg m^{-3} for organic fine sediment (Verkerk et al., 1992).

Previous research has found that the water quality at the surface of the water in a rainwater tank is significantly different from the water quality at the bottom of the rainwater tank (Coombes, 2002). This difference has been attributed to stratification. Stratification occurs in lakes and wetlands (Doods, 2002) and could occur in rainwater harvesting systems, which results in chemical and physical gradients in the water column. In addition, a biofilm can form on the air/liquid interface and the solid/liquid interface (Hermansson & Dahlbäck, 1983; Hermansson et al., 1987; Marshall, 1980; Parker & Barson, 1970; Percival et al., 2000; Sigee, 2005; Spinks et al., 2003). Stratification occurs in a body of water, due to warm water being less dense than cooler water. The mixed layer of warm water is the epilimnion and floats on the cooler hypolimnion (Davis & Cornwell, 2006; Doods, 2002). De-stratification will occur if the temperature of the total water body is 3.9°C (Doods, 2002), or when storms and wind mix the water. The mixing by wind is minimal in a rainwater tank, but mixing by inflow could occur depending on the storm intensity and depth of water in the tank. Australia also has reasonably high temperatures. Cool temperatures resulting in rainwater tank water cooling to 3.9°C are likely to only occur at high elevations or during extreme winters in Western Sydney or in the high altitude areas of New South Wales and Victoria, for example in the Blue Mountains and the Snowy Mountains. The epilimnion promotes Brownian motion, due to the higher temperatures and is likely to be influenced by inflows causing turbulence. In addition, the movement between layers is mainly molecular diffusion, thereby creating chemical and physical gradients (Doods, 2002). These result in higher chemical and biological activity, and higher concentrations in the upper layers of the water column in comparison to the lower layers. The higher concentrations and warmer temperatures in the epilimnion also support the growth of micro-organisms, and therefore biofilm growth on the air/liquid interface is likely to occur (Parker & Barson, 1970).

The biofilm surface layer may be between 0.1-10 μm, but divers significantly from the remainder of the water body (Characklis & Cooksey, 1983; Sigee, 2005). Biofilms are defined as "microbial cells, attached to a substratum and immobilised in a three dimensional matrix of extra cellular polymers enabling the formation of an independent functioning ecosystem homeostatically regulated" (Percival et al., 2000, p. 61). These biofilms consist of 50% to 90% of Extracellular Polymeric Substances (EPS) and can provide a number of benefits, including (but not limited to) the absorption of nutrients, microbes and heavy metals, protection of temperature changes, desiccation and mediation by protozoa (Ancion et al., 2010; Bester et al., 2010; Characklis & Cooksey, 1983; Percival et al., 2000; Sigee, 2005). The biofilms have a continuous growth cycle and have been shown to protect natural systems against pollution, but are very susceptible to the flow and turbulence of the liquid (Blanchard & Syzdek, 1970; Percival et al., 2000). Biofilm formation was shown to be favoured during non-optimal microbial growing conditions (Landini, 2009) and growth often start off as small micro-colonies (Walker et al., 1995) attached to the tank, pipe walls or sediments (Characklis & Cooksey, 1983; Characklis & Marschall, 1990). The biofilms can form on the air/liquid and solid/liquid interfaces (Marshall, 1980), but the formation is also dependent on van der Waals, electrostatic and steric forces (Van Houdt & Michiels, 2010).

It is expected that these biofilms exist on the air/liquid interface of the rainwater tank and the solid/liquid interface within the rainwater tank, pipes and in bioretention systems (Blanchard & Syzdek, 1970; Characklis & Marschall, 1990; Flemming, 1993; Hermansson & Dahlbäck, 1983; Sawyer et al., 2010; Spiers et al., 2003; Spinks et al., 2003). The velocity of flow within the rainwater tank itself is slow, even when it is raining and when the tank is utilised. It is, therefore, expected that the detachment of biofilms and microbes is minimal inside the tank (Spinks et al., 2003). The biofilms formed on the air/liquid interface provide the bacteria access to oxygen from the atmosphere and nutrients from the liquid (Spiers et al., 2003); however, when the tank is full, it is expected that the air/liquid interface biofilm is removed from the tank through the overflow and thereby increasing the pollutants in the overflow. This allows the water to age and therefore ensures a better water quality within the tank. This however, does not take into consideration the impact of this biofilm in discharges on the receiving waters. The air/liquid interface was tested in a recent study by the authors in Western Sydney to determine if a difference was present in water samples from the bottom of tank and the overflow, thereby investigating if notable stratification occurs in the rainwater tanks.

A sample of the water at the top of two rainwater tanks were taken in the Western Sydney study and tested for *E. coli*, TC and TTC, *Enterococcus sp.* and HPC (van der Sterren, 2011; van der Sterren et al., 2010a). All microbiological results for top of the tanks for the same week as the tanks and overflow samples were higher than results from the overflow and tank samples. The water surface was determined to contain an average of 143 cfu 100 mL^{-1} *Enterococci spp.*, which was significantly higher than the 7 cfu 100 mL^{-1} recorded in the tap sample. All these results showed higher readings in all microbes test in the overflow than in the tank samples, which has been theorised by Coombes et al. (2002; 2000b) to support a microlayer. This is further supported by theory of biofilms (Percival et al., 2000). It is considered critical to test the overflow during rainwater tank water quality studies to determine the long-term effect on overflow water quality, because of the biofilm growth on the air/liquid interface or the effect of chemical gradients and investigate their effect on the receiving waters.

4.2 Water quality within the rainwater tank

The heavy metal results of the tank samples can vary significantly between the sites and can be above the ADWG (NHMRC & NRMMC, 2004). Aluminium concentrations were found to be above the ADWG ((NHMRC & NRMMC, 2004)) for galvanised roofs and tanks and below the ADWG for concrete tanks in a yearlong water quality study in Western Sydney (van der Sterren, 2011; van der Sterren et al., 2010b) and this supported the findings of Berdhal et al. (2008), Föster (1996, 1999) and Magyar et al. (2007; 2008).

The aluminium concentrations in water at sites with galvanized roofs and tanks are expected to be higher than those from those sites with plastic or concrete systems (BlueScope Steel, 2010; Wight et al., 2000), because of the zinc or zinc-aluminium alloy protective layers. The aluminium concentration could also be linked to location and usage of the tanks, as a high water use household is more likely to expose the inside of the tank to atmosphere due to the higher rate of drawdown on the tank. This drawdown could expose the inside of the tank more frequently to oxygen, thereby likely causing an increase in the rate of corrosion (Askeland, 1998; Berdhal et al., 2008). The aluminium concentration determined in the Western Sydney research study were lower than those of Sorenson et al. (1974) in the United States of America (USA), but similar results were found by Herngren et al. (2005) on streets in Brisbane.

Roofs made of Zincalume® can also show elevated levels on zinc in the tank water. The coating on Zincalume® is 55% aluminium, 43.5% zinc and 1.3 % silicon (BlueScope Steel, 2010). The zinc concentrations of tanks samples on sites with galvanised materials were significantly higher ($p<0.05$) than those sites without these materials and were above the recommended guidelines of 3 mg L^{-1} (NHMRC & NRMMC, 2004). The higher levels of zinc in tank samples are likely due to the protective layer that contains zinc, which can cause the higher zinc concentrations in these tanks. The risk of zinc pollution can be significantly reduced using a concrete or plastic tank (enHealth Council, 2004; NRMMC et al., 2009a). The zinc concentrations in the Western Sydney study were similar to the roof run-off results found by Thomas and Greene (1993), and Herngren et al. (2005). The Western Sydney study does not support the finding by Duncan (1999) who obtained a zinc concentration of 10.2 mg L^{-1} (±5 mg L^{-1}), which is higher than the Western Sydney results (μ_{all}=2.63 mg L^{-1}).

Rainwater tank catchment roof areas should not contain lead flashing, especially if the harvested water is to be used for consumption purposes (Cuncliffe, 1998; enHealth Council, 2004; NRMMC et al., 2009a). Sites containing some lead flashing on the rainwater catchment area show elevated levels of lead (Magyar et al., 2007; 2008; O'Connor et al., 2009; van der Sterren, 2011; van der Sterren et al., 2010b). Sites without lead flashing, but with Poly Vinyl Chloride (PVC) pipes can also show elevated levels of lead concentrations (Magyar et al., 2007; 2008; O'Connor et al., 2009; van der Sterren, 2011; van der Sterren et al., 2010b). Lead concentrations are of particular concern, when the residents use their tank for all in-house purposes, including consumption. Magyar et al. (2007; 2008) and O'Connor et al. (2009) found similar lead values in rainwater tank water in Victoria as van der Sterren (2011) found in Western Sydney. Duncan (1999) also showed lead concentration in roof run-off to be on average 0.054 mg L^{-1} (± 5.01 mg L^{-1}). The results collected in the Western Sydney study (μ_{Tall}= 0.009 mg L^{-1}) were similar to other studies in Australia (Magyar et al., 2007; 2008; O'Connor et al., 2009) and are therefore, not considered unusually high.

The copper concentrations for all tanks tested in the Western Sydney study (μ_{T-all}=0.221 mg L^{-1}) were well below the ADWG (NHMRC & NRMMC, 2004), but all tank samples had a hardness falling into the soft range, which is likely to result in corrosion of copper pipe lines. Sites containing concrete tiles or a concrete tank can be beneficial for the users, as the higher hardness reduces copper corrosion. A higher hardness would be beneficial to all sites in the Western Sydney study, especially because the ADWG (NHMRC & NRMMC, 2004) suggest that a good hardness is between 60 and 200 mg L^{-1} $CaCO3$, which was not met by any of the samples on any of the sites.

Total Nitrogen is an important pollutant for receiving water bodies, but in respect to drinking water, the ADWG (NHMRC & NRMMC, 2004) have made recommendations in regards to nitrate, nitrite and ammonium. The tank samples in the Western Sydney study indicated nitrate to be as low as 0.001 mg/L in the tank water (van der Sterren, 2011; van der Sterren et al., 2010b). This is significantly different to the roof catchment study conducted by Evans et al. (2006), who found nitrate concentrations between 0.31 and 4.63 mg L^{-1}. The results are similar to the findings by Nicholson et al. (2010), whom studied roof catchment in the United States of America (USA). The long-term recommended trigger value for Total Phosphorous (TP) in collected roof run-off used for irrigation is 0.05 mg L^{-1}. This can be increased to 0.2 mg L^{-1} if algal blooms in their irrigation system are acceptable (NRMMC et al., 2009a). The RWG (NRMMC et al., 2009a) indicates that it is not practical to reduce the TP concentrations in roof water for domestic applications. The average TP concentrations in the tap samples in the Western Sydney study were 0.18 ± 0.20 mg L^{-1}, which is above this guideline. The results were similar to those found by Duncan (1999) (0.15 ± 1.95 mg L^{-1}) and slightly below the roof run-off TP concentrations found by Miguntanna (2009) (1.85 mg L^{-1}).

The microbiological contamination in tanks used for drinking water must have minimal pathogenic contaminants (enHealth Council, 2004). The ADWG (NHMRC & NRMMC, 2004) states that E. coli concentrations should not be detected and the enHealth guidelines indicate that the microbial quality of rainwater tank water is 'not as good as urban water supplies' (enHealth Council, 2004, p. 2). E. coli was detected above 1 cfu 100 mL^{-1} for at least 75% of the tank samples in the Western Sydney study. High enumeration of E. coli was attributed to low water levels and long antecedent dry periods, resulting in an increased concentration of faecal contamination levels in the roof run-off and tank water. Lower water levels reduce the effect of dilution and the longer antecedent dry period increases the likelihood and build-up of faecal contamination from wildlife deposited onto the roof. In addition, the low water levels can potentially cause significant mixing of the water column as a result of a rainfall event. It is possible that a biofilm or the air/liquid interface are broken up and mixed throughout the small water column during low water levels and a rainfall event.

The mixing of the water column could potentially increase faecal contamination at the bottom of the tank. It has been suggested that a rainwater tank should not be grossly over-designed to ensure that the biofilm is removed from the tank on a regular basis (van Olmen, 2009). This could be done by ensuring that the roof size and usage of the water is sufficient for the site, thereby finding a balance between the formation and cleaning out of the biofilm or air/liquid interface. The results of the enumeration of E. coli in the Western Sydney study showed lower concentrations than reported by McCarthy et al. (2008) in stormwater run-off (50 to 34,770 cfu 100 mL^{-1}). The difference is attributed to the different surface areas contributing to the stormwater run-off tested by McCarthy et al. (2008). Roofs are considered to have lower concentrations of E. coli than other impervious areas. The enHealth guidelines (2004, p. 12) also indicate that there is 'no measurable difference in rates of gastrointestinal illness in children who drank rainwater [sic. harvested roofwater] compared to those who drank mains water', but also indicates that those who are immunocompromised are at a much greater risk.

The turbidity in tank water varies with location and some sites show a greater variation in turbidity due to the location and size of the tank (Thomas & Greene, 1993). The Western Sydney study showed exceedance of the turbidity guidelines from 2% to 33% of the samples taken from the tanks. This is in contrast with the results of Mobbs (1998), where the rainwater tank did not exceed the ADWG (NHMRC & NRMMC, 2004) at all throughout a year of monthly sampling. It should be considered that testing was conducted on a weekly basis for the Western Sydney study, instead of monthly as in Mobbs (1998) and as settlement of solids is often associated with turbidity, this could give different results than monthly grab samples. The increases in turbidity from the tank samples can be attributed to the tanks not being used. When users in the Western Sydney study were on extended leave, the water quality test results showed outliers and when irrigation was not required, an increase in turbidity levels was noted. The second attribute influencing the turbidity levels is the impact of rain on the volume remaining in the tank. Han and Mun (2008) suggest using a 3 m deep tank to reduce the impact of rainfall on the mixing in the tank. Smaller tanks having a high water use and drawdown can result in low water levels on a regular basis. During rainfall events, the rain is likely to disturb the sediment in the bottom of the tank if the water levels are low. Other contributors to elevated turbidity levels have been the dust storm of 23 September 2009 in Sydney and the impact of runoff containing leaf litter (see Section 4.3).

The Dissolved Oxygen in the Western Sydney study was in the acceptable range for the ADWG (NHMRC & NRMMC, 2004), but for one tank in the study, the DO was mainly below the recommended guideline of 85% saturation (HNMRC & NRMMC, 2004). This could be an indication of more oxygen consuming processes taking place in the tank, for example, corrosion and microbial growth, but Chemical Oxygen Demand and Biological Oxygen Demand were not tested for in the Western Sydney study and therefore this hypothesis cannot be confirmed.

The pH was expected to be approximately 6.5 for the tank water (Coombes et al., 2000b; Duncan, 1999; Herngren et al., 2005; Thomas & Greene, 1993) and the average results for the sites containing galvanised materials in the Western Sydney study were not statistically different from this pH ($\mu \neq 6.5$, p>0.05). The sites containing concrete were slightly higher than this, but still within the ADWG (2004) (HNMRC & NRMMC, 2004). The effect of materials on the pH of the tank samples was also presented by Thomas and Greene (1993) and Mendez et al. (2011). The main cause for the pH increase in concrete roofs was attributed to efflorescence (Berdhal et al., 2008). Calcium hydroxide reacts with carbon dioxide in the air producing calcium carbonate, which is transported into the tank during a rain event. The calcium carbonate in the tank dissolves and neutralises the pH. The results for pH found in the Western Sydney study were higher than Duncan (1999) (5.7 ±1.1), but similar to Thomas and Greene (1993) (6.8 to 7.0). Duncan (1999) analysed values from around the world, which could have brought the mean result down due to acid rain in the Northern Hemisphere. Bridgman et al. (1989; 1988) indicated that severity of acid rain in Australia and New Zealand is significantly lower than other parts of the world, which could explain the higher readings in the Western Sydney study and by Thomas and Greene (1993) in comparison to Duncan (1999).

The conductivity concentrations were expected to be between 15 and 297 μS cm^{-1} (Camp Scott Furphy Pty Ltd, 1991; Herngren et al., 2005; Thomas & Greene, 1993). The recorded results from the Western Sydney study fell mostly within this range and any outliers were attributed to extremely low water level. Thomas and Greene (1993) discussed that the higher conductivity was likely to be the result of concrete on their sites, which is supported by the results from the Western Sydney Study.

In summary, the materials of the rainwater tank and roof largely govern the water quality within the rainwater tank. The galvanised steel (and Zincalume® or Colourbond®) roofs and rainwater tank are likely to add significant concentrations of aluminium and zinc to the water supply, as the aluminium and zinc coating is the sacrificial layer. The polypropylene rainwater tank, PVC piping and lead flashing increase the lead concentrations within the rainwater tank, whilst concrete increases the hardness and conductivity in the rainwater tank. A lower hardness can also increase the scaling and therefore the concentration of copper in the rainwater tank and plumbing.

4.3 Water quality of the overflow

A very limited amount of data is available for overflow water quality in the literature. Most rainwater tank water quality testing has focused on the potable or non-potable quality of the harvested tank water. In regards to stormwater management, the overflow quality is of a much greater concern, as it is often directly connected to the receiving drainage system. A study on rainwater tank water quality and quantity discharges has been conducted in Australia, in which the first 2500 mL of overflow was tested (van der Sterren, 2011; van der Sterren et al., 2010a). The parameters analysed were DO, pH, turbidity, conductivity, temperature, TN, TP, aluminium, copper, lead, zinc, *E.coli*, *Enteroccocus spp.*, TC and TTC, which give an indication of the potential water quality of overflows from rainwater tanks. This section summarises the results of the recent study in Western Sydney(van der Sterren, 2011; van der Sterren et al., 2010a) and contrasts the results to the tank samples taken at the same time from the same sites. The overflows tested in the Western Sydney study had significantly greater pollutant loadings than the tank sample itself, which is hypothesised to be the result of the epilimnion characteristics of the air/liquid interface, the potential of biofilm growth or the microlayer on the air/liquid interface, as discussed in Section 4.1.1

Statistically the mean values and distributions of conductivity, copper hardness, lead, pH and turbidity were found to be similar ($\mu_T \neq \mu_O$, $p > 0.05$) between the tank and overflow samples, but aluminium and zinc standard deviations of the tank and overflow samples were significantly different ($\sigma_T \neq \sigma_O$, $p < 0.05$). The concentrations for aluminium ($min_{O\text{-all}}$=0.020 mg L^{-1}) were all above 0.8 µg L^{-1}, which is above the guideline for freshwater of 55 µg L^{-1} (ANZECC & ARMCANZ, 2000a, b). Sites containing galvanised roofs recorded higher readings for aluminium than those without. The majority of the lead results for the overflow ($min_O\text{-all}$= 0.001 mg L^{-1}) were above the AFWG (ANZECC & ARMCANZ, 2000a, b) guideline of 34 µg L^{-1}. A greater variation was observed on sites that contained lead flashing on the roof and conventional PVC downpipes.

The zinc concentrations for the overflow samples had higher concentrations for two sites as a result of the galvanised roofs and tank. All recorded zinc ($min_{O\text{-all}}$=0.050 mg L^{-1}) results were above the AFWG (ANZECC & ARMCANZ, 2000a, b) of 8 µg L^{-1}. The RWG (NRMMC et al., 2009a) indicate that the zinc concentrations in water stored from zinc coated roofs exceeded the long and short term trigger values in the AFWG (ANZECC & ARMCANZ, 2000a, b) and can be toxic to species. The RWG (NRMMC et al., 2009a) recommend low irrigation rates to protect sensitive plants and to minimise contamination of the soils. This would mean that the overflow discharge should be treated using best management stormwater treatment techniques prior to discharging it into the existing drainage system.

The copper concentrations in the overflow also varied, but showed a narrower distributions and similar results between overflows. A comparison of the mean copper concentrations indicated that at 95% confidence interval of the mean values (μ_{all}= 0.081 mg L^{-1}) were similar in the Western Sydney study, but all results were above the AFWG (ANZECC & ARMCANZ, 2000a, b). According to the RWG (NRMMC et al., 2009a) copper concentrations greater than 0.2 mg L^{-1} can be toxic to plants, which occurred occasionally in the overflow samples, thereby posing a potential threat to sensitive plants growing in the overflow discharge path. The high values for the copper concentration occurred on the same day as those outliers for lead and zinc concentrations in the results. It is hypothesised in the Western Sydney study (van der Sterren, 2011) that the humidity and warmth of summer increases the degradation, as identified by Berdahl et al. (2008), thereby increasing the pollutant wash-off.

In addition, a high rate of drawdown can also increase the corrosion, because of wetting, drying and exposure to oxygen on the inside of the tank. It should be noted that the hardness of overflows tested in the Western Sydney study were below the AFWG (ANZECC & ARMCANZ, 2000a, b) criteria of 30 mg L^{-1} $CaCO_3$, which allows the heavy metals guidelines to be applied to the test results. All the hardness results were also below the recommended ambient in-stream guideline of lowland rivers (ANZECC & ARMCANZ, 2000a, b). Sites containing concrete tanks and tiled roofs showed higher hardness concentrations in the overflow, just like the tank samples in the Western Sydney study.

The pH values of the overflows can have an impact on receiving waters, however, dilution of the overflows through the stormwater system, which is mostly concrete, could increase the pH and reduce the risk to the receiving waters. The AFWG (ANZECC & ARMCANZ, 2000a, b) requires the pH to be in between 6.5 and 9.0 for inland rivers, but results indicate average higher values than the minimum guideline for sites containing concrete. This is supported by the findings from Thomas and Greene (1993), whom also found elevated pH results when concrete was used as a catchment area. The conductivity of the overflows is recommended to be below 300 μS cm^{-1} (ANZECC & ARMCANZ, 2000a, b). The conductivity results in the overflow samples indicate that the conductivity was higher in those samples from sites with concrete than those without concrete.

E. coli, TTC, TC and Enterococci spp. counts can vary significantly between different sources and values above the guideline can be expected in rainwater tanks (Evans et al., 2006; Richardson et al., 2009). The E.coli enumeration from the overflows in the Western Sydney study showed a high variation. The enumeration of E.coli exceeded the primary contact standard for faecal coliforms (150 cfu 100 mL^{-1}) and one sample exceeded the secondary contact guideline (150-1000 cfu 100 mL^{-1}) (ANZECC & ARMCANZ, 2000a, b). The results of Enterococcus spp. enumeration also exceeded the secondary contact guidelines in the AFWG (ANZECC & ARMCANZ, 2000a, b).

The variation in the enumeration of the E. coli and Enterococcus spp. may be due to discharge volume, time elapsed between different storm events, and environmental condition, and the total indicator organisms in the run-off entering the tank during an event. The higher readings were found to occur in summer, which was similar to the findings by van Olmen (2009), who indicated that higher temperatures in the water may potentially increase the enumeration of indicator organisms.

The microbiological and nutrient analysis showed that an epilimnion, microlayer or biofilm could exist. As for the microbial and nutrient concentrations, the tank samples were significantly lower than the overflow samples. The E. coli and Enterococcus sp. counts show significantly higher counts in the overflows and were considerably higher than the outliers for the tank samples. In addition, the tank outliers were all within the variation of the overflows and occur after a significant rainfall event or during low water levels. This further supports the possibility of mixing of the water column and biological gradients discharging through the overflow. It should be noted that 50% of the overflows are above the secondary contact guideline of 230 Enterococcus sp cfu 100 mL^{-1} (ANZECC & ARMCANZ, 2000a, b).

All mean overflow temperatures were within the range of 15°C to 23°C with some outliers in summer and some lower temperatures in winter, which were slightly higher than the tank samples. Thermal pollution is, therefore, considered not to be a fundamental issue for the overflow from rainwater tank; however, if all urban areas are considered, thermal pollution of receiving waters may be significant. The heat exchange laws support the growth of biofilms or microlayer at the air/liquid interface, as the direct contact between the water surface and the air is likely to warm up the water at the surface more than the main water body within the tank (Doods, 2002). There is direct contact between the air and liquid at the top of the tank, whilst the main body of the tank has indirect contact with the air. More heat is transferred through air/liquid interface than through the air-solid-liquid interface (tank wall). Furthermore the first 2500 mL of overflow was collected, which is a smaller volume than the tank and therefore after collection, more than likely to have a greater variation in temperature until removed from site as the thermal exchange laws are dependent on mass and thermal conductivity (Askeland, 1998; Young et al., 2004). The temperature and volume in the collection bottle would also affect the DO, as the re-aeration rate of oxygen is dependent on temperature, velocity and water depth (Davis & Cornwell, 2006).

The DO concentrations were highly variable and mostly above the guideline (ANZECC & ARMCANZ, 2000a, b). Some of the results were below the 7.6 mg L^{-1} minimum limit. This indicates that DO in the overflow should be increased prior to discharge into receiving rivers, which often occurs as a result of flow velocity and turbulence through the drainage system. The top of the tank had a higher DO content than the bottom because of the re-aeration is dependent on the depth which causes a temperature gradient (Davis & Cornwell, 2006; Doods, 2002). This higher DO concentration at the air/liquid interface and in the overflow, as a result of the gradients, combined with increases in water temperature could potentially sustain the microbes, thereby resulting in potential higher counts. In addition, when overflows occur, the rainfall coming into the rainwater tank also contains microbes, further increasing the potential survival of indicator organisms and therefore pathogens in the rainwater tank.

The TP concentrations (μ_{o-all}=0.976 mg L^{-1}) however, were all above the AFWG (ANZECC & ARMCANZ, 2000a, b) (0.050 mg L^{-1}) and RWG (NRMMC et al., 2009a). Outliers were attributed to the eucalyptus blossom and other plants in the area. According to the RWG (NRMMC et al., 2009a), the short-term impact on soils for phosphorus contamination is low. It is suggested that phosphorous levels be reduced by using bio-retention and filtration systems to minimise the effect of phosphorus contamination on receiving waters. The overflow from these rainwater tanks should, therefore, be directed to a filtration or retention system to minimise wash-off of TP to the receiving water bodies. All of the nitrate concentrations were well below the AFWG (ANZECC & ARMCANZ, 2000a, b) (600 $\mu g\ L^{-1}$) and the RWG (NRMMC et al., 2009a) (30 mg L^{-1}). This further supports the hypothesis that there is a microlayer or biofilm at the air liquid interface, as these layers are often formed as survival mechanisms in low nutrient environments (Landini, 2009).

Biofilm formation is promoted by non-optimal growing temperature or limited nutrients (Bester et al., 2010; Blanchard & Syzdek, 1970; Landini, 2009). The rainwater tank has been shown to be low in nutrients and is influenced by the similar chemical gradients as a lake. The epilimnion has a limited source of DO and temperature (Chapra, 1997), thereby most likely promoting the formation of a biofilm at the air/liquid interface. The atmosphere provides another additional source reducing the strain on the microbes and allowing the formation of new cells (Bester et al., 2010; Spiers et al., 2003). The chemical and biological gradients (Chapra, 1997), the possible biofilm, as well as the pollutant run-off from the roof water can cause the overflow to have higher pollutants than the bottom of the rainwater tank. This indicates that the rainwater tank, especially larger rainwater tank, can have stratification and therefore should not be modelled as completely mixed tanks (or continuously stirred tanks). On the other hand, when the rainwater tank has a low volume of water, the rainfall can have a significant stirring effect on the minimal volume thereby creating a mixed tank (Chapra, 1997). There is a further need to examine the overflow from rainwater tank and their effects on the stormwater quality to understand its impact. Furthermore, detailed analysis is also required on the stratification effects in the rainwater tank and the potential of biofilm growth. These effects are considered to age and clean the water, but could have a detrimental effect on the outflow water quality of the site. It clearly highlights the need for more than one source control and using a treatment train approach, including UV treatments to control micro-organisms. Furthermore, recent research in Europe indicated that the water quality of rainwater tank improves if an overflow occurs on a regular basis, therefore indicating that a rainwater tank should not be over-designed (van Olmen, 2009).

4.4 Impact on quality from climatic events

Different climatic events can have impacts on the water quality of the rainwater tank. A dust storm covered Sydney and surrounds in a haze on the 23 September 2009 and deposited large amounts of fine particles onto all surfaces, which includes all roofs (Leys et al., 2009). A low intensity rainfall event occurred on the 24 September 2009 and the research conducted in Western Sydney showed high turbidity results (14.8 NTU) in the first-flush after this event. The increase of turbidity levels in the tested tanks were not detected until 6 October 2009, after a more intense rainfall event as well as settling time in the rainwater tank. The overflows also showed higher than average turbidity levels on 6 October 2009. In addition to the elevated turbidity levels, copper, TN, conductivity and hardness also showed significantly high values as a result of the storm event of 2 October 2009. Testing of the deposited dust was conducted by Radhi et al. (2010), who found that the dust contained salt particles and could be traced back to Lake Eyre Basin. The dust storm event is therefore the most likely contributor to these elevated levels. This clearly shows the needs for cleaning and maintenance of the roof after an event like this. Heavy rainfall, is the primary cause of pollutant run-off from the roof and was therefore examined here. All high values were compared to rainfall events and antecedent periods. The tank samples were more likely to show elevated levels when antecedent periods were longer, whilst overflows were more likely to show high values when there is a prolonged period between overflow events (van der Sterren, 2011).

Elevated levels and extremes were more likely to occur in the tank water quality when the antecedent rainfall event exceeded two days. The overflow, on the other hand, was more likely to contain elevated levels when the time to previous overflow exceeded 8.5 days. This is of course a guideline as events such as the dust storm or the intensity of the rainfall event can alter the high values and concentrations after a smaller antecedent period. The focus was on long term sampling (i.e. a year). The benefit of the long term testing is that the seasonal changes as well as long-term (in other words, year) impact can be examined. The antecedent time period can affect the build-up on roofs and wash-off into tanks, as high values are more likely to occur during long antecedent periods. This was previously shown to be important in pollutant build-up by other studies (Egodawatta & Goonetilleke, 2008; Egodawatta et al., 2009; Miguntanna et al., 2010b; Sartor & Boyd, 1972). Overflows are also more likely to contain extreme values when time intervals between overflows are longer.

5. Government subsidies and financial viability

The Australian, State and local governments have a number of subsidies and grants that are provided to citizens to install a rainwater tank on their property. This section looks at the financial viability of rainwater harvesting system and these government subsidies for rainwater tanks. The amount of rebate varies from state to state and also on the size of the rainwater tank and where the water is utilised. For New South Wales, the maximum rebate is $1500. For Australian Capital Territory and Victoria, the maximum rebate is $1000.

A number of studies have examined the financial viability of rainwater harvesting system. These studies have looked at the financial viability of rainwater tanks by themselves, or rainwater tanks as part of an integrated catchment management method. The financial benefit of rainwater tanks as a single entity has been shown to be minimal; however, commonly reduction of cost associated with a reduction in on site detention volume and environmental benefits are often not included in these studies.

A cost benefit analysis of the financial viability of rainwater tanks is dependent on the cost of water, any subsidies and at what time the net present value is computed. In addition, the design and country of origin are also important variables in conducting a life cycle or cost – benefit analysis. Roebuck and Ashley (2006) examined the life cycle costing of a rainwater tank for a school building in United Kingdom (UK) with a project life of 65 years. The long-term savings were estimated to be £18,370 opposed to £122,230 estimated by the tank supplier. They found that capital cost, maintenance cost and mains top-up cost were 31%, 26% and 40% respectively of the total cost and argued that many of the methods of rainwater tank analysis overestimated the cost savings of rainwater tanks. Domenech and Sauri (2011) examined the financial benefit of rainwater tanks in single and multi-unit buildings in Barcelona, Spain. They found minimum pay back periods of 33 to 43 years for a single-family household and 61 years for a 20 m³ tank for multi-unit buildings. Grant and Hallmann (2003) conducted the financial viability of a 600 and a 2250 L rainwater tank in Australia. They found that neither tanks would create a positive return over 30 years using the 2003 water pricing. Life cycle costing of rainwater tanks for multi-storied building in Sydney using a 75 kL tank and for toilet, laundry and outdoor use, the cost benefit ration has been found to range from 0.64 to 1.15 for discount rates of 0% to 7.5% (Rahman et al., 2010)

Although both Spain (Barcelona) and Sydney are considered a mediteranian climate, there is a significant difference between the benefits of rainwater tanks and the pay back periods. The UK life cycle costing can expected to be very different, as the chance of rain and annual volume in the UK is higher than in Spain and Australia. These rainwater tanks are designed according to the climate and therefore affect the size, cost and use of the rainwater tank. It is therefore difficult to compare Australian data with other countries, but overall, there is a significant pay back period of a single rainwater tank and the cost-benefit analysis should take all aspects of the benefits and the cost into account.

When the financial viability of rainwater tanks are considered as part of an integrated water cycle management plan, the rainwater tanks are shown to have significant benefits to authorities. The demand on dam and mains water is significantly reduced when rainwater tanks are implemented throughout the catchment. Various studies have shown that the cost reduction by implementing rainwater tanks throughout the catchment is significant

(Coombes & Kuczera, 2003; Coombes et al., 2000a; Davis & Birch, 2008; Lucas, 2009; Lucas et al., 2009). This clearly indicates that any cost-benefit analysis should not only investigate the cost and benefits of a single rainwater tank, but the cost and benefits of the whole system, including water mains upgrade, cost to the environment and reduction in cost of treating and managing stormwater.

6. Conclusion

Rainwater tanks are popular in Australia as a source of alternative water supply and a means of stormwater management. From a literature review and findings from a field study conducted in Western Sydney in Australia, it has been found that materials of the rainwater tanks and roof largely govern the water quality within the rainwater tanks. The overflow and first flush water quality from the rainwater tanks exceeded water quality guidelines, highlighting the needs for further treatment before discharging the water to urban stormwater systems. The water quality in the rainwater tanks does not meet the drinking water guidelines on many instances, especially during low water levels and the days immediately after a rainfall event. Under these circumstances, the tank water should be disinfected before drinking. Rainwater tanks are found to be not financially viable at lot scale given the current water price, but if all the associated benefits of rainwater tanks at catchment scales are accounted for their financial viability is likely to increase significantly.

7. Acknowledgements

The work presented in this Chapter would not have been completed without the assistance of Jonathan Barnes, Solomon Donald and Ian Turnbull. Thanks also to Surendra Shrestha, John Bavor, Jeff Scott, Sharon Armstrong, Rhonda Gibbons, Bert Aarts, Wayne Higgenbotham, Paul Roddy, Heidi Fitzpatrick from the University of Western Sydney. Thanks also to Turnbull Electrical Contracting, who contributed time and resources for the installations of the systems on site and the funding from the UWS office of Disabilities.

8. References

Ancion, P.-Y., Lear, G., & Lewis, G. D. (2010). Three common metal contaminants of urban runoff (Zn, Cu & Pb) accumulate in freshwater biofilm and modify embedded bacterial communities. *Environmental Pollution, 158*(8), 2738-2745.

Askeland, D. R. (1998). *The science and engineering of materials.* Cheltenham: Stanley Thornes Ltd.

Australian and New Zealand Environment and Conservation Council, & Agriculture and Resource Management Council of Australian and New Zealand (ANZECC & ARMCANZ) (2000a). Australian and New Zealand guidelines for fresh and marine water quality - volume 1. *National Water Quality Management Strategy.* Artarmon: Australian Water Association.

Australian and New Zealand Environment and Conservation Council, & Agriculture and Resource Management Council of Australian and New Zealand (ANZECC & ARMCANZ) (2000b). Australian and New Zealand guidelines for fresh and marine water quality - volume 2 - aquatic ecosystems - rationale and background information. Artarmon: Australian Water Association.

Australian and New Zealand Environment and Conservation Council, & Agriculture and Resource Management Council of Australian and New Zealand (ANZECC & ARMCANZ) (2000c). Australian guidelines for urban stormwater management. *National Water Quality Management Strategy.* Artarmon: Australian Water Association.

Australian Bureau of Statistics (2010). Australia's Environment: Issues and Trends. Canberra: Australian Bureau of Statistics.

Australian Bureau of Statistics (ABS) (2011). Regional Population Growth, Australia. Canberra: Australian Bureau of Statistics.

Bach, P. M., MacCarthy, T. D., & Deletic, A. (2010). Redefining the stormwater first flush phenomenon. *Water Research,* 44(8), 2487-2498.

Barry, M. E., & Coombes, P. J. (2006). Optimisation of mains trickle topup supply to rainwater tanks in an urban setting. *Australian Journal of Water Resources,* 10(3), 269-275.

Barry, M. E., & Coombes, P. J. (2007). Optimisation of mains trickle topup volumes and rates supplying rainwater tanks in the Australian urban setting. *Rainwater and Urban Design Conference.* Sydney: The Institution of Engineers Australia.

BASIX Sustainability Unit (2009). Building Sustainability Index. Sydney: New South Wales Department of Planning.

Berdhal, P., Akbari, H., Levinson, R., & Miller, W. A. (2008). Weathering of roofing materials - an overview. *Construction and Building Materials,* 22(4), 423-433.

Bester, E., Kroukamp, O., Edwards, E. A., & Wolfaardt, G. M. (2010). Biofilm form and function: carbon activity affects biofilm architecture, metabolic activity and planktonic cell yield. *Journal of Applied Microbiology,* 110(2), 387-398.

Blackburn, N., Morison, P., & Brown, R. (2010). A review of factors indicating likelihood of and motivations for household rainwater tank adoption. *Stormwater 2010: National Conference of the Stormwater Industry Association.* Sydney, pp. 1-10.

Blacktown City Council (2006). Blacktown development control plan 2006. Blacktown: Blacktown City Council.

Blanchard, D. C., & Syzdek, L. (1970). Mechanisms for the water-to-air transfer and concentration of bacteria. *Science, New Series,* 170(3958), 626-628.

Blue Mountains City Council (2005). Better Living Development Control Plan. Katoomba: Blue Mountains City Council.

BlueScope Steel (2010). Benefits of ZINCALUME® steel.

Bridgman, H. A. (1989). Acid rain studies in Australia and New Zealand. *Archives of Environmental Contamination and Toxicology,* 18(3), 137-146.

Bridgman, H. A., Rothwell, R., Tio, P.-H., & Pang-Way, C. (1988). The Hunter region (Australia) acid rain project. *Bulletin of the American Meteorological Society,* 69(3), 266-271.

Brisbane City Council (2008). Subdivision and development guidelines - Part C water quality management guidelines. Brisbane: Brisbane City Council.

Brodie, I. M. (2007). Investigation of stormwater particles generated from common urban surfaces. *Faculty of Engineering and Surveying:* University of Southern Queensland.

Bucheli, T. D., Müller, S. R., Heberle, S., & Schwarzenbach, R. P. (1998). Occurrence and behaviour of pesticides in rainwater, roof runoff, and artificial stormwater infiltration. *Environmental Science and Technology,* 32(22), 3457-3464.

Camp Scott Furphy Pty Ltd (1991). *Urban runoff study*: Nepean-Hawkesbury Catchment Management Council.

Chapra, S. C. (1997). *Surface water-quality modeling*: WCB/McGraw-Hill.

Characklis, W. G., & Cooksey, K. E. (1983). Biofilms and microbial fouling. *Advances in Applied Microbiology*, 29, 93-138.

Characklis, W. G., & Marschall, K. C., eds. (1990). *Biofilms*. New York: John Wiley & Sons.

Coombes, P. J. (2002). Rainwater tanks revisited: new opportunities for Urban Water Cycle Management. Newcastle: University of Newcastle.

Coombes, P. J., & Kuczera, G. (2003). A sensitivity analysis of an investment model used to determine the economic benefits of rainwater tanks. *28th International Hydrology and Water Resources Symposium*. Wollongong, Australia: Institution of Engineers Australia, pp. 243-250.

Coombes, P. J., Kuczera, G., & Kalma, J. D. (2000a). Economic benefits arising from use of Water Sensitive Urban Development source control measures. *Hydro 2000 - 3rd International Hydrology and Water Resources Symposium of the Institution of Engineers Australia*. Perth, Western Australia: The Institution of Engineers Australia, pp. 152-157.

Coombes, P. J., Kuczera, G., & Kalma, J. D. (2000b). Rainwater quality from roofs, tanks and hot water systems at Figtree Place. *Hydro 2000 - 3rd International Hydrology and Water Resources Symposium of the Institution of Engineers Australia*. Perth, Western Australia: The Institution of Engineers Australia, pp. 1042-1047.

Cuncliffe, D. A. (1998). Guidance on the use rainwater tanks. *National Environmental Health Forum Monographs, Water Series Number 3*. Rundle Mall: Department of Human Services.

Davis, B. S., & Birch, G. F. (2008). Catchment-wide assessment of the cost-effectiveness of stormwater remediation measures in urban areas. *Environmental Science & Policy*, 12(1), 84-91.

Davis, M. L., & Cornwell, D. A. (2006). *Introduction to environmental engineering*. New York: McGraw-Hill.

Domènech, L., & Saurí, D. (2011). A comparative appraisal of the use of rainwater harvesting in single and multifamily buildings of the Metropolitan Area of Barcelona (Spain): social experience, drinking water savings and economic costs. *Journal of Cleaner Production*, 19(1), 598-608.

Doods, W. (2002). Freshwater ecology - concepts and environmental applications. Elsevier.

Duncan, H. P. (1999). Urban stormwater quality: A statistical overview. Cooperative Research Centre for Catchment Hydrology.

Egodawatta, P., & Goonetilleke, A. (2008). Modelling pollutant build-up and wash-off in urban road and roof surfaces. *31st Hydrology and Water Resources Symposium and 4th International Conference on Water Resources and Environmental Research: Water Down Under 2008*. Adelaide, Australia, pp. 418-427.

Egodawatta, P., Thomas, E., & Goonetilleke, A. (2009). Understanding the physical processes of pollutant build-up and wash-off on roof surfaces. *Science of the total Environment*, 407(6), 1834-1841.

enHealth Council (2004). Guidance on use of rainwater tanks. In: Australian Government, ed: Australian Government Department of Health and Ageing, pp. 1-72.

Environmental Health Branch (2008). Rainwater tanks where a public water supply is available - use of. North Sydney: Department of Health, New South Wales Government.

Evans, C. A., Coombes, P. J., & Dunstan, R. H. (2006). Wind, rain and bacteria: The effect of weather on the microbial composition of roof-harvested rainwater. *Water Research*, 40(1), 37-44.

Flemming, H.-C. (1993). Biofilms and environmental protection. *Water Science and Technology*, 27(7-8), 1-10.

Föster, J. (1996). Patterns of roof runoff contamination and their potential implications on practice and regulations of treatment and local infiltration. *Water Science and Technology*, 33(6), 39-48.

Föster, J. (1999). Variability of roof runoff quality. *Water Science and Technology*, 39(5), 137-144.

Goonetilleke, A., Thomas, E., Ginn, S., & Gilbert, D. (2005). Understanding the role of land use in urban stormwater quality management. *Journal of Environmental Management*, 74(1), 31-42.

Goyen, A. G., Lees, S. J., & Phillips, B. C. (2002). Analysis of allotment based storage, infiltration and reuse drainage strategies to minimize urbanization effects. *9th International Conference on Urban Drainage, Global Solutions for Urban Drainage:9ICUD 2002*. Portland, Oregon USA: American Society of Civil Engineers, pp. 1-17.

Grant, T., & Hallmann, M. (2003). Urban domestic rainwater tanks: Life cycle assessment. *Water*, 30(5), 36-41.

Griffin, D. M. J., Randall, C., & Grizzard, T. J. (1980). Efficient design of stormwater holding basins used for water quality protection. *Water Research*, 14(10), 1540-1554.

Han, M. Y., & Mun, J. S. (2008). Particle behaviour consideration to maximize the settling capacity of rainwater storage tanks. *Water Science and Technology*, 56(11), 73-79.

Hawkesbury City Council (2000). Development control plan 2000. Windsor: Hawkesbury City Council.

Hermansson, M., & Dahlbäck, B. (1983). Bacterial activity at the air/water Interface. *Microbial Ecology*, 9(4), 317-328.

Hermansson, M., Jones, G. W., & Kjelleberg, S. (1987). Frequency of antibiotic and heavy metal resistance, pigmentation and plasmids in bacteria of the marine air-water interface. *Applied and Environmental Microbiology*, 53(10), 2338-2342.

Herngren, L., Goonetilleke, A., & Ayoko, G. A. (2005). Understanding heavy metal and suspended solids relationships in urban stormwater using simulated rainfall. *Journal of Environmental Management*, 76(2), 149-158.

Heyworth, J. S., Maynard, E. J., & Cuncliffe, D. A. (1998). Who drinks what? Potable water use in South Australia. *Water (Melbourne)*, 25(1), 9-13.

Horstman, M. (2007). 1000 year drought. In: Hiscox, M., ed. *Catalyst*. Australia: Australian Broadcasting Company, p. 9:38.

Landini, P. (2009). Cross-talk mechanisms in biofilm formation and responses to environmental and physiological stress in *Escherichia coli*. *Research in Microbiology*, 160(4), 259-266.

Leys, J., Heidenreich, S., & Case, M. (2009). DustWatch report for week ending 21 September 2009. 1-6.

Lucas, S. A. (2009). Mains water savings and stormwater management benefits from large architecturally - designed under-floor rainwater storages. *32nd Hydrology and Water Resources Symposium*. Newcastle: Engineers Australia, pp. 628-640.

Lucas, S. A., Coombes, P. J., & Sharma, A. K. (2009). Residential diurnal water use patterns and peak demands - implications for integrated water infrastructure planning. *32nd Hydrology and Water Resources Symposium*. Newcastle Australia: Engineers Australia, pp. 1081-1091.

Magyar, M. I., Mitchell, V. G., Ladson, A. R., & Diaper, C. (2007). An investigation of rainwater tanks quality and sediment dynamics. *Water Science and Technology*, 56(9), 21-28.

Magyar, M. I., Mitchell, V. G., Ladson, A. R., & Diaper, C. (2008). Lead and other heavy metals: common contaminants of rainwater tanks in Melbourne. *31st Hydrology and Water Resources Symposium and 4th International Conference on Water Resources and Environmental Research: Water Down Under 2008*. Adelaide, Australia: Institute of Engineers Australia, pp. 409-417.

Marshall, K. C. (1980). Microorganisms and interfaces. *BioScience*, 30(4), 246-249.

McCarthy, D. T., Deletic, A., Mitchell, V. G., Fletcher, T. D., & Diaper, C. (2008). Uncertainties in stormwater *E.coli* levels. *Water Research*, 42(-), 1812-1824.

Melbourne Water (2005). *Water Sensitive Urban Design engineering procedures: stormwater*. Melbourne Australia: CSIRO Publishing.

Mendez, C. B., Klenzendorf, J. B., Afshar, B. R., Simmons, M. T., Barrett, M., Kinney, K. A., & Kirisits, M. J. (2011). The effect of roofing material on the quality of harvested rainwater *Water Research*, Article in Press, 1-11.

Miguntanna, N. (2009). Determining a set of surrogate parameters to evaluate urban stormwater quality. *Faculty of Built Environment and Engineering*: Queensland University of Technology.

Miguntanna, N., Egodawatta, P., Kokot, S., & Goonetilleke, A. (2010a). Determination of a set of surrogate parameters to assess urban stormwater quality. *Science of the total Environment*, 408(24), 6251-6259.

Miguntanna, N., Goonetilleke, A., Egodawatta, P., & Kokot, S. (2010b). Understanding nutrient build-up on road surfaces. *Journal of Environmental Sciences*, 22(6), 806 - 812.

Mobbs, M. (1998). *Sustainable house*. Marrickville, Australia: CHOICE Books.

Moeller, T., Bailar, J. C., Jr., Kleinberg, J., Guss, C. O., Castellion, M. E., & Metz, C. (1984). *Chemistry with inorganic qualitative analysis*. Orlando, Florida: Academic Press Inc.

National Health and Medical Research Council & Natural Resource Management Ministerial Council (NHMRC & NRMMC) (2004). National water quality management strategy - Australian Drinking Water Guidelines. Australian Government.

Natural Resource Management Ministerial Council, Environment Protection and Heritage Council, & National Health and Medical Research Council (NRMMC / EPHC & HNMRC) (2009a). National water quality management strategy - Australian guidelines for water recycling: Stormwater harvesting and reuse. Australian Government.

Natural Resource Management Ministerial Council, Environment Protection and Heritage Council, & National Health and Medical Research Council (NRMMC / EPHC & HNMRC) (2009b). National water quality management strategy - Australian Guidelines for Water Recylcing (Phase 2): Augmentation of drinking water supplies. Australian Government.

Natural Resource Management Ministerial Council, Environment Protection and Heritage Council, & National Health and Medical Research Council (NRMMC / EPHC & HNMRC) (2009c). National water quality management strategy - Overview of the Australian guidelines for water recycling: Managing health and environmental risks 2006. Australian Government.

Nicholas, D. I. (1995). On-site Stormwater Detention: improved implementation techniques for runoff quantity and quality management in Sydney. *Water Science and Technology*, 32(1), 85-91.

Nicholson, N., Clark, S. E., Long, B. V., Siu, C. Y. S., Spicher, J., & Steele, K. A. (2010). Roof runoff water quality - a comparison of traditional roofing materials. *World Environmental and Water Resources Congress 2010: Challenges of Change*. Providence, Rhode Island: American Society of Civil Engineers, pp. 3349-3355.

O'Connor, J. B., Mitchell, V. G., Magyar, M. I., Ladson, A. R., & Diaper, C. (2009). Where is the lead in rainwater tanks coming from? *32nd Hydrology and Water Resources Symposium*. Newcastle, Australia: Institute of Engineers Australia, pp. 653-660.

Parker, B., & Barson, G. (1970). Biological and chemical significance of surface microlayers in aquatic ecosystems. *BioScience*, 20(2), 87-93.

Penrith City Council (2008). *Exhibition of stage 1 planning documents*. Penrith, Australia: Penrith City Council.

Penrith City Council (2010). Penrith City Council Development Control Plan. Penrith.

Percival, S. L., J. T. Walker, et al. (2000). Microbiological aspects of biofilms and drinking water. Vreeland, R. H. (ed). Boca Raton: CRC Press.

Radhi, M., Box, M. A., Box, G. P., & Cohen, D. D. (2010). Size-resolved chemical composition of the September 2009 Sydney dust storm. *Air Quality and Climate Change*, 44(3), 25-30.

Rahman, A., Dbais, J., & Imteaz, M. A. (2010). Sustainability of RWHSs in Multistorey Residential Buildings. *American Journal of Engineering and Applied Sciences*, 1(3), 889-898.

Richardson, H. Y., Nichols, G., Lane, C., Lake, I. R., & Hunter, P. R. (2009). Microbiological surveillance of private water supplies in England - the impact of environmental and climate factors on water quality. *Water Research*, 43(8), 2159-2168.

Roebuck, R. M., & Ashley, R. (2006). Predicting the hydraulic and life-cycle cost performance of rainwater harvesting systems using a computer based modeling tool. *4th International Conference on Water Sensitive Urban Design*. Melbourne: Engiener Australian, pp. 699-709.

Sartor, J. D., & Boyd, G. B. (1972). Water pollution aspects of street surface contaminants. In: monitoring, O. o. r. a., ed. Washington: US Environmental Protection Agency, p. 236.

Sawyer, C. B., Hayes, J. C., & English, W. R. (2010). Characterization of *Escherichia coli* for sediment basin systems at construction sites. *World Environmental and Water Resources Congress 2010: Challenges of Change*. Providence, Rhode Island: American Society of Civil Engineers, pp. 3198-3208.

Sigee, D. C. (2005). *Freshwater microbiology - biodiversity and dynamic interactions of microorganisms in the aquatic environment.* Chichester: John Wiley & Sons.

Sorenson, J. R. J., Campbell, I. R., Tepper, L. B., & Lingg, R. D. (1974). Aluminium in the environment and human health. 8, 3-95.

Spiers, A. J., Bohannon, J., Gehrig, S. M., & Rainey, P. B. (2003). Biofilm formation at the air-liquid interface by the *Pseudomonas fluroscens* SBW25 wrinkly spreader requires an acetylated form of cellulose. *Molecular Microbiology,* 50(1), 15-27.

Spinks, A. T., Coombes, P. J., Dunstan, R. H., & Kuczera, G. (2003). Water quality treatment processes in domestic rainwater harvesting systems. *28th International Hydrology and Water Resources Symposium.* Wollongong, Australia: Institution of Engineers Australia, pp. 227-234.

Standards Australia (2006). Plumbing and drainage part 3: Stormwater drainage. In: Australian and New Zealand Standards, ed. Sydney: Australian Standards.

Standards Australia (2008). HB 230-2008 Rainwater tank design and installation handbook. In: Australian and New Zealand Standards, ed. Sydney, Australia: Standards Australia.

Sydney Water (2007). Water analysis - Typical water analysis for Sydney Water's drinking water supply. Sydney: Sydney Water.

Thomas, P. R., & Greene, G. R. (1993). Rainwater quality from different roof catchments. *Water Science and Technology,* 28(3-5), 291-299.

Upper Parramatta River Catchment Trust (UPRCT), Cardno Willing (NSW) Pty Ltd, Haddad Khalil Mance Arraj Partners, & Brown Consulting (NSW) Pty Ltd (2005). On-site stormwater detention handbook. Upper Parramatta River Catchment Trust.

van der Sterren, M. (2011). Assessment of the impact of rainwater tanks on urban run-off quantity and quality characteristics (under review). *School of Engineering.* Penrith: University of Western Sydney, p. 604.

van der Sterren, M., Dennis, G. R., Chuck, J., & Rahman, A. (2010a). Rainwater tank water quality testing in Western Sydney, Australia. *World Environmental and Water Resources Congress - Challenges of Change.* Providence, USA, pp. 4048-4058.

van der Sterren, M., Rahman, A., & Dennis, G. R. (2009). A case study of rainwater tank quality testing in Western Sydney, Australia. *The 7th International Symposium on Ecohydraulics and 8th International Conference on Hydroinformatics* Concepcíon, Chile.

van der Sterren, M., Rahman, A., & Ryan, G. (2010b). Investigation of water quality and quantity of five rainwater tanks in Western Sydney, Australia. *World Environmental and Water Resources Congress - Challenges of Change.* Providence, USA, pp. 3933-3941.

Van Houdt, R., & Michiels, C. W. (2010). Biofilm formation and the food industry, a focus on the bacterial outer surface. *Journal of Applied Microbiology,* 109(4), 1117-1131.

van Olmen, R. (2009). Gebruik van hemelwater in woningen - economische en ecologische analyse. Belgium: Universiteit Hasselt.

Verkerk, G., Broens, J. B., Kranendonk, W., van der Puijl, F. J., Sikkema, J. L., & Stam, C. W. (1992). *BINAS.* Groningen: Wolters-Noordhoff.

Walker, J. T., Mackerness, C. W., Rogers, J., & Keevil, C. W. (1995). Heterogeneous mosaic biofilm - a haven for waterborne pathogens. In: Lappin-Scott, H. M., & Costerton, J. W., eds. *Microbial Biofilms.* Cambridge: Cambridge University Press.

White, I. (2010). Rainwater harvesting: theorising and modelling issues that influence household adoption. *Water Science and Technology,* 62(2), 370-377.

Wight, I., Huddle, L., & Mullens, R. (2000). Why any old galvanised iron won't do... *The Heritage Advisor*.

Wong, T. H. F., ed. (2006). *Australian runoff quality - A guide to Water Sensitive Urban Design*. Crows Nest: Engineers Australia

Young, H. D., Freedman, R. A., & Ford, L. A. (2004). *Sears and Zemansky's university physics with modern physics*. Sydney: Pearson Education.

Permissions

The contributors of this book come from diverse backgrounds, making this book a truly international effort. This book will bring forth new frontiers with its revolutionizing research information and detailed analysis of the nascent developments around the world.

We would like to thank Kostas Voudouris and Dimitra Voutsa, for lending their expertise to make the book truly unique. They have played a crucial role in the development of this book. Without their invaluable contribution this book wouldn't have been possible. They have made vital efforts to compile up to date information on the varied aspects of this subject to make this book a valuable addition to the collection of many professionals and students.

This book was conceptualized with the vision of imparting up-to-date information and advanced data in this field. To ensure the same, a matchless editorial board was set up. Every individual on the board went through rigorous rounds of assessment to prove their worth. After which they invested a large part of their time researching and compiling the most relevant data for our readers. Conferences and sessions were held from time to time between the editorial board and the contributing authors to present the data in the most comprehensible form. The editorial team has worked tirelessly to provide valuable and valid information to help people across the globe.

Every chapter published in this book has been scrutinized by our experts. Their significance has been extensively debated. The topics covered herein carry significant findings which will fuel the growth of the discipline. They may even be implemented as practical applications or may be referred to as a beginning point for another development. Chapters in this book were first published by InTech; hereby published with permission under the Creative Commons Attribution License or equivalent.

The editorial board has been involved in producing this book since its inception. They have spent rigorous hours researching and exploring the diverse topics which have resulted in the successful publishing of this book. They have passed on their knowledge of decades through this book. To expedite this challenging task, the publisher supported the team at every step. A small team of assistant editors was also appointed to further simplify the editing procedure and attain best results for the readers.

Our editorial team has been hand-picked from every corner of the world. Their multi-ethnicity adds dynamic inputs to the discussions which result in innovative outcomes. These outcomes are then further discussed with the researchers and contributors who give their valuable feedback and opinion regarding the same. The feedback is then collaborated with the researches and they are edited in a comprehensive manner to aid the understanding of the subject.

Apart from the editorial board, the designing team has also invested a significant amount of their time in understanding the subject and creating the most relevant covers. They scrutinized every image to scout for the most suitable representation of the subject and create an appropriate cover for the book.

The publishing team has been involved in this book since its early stages. They were actively engaged in every process, be it collecting the data, connecting with the contributors or procuring relevant information. The team has been an ardent support to the editorial, designing and production team. Their endless efforts to recruit the best for this project, has resulted in the accomplishment of this book. They are a veteran in the field of academics and their pool of knowledge is as vast as their experience in printing. Their expertise and guidance has proved useful at every step. Their uncompromising quality standards have made this book an exceptional effort. Their encouragement from time to time has been an inspiration for everyone.

The publisher and the editorial board hope that this book will prove to be a valuable piece of knowledge for researchers, students, practitioners and scholars across the globe.

List of Contributors

Rouzbeh Nazari and Reza Khanbilvardi
City University of New York, USA

Saeid Eslamian
Isfahan University of Technology, Iran

Flávia Vieira da Silva-Medeiros, Flávia Sayuri Arakawa, Gilselaine Afonso Lovato, Célia Regina Granhen Tavares and Rosângela Bergamasco
Universidade Estadual de Maringá, Maringá, PR, Brazil

Maria Teresa Pessoa Sousa de Amorim
Universidade do Minho, Campus Azurém, Guimarães, Brazil

Miria Hespanhol Miranda Reis
Universidade Federal de Uberlândia, Uberlândia, MG, Portugal

Rosângela Bergamasco, Angélica Marquetotti Salcedo Vieira and Letícia Nishi
Universidade Estadual de Maringá, Brazil

Álvaro Alberto de Araújo and Gabriel Francisco da Silva
Universidade Federal de Sergipe, Brazil

Davi Gasparini Fernandes Cunha, Maria do Carmo Calijuri, Doron Grull and Pedro Caetano Sanches Mancuso
Universidade de São Paulo, Brazil

Daniel R. Thévenot
LEESU, Université Paris-Est, France

Edyta Kiedrzyńska and Maciej Zalewski
International Institute of the Polish Academy of Sciences, European Regional Centre for Ecohydrology Under the Auspices of UNESCO, Lodz, Poland
University of Lodz, Department of Applied Ecology, Lodz, Poland

Chad S. Boyd, Tony J. Svejcar and Jose J. Zamora
USDA-Agricultural Research Service, Eastern Oregon Agricultural Research Center, Burns, OR, USA

Y.O. Yusuf and M.I. Shuaib
Department of Geography, Ahmadu Bello University, Zaria, Nigeria

Medjram Mohamed Salah and Boussaa Zehou El-Fala Mohamed
Laboratoire LARMACS, Université de Skikda Algérie, France

Bachir Meghzili
Laboratoire LARMACS, Université de Skikda Algérie, France
Université de Biskra, France

Michel Soulard
Equipe Matériaux à Porosité Contrôlée, IS2M, LRC CNRS 7228, UHA, ENSCMu, France

Binghui He and Tian Guo
Southwest University, China

Munjed A. Maraqa, Hassan D. Imran and Waleed Hamza
United Arab Emirates University, United Arab Emirates

Ayub Ali
Griffith University, Australia

Saed Al Awadi
Ports, Customs and Free Zone Corporation, United Arab Emirates

Florentina Bunea
National Institute for R&D in Electrical Engineering ICPE-CA, Romania

Diana Maria Bucur and Gabriela Elena Dumitran
Politehnica University of Bucharest, Romania

Gabriel Dan Ciocan
Université Laval, Laboratoire de Machines Hydrauliques, Canada

W.B. Wan Nik and M.F. Ahmad
Dept. of Maritime Technology, Universiti Malaysia Terengganu, Malaysia

M.M. Rahman
Dept. of Pharm. Chemistry, Faculty of Pharmacy, Int. Islamic University Malaysia, Malaysia

J. Ahmad
Dept. of Engineering Science, Universiti Malaysia Terengganu, Malaysia

A. M Yusof
Department of Chemistry, Faculty of Science, Universiti Teknologi Malaysia, Malaysia

M. Mosharraf Hossain and K.M. Nazmul Islam
Institute of Forestry and Environmental Sciences, University of Chittagong, Chittagong, Bangladesh

Ismail M.M. Rahman
Department of Applied and Environmental Chemistry, University of Chittagong, Chittagong, Bangladesh

Y.O. Yusuf, E.O. Iguisi and A.M. Falade
Department of Geography, Ahmadu Bello University, Zaria, Nigeria

A. Rahman and G.R. Dennis
University of Western Sydney, Australia

M. van der Sterren
University of Western Sydney, Australia
NSW Office of Environment and Heritage, Australia

Printed in the USA
CPSIA information can be obtained
at www.ICGtesting.com
JSHW011446221024
72173JS00004B/967